Optoelectronics

TEXAS INSTRUMENTS ELECTRONICS SERIES

Optoelectronics

Theory and Practice

Edited by Alan Chappell
Texas Instruments Ltd.
Original German Version: Volkmar Härtel
assisted by Eilhard Haseloff, Gerhard Jahn,
and Günther Suhrke
Texas Instruments Deutschland GmbH

McGRAW-HILL BOOK COMPANY

New York St. Louis San Francisco Aukland Bogotá
Düsseldorf Johannesburg London Madrid Mexico
Montreal New Delhi Panama Paris São Paulo
Singapore Sydney Tokyo Toronto

Library of Congress Cataloging in Publication Data
Main entry under title:

Optoelectronics.

(Texas Instruments electronics series)
Includes index.
1. Optoelectronics.   I. Chappell, Alan.
TA1750.068   621.38′0414   78-8021
ISBN 0-07-063755-5

The circuits, sub-assemblies and procedures
described in this book have been tested by
Texas InstrumentsLtd. (T.I.L.); but this
does not guarantee their reliability in
operation.

Neither can T.I.L. give any warranty that
these circuits etc. are not covered by patent
rights of third parties.

T.I.L. reserves all rights in this book.
Without express permission from T.I.L.
neither the book nor parts of it may be
duplicated or distributed in any manner by
photocopies, microfilms or any other
process. This applies also to the right of
reproduction in public.

Texas Instruments reserves the right to make changes at any time
in order to improve design and supply the best product possible.
Information contained in this publication is believed to be
accurate and reliable. However, responsibility is assumed neither
for its use nor for any infringement of patent or rights of others
which may result from its use. No license is granted by implication
or otherwise under any patent or patent right of Texas Instruments
or others.

Design/typography
David Muriel Presentation Unit

Foreword

Over the years progress in electronics has
given birth to an ever increasing number of
circuit components for a great variety of
uses. The so called optoelectronic devices in
particular, which make use of the mutual
interaction of radiation and the electronic
structure of materials, have become widely
used in recent years. This has been brought
about chiefly through improvements in the
semiconductor manufacturing process
which have enabled optoelectronic devices
to be used economically in many and varied
applications.

The purpose of this book is to provide both
a theoretical and practical introduction to
these optoelectronic components. It is
written primarily for engineers, technicians
and students who need to acquire the
background knowledge and the ability to
use these devices. With this in mind
particular attention has been paid to
providing an abundance of practical hints
and suggestions as well as the necessary
theoretical background to enable users to
develop their own circuits and applications.

The Editors.

Contents

ix

Introduction

The term "Optoelectronics" is today understood to mean the production, utilisation and evaluation of electromagnetic radiation in the optical wavelength range and its conversion into electrical signals. Two basic components are needed – a radiation source as a transmitter and a photoelectric converter as a receiver. Components which emit or are sensitive to radiation in the UV, IR and visible range are defined as "optoelectronic components". Optoelectronics is today classified as a subsidiary area of telecommunications, while its origins go back into the last century. Thus, in 1873, Smith discovered the change in conductivity when selenium is irradiated, and in 1887 Hertz discovered the effect named after him, by which the spark discharge of a spark gap starts at lower voltages under UV radiation. In 1888, Hallwachs found the effect which bears his name, that negatively-charged metal plates lose their charge under UV radiation, but positively-charged plates do not. Numerous further examples are named in the literature.

With the introduction of transistors in the fifties, semiconductor optoelectronics first made use of the radiation-dependence, which was undesirable in normal use, of ordinary diodes and transistors, for radiation receivers. After further research work, semiconductor radiation sources were also manufactured in the early sixties.

As the largest semiconductor manufacturer in the world, Texas Instruments recognised at an early stage that optoelectronic semiconductor products will find widespread application. TI carried out pioneer work in the research, development and manufacture of optoelectronic components. Thus, as long ago as 1957, Texas Instruments introduced the first solar cells and in 1959 the 1N2175 silicon photodiode. Through continuous further developments, TI have also provided the largest range of optoelectronic components, as is shown by a range of over 125 standard products, which take account of the most widely-varying needs of the user.

As the following list shows, optoelectronic components are used today in most branches of industry and areas of daily life.

Power engineering, data-processing, machine control, heat engineering, the automobile industry, space systems, sound and video recording, photography, consumer electronics, medicine, telecommunications, environmental protection, domestic appliances etc.

The fields of application can be divided into six groups:

1
Measuring, monitoring, evaluation, control and testing of given light sources.

2
Optoelectronic devices with unmodulated optical radiation.

3
Optoelectronic devices with modulated optical radiation.

4
Optoelectronic devices for alphanumeric displays.

5
Optoelectronic devices for recording or transmission of images.

6
Optoelectronic devices for image reproduction.

In group 1, a radiation source is defined as a natural, existing light source, if its spectral emission distribution lies mainly in the visible range. Such light sources include, for example, the sun, artificial light or naked flames. Fields of application for optoelectronic components include illumination meters in photometry, exposure meters, illumination switches, flash time controllers and flash release in photography, dust density meters for environmental protection, fog density meters and parking-light switches in transport, flame monitors and combustion monitoring devices in heating engineering, monitoring devices on sanitary installations, twilight switches for monitoring the lighting of streets and shop windows, brightness controls on television receivers, in relation to the ambient lighting, brightness regulators in lighting, inspection and control equipment, machine tool controls for the measurement of lengths, positions and angles, revolution counters for the control of motors and converters to transform radiation and light energy into electrical energy.

Group 2 comprises optoelectronic devices with unmodulated optical radiation. In these, the non-directional emission from an optical radiation source is usually converted into a parallel beam and directed onto a photosensitive receiver. The receiver converts the incident radiant energy into a DC signal which is processed further and evaluated in DC amplifiers. This describes the principle of the direct light-beam or optoelectronic coupler. If the receiver receives the transmitted radiation through a mirror or reflector, then we have a reflection coupler. This group includes couplers for production control, the counting of individual items, as safety devices in machines, as punched card and tape readers and code and text readers, as distance and angle detectors in control devices and for the detection of pointer positions in measuring instruments.

Group 3 is concerned with radiation which is modulated at the transmitting end. The receiver converts this modulated radiation into an AC signal. In other respects, the principle of Group 2 is retained. The receivers are designed as AC amplifiers and can therefore be considerably more sensitive and have greater temperature-stability than DC amplifiers.

The applications cover those of Group 2 and additional new areas such as optical sound recording and reproduction, optical telephony, remote controls, garage door openers, range-finders, alarm systems, signal transmission with electrical isolation and AC telegraphy.

Optoelectronic indicating components, indicating units and displays, which radiate in the visible range, are classified in Group 4. In the simplest form, filament lamps, discharge lamps or light-emitting diodes (LEDs) are used for operational and warning indications. In general terms, a display is a device for the representation of numerical and alphanumeric information. These include, among others, mechanical indicating units, cold cathode indicator tubes, seven- or fourteen-segment indicator units, projection displays and alphanumeric displays on image tubes. The segment displays exist with different light sources, e.g., with gallium arsenide phosphide (GaAsP) diodes, incandescent filaments, discharge lamps, AC luminescence lamps, fluorescent displays by means of accelerated electrons and translucent or reflecting liquid-crystal display units.

Group 5 concerns itself with optoelectronic devices for image recording. By means of a lens system, an image is projected either onto a photocathode or on a target. Image converters convert the image appearing on the photocathode, by means of an electron beam, into a visible fluorescent image on the anode or camera tubes scan the image projected on the target. On the target load resistance, every point on the image is converted into an electrical signal corresponding to its brightness.

Group 6 includes optoelectronic devices for picture reproduction. This is understood to mean the electronic reproduction of images on a screen. They include the conventional television picture tubes as well as radar and oscilloscope tubes and the more recent flat picture displays.

On the optical side, optoelectronics touches on the fields of geometrical optics, physiological optics, physical optics, quantum optics, applied optics and in future possibly the field of integrated optics. Geometrical optics is concerned with the propagation, refraction and reflection of rays in accordance with geometrical laws, physiological optics is concerned with the process of vision and the evaluation of light by the eye, and physical or wave optics with the propagation of light as waves (diffraction, interference, polarisation, colour dispersion). Quantum optics deals with the laws of temperature radiation, the atomic model, excitation conditions and luminescence and emission phenomena, the field of X-rays and the interaction between radiation and matter. Applied optics is concerned with optical instruments such as magnifying devices, microscopes etc. Integrated optics deals with optical circuits in a manner analoguous to integrated electronic circuits.

On the electronic side, optoelectronics concerns the production of modulation and the control of electrically operated radiation sources, the drive and deflection circuits of image recording and reproducing tubes, and the further processing and evaluation of the electrical signals produced from a photo-electric converter.

As these statements show, the field of opto-electronics, with its applications, is extremely extensive and many-sided; the present book therefore has the following limits:

The first part deals with the fundamental principles of optoelectronics, photometric and radiation units, black bodies and Lambert radiators, laws of radiation, radiation, luminescence and photo-emission phenomena. In the second, technical part, the emphasis lies on the description, calculation and application of optoelectronic semiconductor components. Finally, the third part is concerned with circuits proven in the laboratory and in practice.

Volkmar Härtel

1
Physics of Optical Radiation

Physics of Optical Radiation

1.1
Optical Radiation and Light

Wavelength range	Designation of Radiation
100 nm – 280 nm	UV – C
280 nm – 315 nm	UV – B
315 nm – 380 nm	UV – A
380 nm – 440 nm	Light – violet
440 nm – 495 nm	Light – blue
495 nm – 558 nm	Light – green
580 nm – 640 nm	Light – yellow
640 nm – 750 nm	Light – red
750 nm – 1 400 nm	IR – A
1·4 μm – 3 μm	IR – B
3 μm – 1 000 μm	IR – C

Table 1.1
Subdivision of the optical radiation
spectrum according to DIN 5031

1.1.1
Basic Definitions

Optical radiation is understood to mean electromagnetic radiation in the range of wavelengths between 10 nm and 1 mm. This range is illustrated in Figure 1.1 as part of the whole electromagnetic spectrum. The optical radiation band consists of the sub-ranges *UV* (Ultraviolet), *visible* radiation (Light) and *IR* (infra-red). The transitions between the individual ranges are fluid. According to DIN 5031, Part 7, the UV range starts at 100 nm. The UV and IR ranges are divided into sub-groups A, B and C and the visible range into the relevant colours, as shown in Table 1.1. The expression "Light" only relates to the optical radiation perceived and evaluated by the human eye.

1.1.2
The quantum nature of radiation

Until the beginning of this century, electromagnetic radiation, including optical radiation, was considered to be continuous trains of waves.

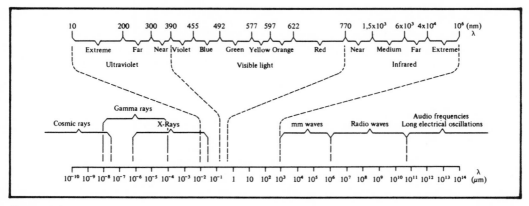

Figure 1.1
Electromagnetic Radiation Spectrum

Physical investigations and considerations on the phenomena of the external photo-electric effect led, however, to the recognition of the fact that radiation does not interact with matter continuously, but in small portions which cannot be further sub-divided, the so-called *quanta*. Such a radiation quantum, also called a *photon*, corresponds to a certain amount of energy, dependent on the frequency of the radiation, which has a minimum value for any given frequency. For the relationship between the energy and the frequency of a quantum of radiation, the equation

$$W_{Ph} = h \cdot \nu \tag{1.1}$$

applies, where $h = 6 \cdot 62 \cdot 10^{-34}$ watt. sec^2 (W.s^2) Planck's constant and ν = the frequency of the radiation in Hz.

In the optical range, the amounts of energy in the individual quanta are so small that the quantum structure of such radiation is beyond the limits of most conventional measurement methods and observations.

All quanta arise from changes in energy in atoms and molecules. The radiations which are of importance for opto-electronics have their origin in the outer electron orbits of the atoms. In the normal state, each electron is in the physically lowest possible level, where it has a certain amount of potential and kinetic energy. By excitation processes, e.g., by the introduction of electrical, thermal or radiant energy, electrons can temporarily leave their basic state and occupy a higher level with a correspondingly higher energy content. This state is not stable; therefore after a very short time the excited electrons fall back to the basic state, while emitting a quantum of radiation, the frequency of which, as shown in formula (1.1), corresponds to the energy difference between the two levels.

1.1.3
The dualism of waves and particles

If the nature of electromagnetic radiation is investigated, then dependent upon the experimental conditions, sometimes it appears to behave like a wave, but at other times like a stream of particles (corpuscles). Therefore, when describing radiation, these two aspects are considered.

Electromagnetic radiation can be demonstrated to exhibit typical wave characteristics such as interference, refraction and polarisation, which can obviously be interpreted by the periodic nature of a wave-train. In theoretical physics, the propagation of electromagnetic radiation is derived from a wave process with the aid of Maxwell's equations.

From the corpuscular viewpoint, radiation takes on the character of a stream of particles. Each photon is then considered as an elementary particle with zero stationary mass, moving at the speed of light. The shorter the wavelength of radiation, the more prominent does the corpuscular nature become in comparison with the wave character. In the range of gamma rays, therefore, their particle nature becomes the predominant characteristic.

1.1.4
Wavelength and propagation speed

The speed of propagation of light in vacuo, and also approximately in the atmosphere, is $c_O = 2 \cdot 998 \cdot 10^8$ m/s.

For practical calculations, the rounded value of $c_O = 3 \cdot 10^8$ m/s is adequate. The relationship between the three fundamental values, speed of light c_O, wavelength λ and frequency ν is given by

$$c_O = \lambda \cdot \nu \tag{1.2}$$

By combining the equations (1.1) and (1.2), the wavelength of a photon can be calculated, if the photon energy W_{Ph} is known.

$$\lambda = \frac{c_O \cdot n}{W_{Ph}} \tag{1.3}$$

The photon energy is usually stated in electron volts (eV). One electron volt corresponds to the kinetic energy received by an electron through acceleration in an electric field with a potential difference of one volt.

For the conversion of eV into the SI unit Ws or J:

$$1 \text{ eV} = 1.602 \cdot 10^{-19} \text{ Watt secs (Ws)}$$
$$1 \text{ eV} = 1.602 \cdot 10^{-19} \text{ Joules (J)}$$

The equation (1.3) can be simplified as a numerical equation, if the numerical values of the natural constants and the conversion factor for eV into Ws are inserted:

$$\lambda = \frac{3 \cdot 10^8 \cdot 6 \cdot 62 \cdot 10^{-34}}{W_{Ph} \cdot 1 \cdot 6 \cdot 10^{-19}}$$

Where λ is measured in μm
        C   is measured in ms^{-1}
and W_{Ph} is measured in eV

This results in the numerical equation

$$\lambda = \frac{1 \cdot 24}{W_{Ph}} \qquad (1.4)$$

From these equations it can be seen, that the greater the energy of the photons, the shorter does the wavelength of the electromagnetic wave becomes.

1.2
Radiation and Luminescence Phenomena

1.2.1
Atomic and Band Model

For the physical explanation of radiation phenomena and also of the photoemission which is described later, an understanding of the atom and band models is necessary. The atomic model describes, in a simplified representation, the spatial structure of an atom, while the band model gives information on the energy content of electrons,

both in single atoms and also in combinations of several atoms, e.g., in a crystal lattice. In *Figure 1.2*, the atomic model for germanium is shown as an example.

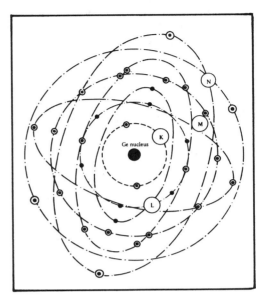

Figure 1.2
Model of germanium atom

Recent work has shown the atom to be extremely complex but for simplicity it can be considered to be composed of negatively charged electrons, orbiting around a charged nucleus. The atomic nucleus itself consists of positively charged protons bound together with neutral particles called neutrons. The number of protons determines which chemical element is concerned. In an electrically neutral atom, the number of electrons and protons is equal. The electrons are located in specific orbits around the nucleus, and these are grouped into so-called shells; in *Figure 1.2* these shells are shown with their normal designations K, L, M and N.

Nearly all atomic nuclei with more than 82 protons and a few smaller nuclei are unstable and disintegrate, through a process known as as nuclear fission, spontaneously over varying periods of time. Through this fission process the basic elements are converted into others. Radium, for example, disintegrates to form the stable element Lead, with the emission of radiation from the nucleus which is characterised into α-, β-, and the short-wave γ-rays. While this radiation originates in the nucleus of the atom, the emission and absorption of optical radiation takes place within the cloud of electrons surrounding the nucleus.

The energy state of the electrons is described by four quantum numbers, The quantum numbers are designated by small letters:

1
n = Principal quantum number (associated with a particular shell)

2
l = Secondary quantum number (sub-group is the shell and shape of the electron orbit)

3
m = Magnetic quantum number (location in space of the angular momentum vector of the orbit)

4
s = Spin quantum number (angular momentum of the electron itself about an imaginary axis)

All electrons associated with an atom or a molecule differ from one another in at least one quantum number. This principle of exclusion of two identical electron states, the so-called Pauli exclusion principle, determines the maximum number of electrons within a given shell, an inter-mediate shell or an energy level. If all the electrons of an atom are at their lowest possible energy level, that is, in the inner-most shells, then the atom is in its basic state.

With the single atom, the electrons adopt exactly defined, discrete energy states. In a crystal, on the other hand, because of the interaction of the electrons belonging to the different atomic nuclei, the previous discrete bands divide into ranges, the *energy bands*. These ranges are illustrated in the so-called band model, while in semi-conductor technology one restricts oneself to the *valency band* and the *conduction band*, which are of interest here, with a "forbidden gap or band" between them. *Figure 1.3* shows a few examples.

In the cases of metals and insulators, the valency band is occupied by electrons, which are fixed in their places. Thus, no movable electrical charges are possible within the valency band. In the case of insulators, the conduction band is unoccupied, i.e., it is free of charge carriers, while in the case of a metal each atom gives up one or more electrons into this band. These electrons are then very loosely bound to a given atomic nucleus and are therefore free to move in the crystal lattice. Their number and mobility in the conduction band determines the conductivity of the substance. In contrast to the metals, which are known to be good electrical conductors, the semiconductor, at low temperatures has almost all electrons in the valency band, so that it is then almost an insulator. With rising temperature, more and more covalent bonds[1] break apart, since through the external supply of energy in the form of heat, a certain number of electrons can leave the valency band and move up into the conduction band. This is possible without difficulty with semiconductors, since the width of the forbidden band is very narrow in comparison with that of insulators. A hole, is then produced in the valency band and behaves like a positively charged particle which can migrate within the band. The *presence of holes in the valency band and electrons in the conduction band causes the conductivity of semiconductor materials to lie somewhere between those of metals and insulators.* The width of the forbidden band, which is

1   *Chemical bond between the atoms of the crystal lattice*

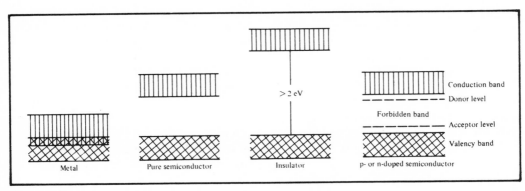

Figure 1.3
Energy bands of a few materials

generally stated in eV, determines the necessary minimum energy, which must be supplied to an electron, in order to raise it from the valency band to the conduction band.

In the preparation of semiconductor devices "impurities" are intentionally added in small, defined quantities to the undoped or "intrinsic" semiconductor material. Silicon and germanium are two quadrivalent or group IV semiconductors which are commonly used for producing useful electrical devices. However, semiconducting compound materials such as gallium arsenide can also have characteristics which can be readily utilised. The conductivity of the basic material is modified as required by adding either trivalent (group III) atoms, so called "acceptors," to give P-material or pentavalent (group V) atoms, so called "donors," to give N-material. Since the donor level is close, in energy, to the conduction band, a very small amount of energy is sufficient to raise an electron of the donor substance into the conduction band. Therefore, the pentavalent donors, with one electron which is only loosely bound to the atom, increase the basic conductivity. The acceptors in turn attract either loosely bound or free electrons, in order to fill the gaps (holes) caused by their addition. The acceptor level is close, in energy, to the valency band.

The conductivity of the doped semiconductor can be altered by the presence of an external energy source.

The excitation energy can be supplied in the form of heat, by photons (light energy) or by the application of an external voltage (electrical energy). If the excitation takes place through photons (irradiation), the term "internal photoeffect" is used. If an excited electron is in the conduction band, this state of excitation is not stable. After a certain time, the electron falls back again and recombines with a hole. Electromagnetic radiation, corresponding to the energy difference liberated, can then be emitted.

Under certain conditions, electrons can be released completely, by energetic excitation, from their parent substance, for example, from alkali metals or certain oxides, and can move freely in space. This process is called emission. If this emission is caused by light quanta, then the term "external photo-effect" is also used.

On the basis of our knowledge of the atomic model, we know that excited electrons are located in orbits with higher energy levels and that on their return to the basic state, electromagnetic radiation can be emitted. A distinction can be made between three kinds of radiation:

a
Radiation, which occurs through the return of the electron to the basic state by direct recombination.

b
Fluorescence radiation.

7

c
Phosphorescence radiation.

From the band model, a distinction is made between four kinds of recombination:

1
Recombination between free electrons of the conduction band and free holes of the valency band.

2
Recombination between free electrons of the conduction band and bonded holes of the acceptor level.

3
Recombination between electrons of the donor level and free holes of the valency band.

4
Recombination between electrons of the donor level and bonded holes of the acceptor level.

The kind of recombination is determined by the characteristics of the semiconductor and its doping.

1.2.2
Luminescence, Fluorescence, Phosphoresence

For all cases of light emission, which do not have their cause solely in the temperature of the material, E. Wiedemann introduced the term "Luminescence" as long ago as 1889. This is understood to mean radiation phenomena in the visible range. In a broader sense, it is also understood, in the literature, to mean the optical radiation range. Depending on the kind of excitation energy, which causes the luminescence phenomenon, distinctions are made, among others, between:

a
Thermoluminescence:
Excitation by raising the temperature of

crystals, in which electrons have been raised in energy by absorbed light do not return at once to the basic state, with emission of the luminescent light, but are stored in energy levels, which are somewhat below the starting energy needed for luminescence.

b
Bioluminescence:
Is part of chemiluminescence. It occurs in nature, e.g., in glow-worms and fireflies.

c
Chemiluminescence:
Occurs through certain chemical reactions, in which energy is liberated and emitted as radiation, e.g., phosphorus glows through oxidation in the air.

d
Cathodoluminescence:
Occurs through accelerated fast electrons, which, on collision with atoms, excite the corresponding valency electrons and cause the emission of radiation or light. Typical examples are television and oscilloscope tubes.

e
A.C. Electroluminescence:
Obtains the excitation energy through an electric field, e.g., in the dielectric of a capacitor (Destriau effect). The luminescent capacitor contains the thin luminescent dielectric and also a transparent electrode.

f
Photoluminescence:
Is caused by fluorescence. The exciting radiation, for example, UV, is more energetic than the radiation emitted in the visible range.

g
Radioluminescence:
Obtains the excitation energy through X-rays or gamma rays.

h
Betaluminescence:
The excitation energy is beta radiation.

i

Crystal luminescence:

Is produced by the deformation of certain crystals.

k

Triboluminescence:

Occurs through the supply of mechanical energy with certain crystals. For example, quartz or zinc sheets glow with a faint light through rubbing, drilling, scratching etc.

As well as classification by the excitation energy, luminescence phenomena are also classified according to the way in which they occur:

a

Fluorescence:

With this type of radiation, the excited electrons fall back, in one or more steps, within about 10^{-8} seconds, to the basic state and light is emitted. In this process, the excitation energy generally has a higher quantum energy than the radiation emitted. Fluorescent substances act, to a certain extent, as frequency converters. In contrast to phosphorescence, fluorescence only gives an emission, as long as an external supply (e.g., radiation) is maintained.

b

Phosphorescence:

A radiation, with which the excited electrons at first remain in a metastable state. This metastable state occurs under the influence of activators (foreign metallic atoms in small concentrations in the basic material), while the electrons fall back into the basic state after a dwell time of varying duration. Phosphorescent materials radiate both during the presence of the excitation energy and also after this excitation energy is switched off, according to the after-glow time.

1.2.3

Luminescence phenomena in semiconductors, Injection luminescence

In a semiconductor diode operated in the forward direction, the junction region is enriched with electrons and holes. These two kinds of charge carrier recombine with one another, and at every recombination an electron is transferred from the conduction band into the valency band. At the same time it gives up the amount of energy, which corresponds to the difference in energy between the conduction band and the valency band.

Depending on the given conditions, the energy thus liberated can be converted into radiant energy (photons) or into heat (lattice vibrations of the crystal, also called phonons). If a photon conversion takes place in the semiconductor materials known up to the present day, radiation in the range from infrared to the visible range appears. Since they are caused injection of charge carriers into a junction region, radiation phenomena of this kind are called *injection luminescence*. The probability of photon radiation taking place depends to a great extend on whether the material used is a "direct" or an "indirect" semiconductor.

Both on the basis of the wave-particle dualism and also according to the theory of wave mechanics, a wave function can be ascribed to a particle of matter. In this process, a moving particle, e.g., either an electron or a hole, can be treated mathematically like a wave. From this wave function, a term which is important for semiconductor considerations can be derived by quantum mechanics which is the wave-number vector $\vec{K}$. This value, which is also known as the propagation vector, is proportional to the momentum (p = m.v) of the moving charge-carrier, as long as the particle can be considered to be "free". In this case it can also be proved that the energy W of the particle is a quadratic function of $\vec{K}$. In a crystal with its periodic three-dimensional lattice the conditions are more complicated, through the interactions of the lattice components with the moving charge carriers. Here, the

9

function W = f($\vec{K}$) is no longer a quadratic function, as before, but the curve which is now produced can contain several maxima and minima. Also, the shape of the curve depends on the geometrical crystal direction, in relation to the major crystal axes, in which the "particle wave" is moving. A few examples are shown in *Figure 1.4*. In these, the holes always have an energy maximum at $\vec{K}$ = 0, while the curve shapes differ for electrons.

The probability that an electron will remain is always highest, where its energy becomes a minimum, while the holes endeavour to reach a level with the maximum possible energy in the valency band. At the points, where a minimum of electron energy is directly opposite a maximum of hole energy, the electron can fill the hole, in a recombination, without a change of the wave-number vector $\vec{K}$ or of its momentum Semiconductors, where this recombination is possible, are called "direct semiconductors". If the electron energy minimum and the hole energy maximum are not directly opposite, then a recombination can only take place with a simultaneous change of $\vec{K}$. In this case, we speak of an "indirect semiconductor".

The physical law of the conservation of momentum in a self-contained system requires, that when a light quantum is either absorbed or emitted from a semiconductor, the momentum of the light quantum causes a corresponding change in momentum in the crystal system. If the momentum values are calculated, both of a moving charge carrier and of a light quantum in the wavelength range which is of interest, it is found, that the momentum of the light quantum is negligibly small in comparison with that of the charge carrier, so that in practice, only the changes of momentum of the charge carriers need to be taken into account, even though light quanta are involved in the process.

Electron transitions in direct semiconductors take place without significant change of momentum, so with recombinations in these materials the probability of the emission of radiation is high. Things are different with indirect semiconductors. In the case of a recombination, here, as well as the energy given up, a change in momentum must also be taken into account. Under these conditions, the production of phonons is again probable, since as well as the energy, these also take up a momentum

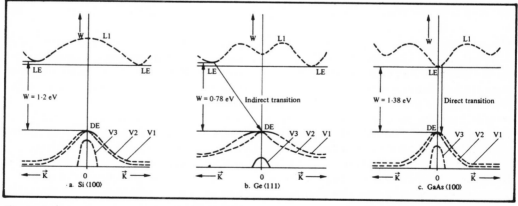

Figure 1.4
Band diagrams in the energy-momentum graph. The abbreviations denote LE = Conduction electron, DE = hole, W = Energy, $\vec{K}$ = wave number vector, L1 = Conduction band, V1, V2, V3 = Valency bands, $\langle 100 \rangle$, $\langle 111 \rangle$ = Miller's indices[1]

1   Miller's indices are the reciprocal values of the points of intersection of the crystal axes with cut surface.

of the same order of magnitude as the electron momentum. However, a small yield of photons is also possible here with a few materials. *Figure 1.4* a shows the W ($\vec{K}$) diagram of silicon. If can be seen from the curve shape, that it is an indirect semiconductor. Injection luminescence does not occur with silicon. The conditions are similar with germanium (*Figure 1.4 b*). Although there is a minimum of electron energy at $\vec{K}$ = 0 here, no radiation emission occurs, since the conduction electrons are to be found in the deeper energy minima at the sides, from where a direct transition is not possible.

In contrast to the semiconductors just described, gallium arsenide (GaAs) shows a deepest electron energy minimum at $\vec{K}$ = 0, directly opposite the maximum in the conduction band (*Figure 1.4 c*). As a direct semiconductor, GaAs is capable of converting the energy liberated on recombination into radiation. From the width of the forbidden band $\Delta W = 1 \cdot 38$ eV, using the equation (1.4), the wavelength $\lambda = 898$ nm is obtained. This radiation lies in the infrared range and is therefore not visible to the human eye. It is, however, important, that silicon photodetectors are particularly sensitive just in this wavelength range, so that GaAs light-emitting or luminescent diodes find many applications in opto-electronics in combination with Si detectors.

By suitable doping of the intrinsic or undoped semiconductor materials the performance in terms of the radiated wavelength or the efficiency, can be varied. With pure GaAs, the radiation efficiency is of the order of a few percent and increases somewhat with silicon doping, for example to 12% with the infra-red emitting diode TIXL 12 (at room temperature), while the wavelength increases at the same time to 925 nm.

A further important material for light-emitting diodes is gallium phosphide (GaP), an indirect semiconductor, from which light-emitting diodes with good quantum yield

can be produced. This is achieved by adding the so-called isoelectronic centres in addition to the donors and acceptors. With nitrogen additions, green light is obtained, while zinc-oxygen produces red light. The best published quantum yields lie around 10% for red and around 1% for green.

A partial replacement of the arsenic by phosphorus in gallium arsenide gives a mixed crystal, which is used as gallium arsenide phosphide (GaAsP) in many light-emitting diodes. Depending on the proportion of phosphorus, radiation in the spectral range red to yellow is achieved. Up to a phosphorus-arsenic ratio of about 4 : 6, GaAsP is a direct semiconductor, which emits red light. With increasing phosphorus content, the material changes to an indirect semiconductor, the radiation wavelength becomes shorter (yellow) and the radiation efficiency falls. Exactly as GaP, however, the decreasing efficiency with shorter-wavelength light is substantially compensated for by the increased sensitivity of the eye. Very often, a mixed crystal with the composition $GaAs_{0.56}P_{0.44}$, which emits radiation at a wavelength of 650 nm with an efficiency of 0.1%, is used. The luminance achieved is still completely adequate, even in brightly lit rooms.

By means of special design measures, it is possible to build semiconductor lasers from GaAs of GaAlAs with a small aluminium concentration. Optical feedback is achieved through the use of parallel cleavage surfaces perpendicular to the plane of the PN junction or alternatively through plane surface grinding. If the threshold current density is exceeded with such devices, they emit coherent beams of radiation. Typical values of the threshold current density lie around $15 \cdot 10^3$ A/cm^2. With special GaAlAs heterostructures, threshold current densities under 10^3 A/cm^2 are achieved; such injection lasers can then run at room temperature in continuous pulse operation.

Efforts are still being made, to search for new materials which will produce practical light-emitting diodes, particularly at either end of the currently achievable spectrum range. For a blue luminescent diode, useful results are expected from the material gallium nitride (GaN), but various technological problems still have to be solved here.

In future developments in this field, the so-called II-VI compounds should also have a certain part to play. These have long been used as phosphorescent materials (e.g., the well-known fluorescent materials for television and oscilloscope tubes). They include, among others: Barium sulphide (BaS), Barium selenide (BaSe), Barium Telluride (BaTe), Cadmium sulphide (CdS), Cadmium selenide (CdSe), Zinc sulphide (ZnS), Zinc selenide (ZnSe). Table 1.2 shows, among other factors, the band energy spacing ΔW, the wavelength and the transition type of various semiconductors, which are of importance for optoelectronics.

The highest efficiencies which have so far been achieved for radiation output lie in the infrared range. Using special phosphorescent materials with lanthanides as doping elements, it is possible, by double excitation of electrons, to convert the infrared radiation of a GaAs diode into visible green light. Although this radiation conversion process has only a low efficiency, the luminance of the phosphor achieved is sufficient, because of the high green sensitivity of the eye. *Figure 1.5* shows a luminescent diode working on this principle.

A characteristic property of luminescent diodes is the relatively narrow-band spectral range of the light quanta emitted. This can also be seen from *Figure 1.6*, in which the emission power of various semiconductor materials, and also the spectral sensitivity of a silicon photo-detector and that of the human eye, are plotted against the wavelength.

It should also be mentioned, that luminescent diodes, depending on their doping and manufacture, can be modulated up to very high frequencies (see later chapter on rise and fall times).

Finally, in *Figure 1.7*, the construction principle of a planar gallium arsenide phosphide luminescent diode is shown. The PN junction must lie as closely as possible under the surface, so that a good

Material	Band energy spacing (eV)	Wavelength at 300 K (μm)	Radiation equivalent (Lm/W)	Radiation range	Transition type
Ge	0·66				Indirect
Si	1·09				Indirect
SiC	2·5	0·496	200	Blue	Indirect
InSb	0·18	6·9		IR − C	Direct
InAs	0·36	3·45		IR − C	Direct
GaSb	0·7	1·77		IR − B	Direct
InP	1·26	0·985		IR − A	Direct
GaAs	1·38	0·898		IR − A	Direct
GaAsP	1·90	0·65	70	Red	Direct
GaP	2·19	0·565	590	Green	Indirect
GaP	1·8	0·69	5·5	Red	Indirect
GaN	3·1	0·4	0·3	Violet	Indirect

Table 1.2
Band energy spacing and wavelength of various semiconductors

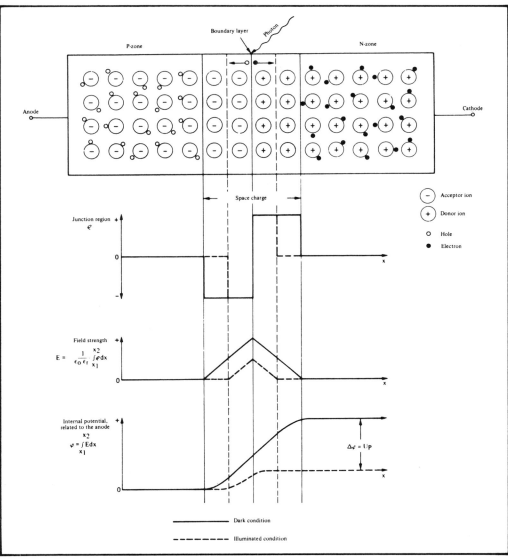

Figure 1.10
Schematic construction of a semiconductor photo-diode

used as radiation detectors. Without any irradiation, only a reverse current flows which is generally negligibly small, this is the so-called dark current. With increasing irradiation of the junction, the number of additional free charge carriers thus produced increases and raises the reverse current quite considerably. The selection of the operating mode — photocell or operation with reverse voltage — mainly depends on the application and on the matching conditions to the subsequent circuit.

A further component in this category is the phototransistor. It is several times more sensitive than the photodiode, but its upper limiting frequency is lower. As with the diode, at first a photocurrent is produced by irradiation of the system, but this is considerably amplified by the transistor effect. Phototransistors with the base connection brought out offer a number of additional possibilities for varying the mode of operation, e.g., adjustment of the working point, increasing the cut-off frequency with a simultaneous reduction of the sensitivity, photo-diode operation by not using either the emitter or collector connection.

As a result of the geometrical dimensions, the collector-base diode generally has a higher photosensitivity than the emitter-base diode. As with every normal silicon-transistor, the reverse emitter-base breakdown voltage is only about 6 V, because of the emitter doping which is higher than that of the base and the collector. Furthermore, both junctions can also be used as photocells.

Finally, as a very sensitive photoelectric component, let us also mention the Darlington phototransistor, which, as well as the actual phototransistor, also in one case contains an additional emitter follower.

2
Principles of calculation in radiation physics and optics

Principles of calculation in radiation physics and optics

The definitions of the physical and optical values have been compiled strictly on the basis of the DIN Standards 1301, 5031, 5033 and 44020, as revised from 1964 to 1972. For the practical worker in formulae listed from the DIN standards have also been shown in simplified form under certain stated conditions.

For convenience the optical spectrum can be divided into the two invisible parts (IR and UV) and the visible part (see Section 1.1). Historically and for practical purposes the visible part was of chief importance. The parameters defining this part of the spectrum are defined as "photometric" units. The definition of an optical (photometric) unit includes the weighting in accordance with the ocular sensitivity $V_{(\lambda)}$ laid down by the CIE (International

In this chapter we also define radiometric units for the optical spectrum, which are by analogy also applicable, to the photometric units. For example: The radiant flux is termed photometrically as luminous flux and is defined as follows: The luminous flux is the radiant flux emitted, which, when falling on the retina of the eye, causes an impression of light, weighted in accordance with the spectral sensitivity $V_{(\lambda)}$ of the eye.

The symbols for the radiometric units are given the index "e" (energetic), those for the photometric values the index "v" (visible).

2.1
Radiant flux – Luminous flux, Radiant energy – Luminous energy

The radiant flux or radiant power ϕ_e is the quotient of the radiant energy or quantity of radiation dW_e, transmitted by radiation, divided by the time dt. The total radiant energy emitted in all direction by a radiation source in one second is called the total radiant flux.

$$\Phi_e = \frac{dW_e}{dt} \qquad (2.1)$$

The radiant energy W_e is obtained by integration of the radiant flux Φ_e over the time t.

$$W_e = \int \Phi_e \cdot dt \qquad (2.2)$$

With constant radiant energy W_e per unit time, the simplified form:

$$\Phi_e = \frac{W_e}{t} \qquad (2.3)$$

is obtained from equation (2.1).

If the radiant flux Φ_e is uniform over the time t, then the simplified form:

$$W_e = \Phi_e \cdot t \qquad (2.4)$$

is obtained from equation (2.2).

The unit of radiant flux Φ_e is the Watt.

$$[\Phi_e] = W \qquad (2.5)$$

The unit of radiant energy W_e is the watt-second.

$$[W_e] = Ws = J \qquad (2.6)$$

The spectral portion of the radiant flux, to which the human eye is sensitive, is called the luminous power or luminous flux Φ_v. The unit of luminous power is the Lumen.

$$[\Phi_v] = lm \qquad (2.7)$$

The unit of luminous energy W_v is the Lumen-second.

$$[W_v] = lms \qquad (2.8)$$

Table 2.1, below, shows a summary, in which the various definitions are listed.

2.2
Units related to the radiation source

By analogy with electronics, the radiation sources can be defined as "transmitters" and radiation detectors as "receivers". In the following sections, the units at the "transmitting end" will be dealt with.

2.2.1
Radiant emittance – Luminous emittance

The radiant emittance M_e is the quotient of the radiant flux $d\Phi_e$ emitted from the surface element dA_S, divided by the area of the surface element dA_S. Thus it is the density of the radiant flux per unit area. A plane surface element can only radiate in a hemisphere or a semi-infinite space. The following formula applies:

dA_S $M_e = \dfrac{d\Phi_e}{dA_S}$ $\qquad (2.9)$

The radiant flux Φ_e is obtained by integration of the radiant emittance M_e over all surface elements dA_S.

$$\Phi_e = \int M_e \cdot dA_S \qquad (2.10)$$

If the radiant flux Φ_e leaving the surface A_S is uniform over all surface elements dA_S, then (2.9) simplifies to:

$$M_e = \frac{\Phi_e}{A_S} \qquad (2.11)$$

If the radiant emittance M_e from every surface element dA_S is equal then the equation becomes, from (2.10):

$$\Phi_e = M_e \cdot A_S \qquad (2.12)$$

The unit of radiant emittance M_e is the watt per m^2

$$[M_e] = \frac{W}{m^2} \qquad (2.13)$$

The unit of luminous emittance M_v is the Lumen per m^2

$$[M_v] = \frac{lm}{m^2} \qquad (2.14)$$

These various definitions are summarised in *Table 2*.

	Parameter	Symbol	Dimension	Unit
radiometric	radiant flux radiant power	Φ_e (P_o)	power	W
photometric	luminous power luminous flux	Φ_v (P_v)	power	lm
radiometric	radiant energy	W_e	power x time	Ws
photometric	quantity of light (luminous energy)	W_v	power x time	lms

Table 2.1
Summary of radiant and luminous flux and radiant and luminous energy parameters.

	Parameter	Symbol	Dimension	Unit
radiometric	radiant exitance (radiant emittance)	M_e	power / area of active region	Wm^{-2} (Wcm^{-2})
photometric	luminous exitance (Luminous emittance)	M_v	power / area of active region	lmm^{-2}

Table 2.2
Summary of radiant emittance and luminous emittance paramters

2.2.2
Radiant intensity – Luminous intensity

The radiant intensity I_e is the quotient of the radiant flux $d\Phi_e$ leaving the radiation source in a given direction, divided by the solid-angle element $d\Omega$ covered by the radiation. Radiant intensity can only be associated with a radiation source which is considered as a point (see Section 3.3).

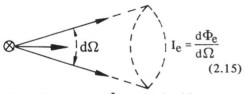

$$I_e = \frac{d\Phi_e}{d\Omega} \qquad (2.15)$$

The radiant power Φ_e obtained by integration of the radiant intensity I_e over the the solid angle elements $d\Omega$.

$$\Phi_e = \int_\Omega I_e \cdot d\Omega \qquad (2.16)$$

If the radiant power Φ_e is uniform in all solid-angle element $d\Omega$, the following simplified form applies:

$$I_e = \frac{\Phi_e}{\Omega} \qquad (2.17)$$

This is obtained from the equation (2.15).

If the radiant intensity I_e is the same in every solid-angle element $d\Omega$, then the simplified form applies:

$$\Phi_e = I_e \cdot \Omega \qquad 2.18)$$

The solid angle Ω is stated in steradians, or sterads, abbreviated as sr (see Sections 3.1 and 10.1):

$$[\Omega_o] = \frac{2\pi sr}{2\pi} = 1\,sr = sr \qquad (2.19)$$

The unit of radiant intensity is the watt per steradian.

$$[I_e] = \frac{W}{sr} \qquad (2.20)$$

The unit of luminous intensity is the Lumen per steradian or candela, abbreviated to cd:

$$[I_v] = \frac{lm}{sr} = cd \qquad (2.21)$$

A "black-body radiator" (see the chapter on radiation laws of the black body), with a temperature of $2042°K$ (since 1969, $2045°K$, the solidification point of platinum, applies) has a radiant intensity per cm^2 of surface in the normal direction of approx. $31\cdot3$ W/sr. Photometrically, this corresponds to a luminous intensity of 60 cd.

The various definitions are summaried in *Table 2.3*.

	Parameter	Symbol	Dimension	Unit
radiometric	radiant intensity	I_e	$\dfrac{power}{solid\ angle}$	Wsr^{-1}
photometric	luminous intensity	I_v	$\dfrac{power}{solid\ angle}$	$lmsr^{-1}$; cd

Table 2.3
Summary of radiant and luminous intensity parameters

2.2.3
Radiance – Luminance

The radiance L_e is the quotient of the radiant intensity dI_e of radiation which leaves or passes through a surface element dA_S in a given direction, divided by the projection $dA_S.\cos\varphi$ of the surface element, if φ is the angle between the direction of radiation and the normal to the surface.

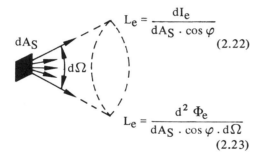

$$L_e = \frac{dI_e}{dA_S \cdot \cos\varphi} \qquad (2.22)$$

$$L_e = \frac{d^2\,\Phi_e}{dA_S \cdot \cos\varphi \cdot d\Omega} \qquad (2.23)$$

For the same given direction, the radiant intensity I_e is obtained by integration of the radiance L_e over the surface element dA_S:

$$I_e = \int_{A_S} L_e \cdot dA_S \cdot \cos\varphi \qquad (2.24)$$

The radiant emittance M_e is obtained by integration of the radiance L_e over the solid angle elements $d\Omega$:

$$M_e = \int_{\Omega} L_e \cdot d\Omega \qquad (2.25)$$

The radiant power Φ_e is obtained by integration of the radiance L_e over the

surface elements dA_S and the solid angle elements $d\Omega$:

$$\Phi_e = \int_{A_S} \int_{\Omega} L_e \cdot dA_S \cdot \cos\varphi \cdot d\Omega \qquad (2.26)$$

The limits of the integral of the radiance L_e over the solid angle Ω in equation (2.26) or of the radiant intensity I_e over the solid angle in equation (2.16) depend on the radiation angle of the radiation source (see radiant intensity distribution curve). With plane surface radiators, the integration is carried out over the semi-infinite space from 0 to 2πsr. In the case of sources radiating in all spatial directions, the integration is carried out over the infinite space from 0 to 4πsr. (See Sections 3.1 and 10.1).

If the radiant flux Φ_e leaving the surface A_S is uniform over all surface elements dA_s and in all solid angle elements $d\Omega$, then the simplified forms of equation (2.22) are obtained:

$$L_e = \frac{I_e}{A_S \cdot \cos\varphi} \qquad (2.27)$$

equation (2.23):

$$L_e = \frac{\Phi_e}{A_S \cdot \cos\varphi \cdot \Omega} \qquad (2.28)$$

If the radiance from all surface elements dA_S is the same in every solid angle element $d\Omega$ then the simplified forms are obtained from equation (2.24):

24

	Parameter	Symbol	Dimension	Unit
radiometric	radiance	L_e	$\dfrac{\text{power}}{\text{area of active region . solid angle}}$	$\dfrac{\text{Watt}}{\text{m}^2 \text{. sr}}$ $\left(\dfrac{\text{Watt}}{\text{cm}^2 \text{. sr}}\right)$
photometric	luminance	L_v	$\dfrac{\text{power}}{\text{area of active region . solid angle}}$	ftLa $\dfrac{\text{Candela}}{\text{m}^2}$

Table 2.4
Summary of radiance and luminance parameters

$$I_e = L_e \cdot A_S \cdot \cos \varphi \qquad (2.29)$$

equation (2.25)

$$M_e = L_e \cdot \Omega \qquad (2.30)$$

and equation (2.26):

$$\Phi_e = L_e \cdot A_S \cdot \cos \varphi \cdot \Omega \qquad (2.31)$$

The unit of radiance is the Watt per (m² steradian):

$$[L_e] = \frac{W}{m^2 \cdot sr} \qquad (2.32)$$

The unit of luminance is the Lumen per (m² steradian) or Candela per m²:

$$[L_v] = \frac{lm}{m^2 \cdot sr} = \frac{cd}{m^2} \qquad (2.33)$$

All important parameters are summarised in *Table 2.4*

Unit	cd . m⁻²	asb	cd . cm⁻², sb	L	cd . ft⁻²	fL	cd . in⁻²	Remarks
1 cd . m⁻²	1	π	10^{-4}	$\pi \cdot 10^{-4}$	$9 \cdot 29 \cdot 10^{-2}$	$0 \cdot 2919$	$6 \cdot 45 \cdot 10^{-4}$	Instead of cd . m⁻² also Nit
1 asb (Apostilb)	$\dfrac{1}{\pi}$	1	$\dfrac{1}{\pi} \cdot 10^{-4}$	10^{-4}	$2 \cdot 957 \cdot 10^{-2}$	$0 \cdot 0929$	$2 \cdot 054 \cdot 10^{-4}$	
1 sb = 1 cd . cm⁻²	10^4	$\pi \cdot 10^4$	1	π	929	2919	$6 \cdot 452$	1 in = 2·54 cm 1 ft = 0·3048 m
1 L (Lambert) = cd . cm⁻² . π	$\dfrac{1}{\pi} \cdot 10^4$	10^4	$\dfrac{1}{\pi}$	1	$2 \cdot 957 \cdot 10^2$	929	$2 \cdot 054$	12 in = 1 ft
1 cd . ft⁻²	$10 \cdot 764$	$33 \cdot 82$	$1 \cdot 076 \cdot 10^{-3}$	$3 \cdot 382 \cdot 10^{-3}$	1	π	$6 \cdot 94 \cdot 10^{-3}$	1 yd = 3 ft
1 fL (Footlambert) cd . ft⁻² . π^{-1}	$3 \cdot 426$	$10 \cdot 764$	$3 \cdot 426 \cdot 10^{-4}$	$1 \cdot 0764 \cdot 10^{-3}$	$\dfrac{1}{\pi}$	1	$2 \cdot 211 \cdot 10^{-3}$	ft² = 929 cm² ft² = 0·0929 m² 1m² = 10·764 ft²
1 cd . in⁻²	1550	4869	$0 \cdot 155$	$0 \cdot 4869$	144	$452 \cdot 4$	1	1 in² = 6·4516 cm² 1 ft² = 9·2903 dm²

Example: 1 asb (left-hand column) = (right-hand columns) $\dfrac{1}{\pi} \cdot 10^{-4}$ sb = $\dfrac{1}{\pi} \cdot 10^{-4} \dfrac{cd}{cm^2} = \dfrac{1}{\pi} \cdot \dfrac{cd}{m^2}$

Table 2.5
Conversion factors for the various common units of luminance

2.2.4
Units of luminance

In practice, the need for mathematical determination of the luminance arose at an early date. Therefore consequently there are still several commonly-used units of luminance, which originated in past history. The most common will be explained below and the conversion factors will be stated in *Table 2.5*.

The SI Unit
is now specified by law and it is calculated

in $\dfrac{\text{lm}}{\text{m}^2 \cdot \text{sr}}$ or $\dfrac{\text{cd}}{\text{m}^2}$.

In the literature and data-sheets, however, the following time-honoured measurement units are often to be found.

The Stilb (sb)
is obtained from the unit $\dfrac{\text{cd}}{\text{m}^2}$ by means of the conversion factor 10^4 and relates to the unit of area 1 cm^2.

$$1 \text{ sb} = 1 \frac{\text{cd}}{\text{cm}^2} = 10^4 \frac{\text{cd}}{\text{m}^2}$$

The Apostilb (asb)
takes direct account, with the factor π, of the radiation conditions from a plane surface element of the Lambertian radiator. The luminance of reflecting surfaces, among others, was previously measured in "asb".

$$1 \text{ asb} = \frac{1}{\pi} \cdot \frac{\text{cd}}{\text{m}^2}$$

The Lambert (L)
also takes account, with the factor π, of the radiation conditions of the Lambert radiator. It is greater than 1 asb by the factor 10^4.

$$1 \text{ L} = \frac{1}{\pi} \cdot 10^4 \frac{\text{cd}}{\text{m}^2}$$

cd/ft²
belongs to the same group and the Stilb,

while 1 cd/ft^2 relates to $1 \text{ ft}^2 = 929 \text{ cm}^2$ and the Stilb relates to 1 cm^2 of radiator surface.

$$1 \frac{\text{cd}}{\text{ft}^2} = 10 \cdot 764 \frac{\text{cd}}{\text{m}^2}$$

The Foot-Lambert (fL)
also relates to 1 ft^2 and also, like the Apostilb, has the solid angle $\Omega = \pi$ sr included.

$$1 \text{ fL} = 3 \cdot 426 \frac{\text{cd}}{\text{m}^2}$$

cd/in²
belongs to the same group as the Stilb, while it relates to $1 \text{ inch}^2 = 6 \cdot 4516 \text{ cm}^2$ of radiator area.

$$1 \frac{\text{cd}}{\text{in}^2} = 1550 \frac{\text{cd}}{\text{m}^2}$$

2.2.5
Radiant efficiency and luminous efficiency

The radiant efficiency η_e of a radiation source is the quotient of the total radiant power Φ_e produced, divided by the input power P.

The luminous efficiency η_v of a light source is the quotient of the total luminous power Φ_v produced (total luminous flux) divided by the input power P.

$$\eta_e = \frac{\Phi_e}{P} \tag{2.34}$$

$$\eta_v = \frac{\Phi_v}{P} \tag{2.35}$$

For the input power P and the radiant power Φ_e, the watt is the unit used. The luminous power is measured in Lumens. The radiant efficiency is usually stated as a percentage. The luminous efficiency is quoted in the unit Lumens per watt.

26

$$[\eta_v] = \frac{Lm}{W} \tag{2.36}$$

In the case of a black-body radiator, the luminous efficiency corresponds to the photometric radiant equivalent of the black body (see conversion of radiometric into photometric values).

The radiant efficiency η_e of a radiation source in the wavelength range from λ_1 to λ_2 is the ratio of the radiant flux emitted in this range to the power needed for its generation (for spectral values, see Section 2.3.4).

$$\eta_e = \frac{\int\limits_{\lambda_1}^{\lambda_2} \Phi_{e,\lambda} \cdot d\lambda}{P} \tag{2.37}$$

2.3
Units related to the receiver

Radiation detectors acting as receivers of radiation will be termed "receivers" for brevity. The units related to them will be explained below.

2.3.1
Irradiance – Illuminance

The irradiance E_e is the quotient of the radiant flux $d\Phi_e$ divided by the irradiated surface element dA_E

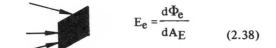

$$E_e = \frac{d\Phi_e}{dA_E} \tag{2.38}$$

The radiant power Φ_e falling on the receiver area A_E is obtained by integration of the irradiance E_e over all surface elements dA_E.

$$\Phi_e = \int\limits_{A_E} E_e \cdot dA_E \tag{2.39}$$

If the incident radiant power Φ_e is equal on all surface elements dA_E, then the simplified form:

$$E_e = \frac{\Phi_e}{A_E} \tag{2.40}$$

applies for equation (2.38).

If the irradiance E_e is equal on all surface elements dA_E, then the simplified form:

$$\Phi_e = E_e \cdot A_E \tag{2.41}$$

applies for equation (2.39).

The unit of irradiance is the watt per m^2:

$$[E_e] = \frac{W}{m^2} \tag{2.42}$$

The irradiance E_e must not be confused with the radiant emittance M_e, since both values have the same unit.

Unit		lx	lm . cm^{-2}	fc	Remarks
1 lx (Lux)	=	1	10^{-4}	0·0929	1 ft^2 = 0·0929 m^2
1 lm.cm^{-2}	=	10^4	1	$0·0929 . 10^4$	Previously alsp Phot (ph) instead of lm . cm^{-2}
1 fc (footcandle)	=	10·764	$10·764 . 10^{-4}$	1	1 m^2 = 10·764 ft^2

Table 2.6
Conversion factors for the various common irradiance units

	Parameter	Symbol	Dimension	Unit
radiometric	irradiance	E_e	$\dfrac{\text{power}}{\text{area}}$	Wm^{-2} (Wcm^{-2})
photometric	illuminance	E_v	$\dfrac{\text{power}}{\text{area}}$	foot candela $lm \cdot m^{-2}$

Table 2.7
Summary of irradiance and illuminance parameters

The unit of illuminance is the Lumen per m^2 or Lux:

$$[E_v] = \frac{lm}{m^2} = lx \tag{2.43}$$

In *Table 2.6*, the conversion factors are listed for the various common units, while in *Table 2.7*, the most important terms are again summarised.

2.3.2
Irradiation – Light exposure

The irradiation H_e is the integral of the irradiance E_e over the time t.

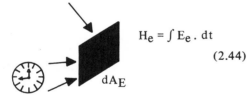

$$H_e = \int E_e \cdot dt \tag{2.44}$$

If the irradiance E_e is constant during time t, then the simplified form:

$$H_e = E_e \cdot t \tag{2.45}$$

applies for equation (2.44).

The measurement unit for irradiation is the watt-second per m^2:

$$[H_e] = \frac{Ws}{m^2} \tag{2.46}$$

The measurement unit for light exposure is the Lumen-second per m^2 or the Lux-second.

$$[H_v] = \frac{lms}{m^2} = lxs \tag{2.47}$$

This data is summarised in *Table 2.8*.

	Parameter	Symbol	Dimension	Unit
radiometric	radiant exposure (irradiation)	H_e	$\dfrac{\text{energy}}{\text{area}}$	Wsm^{-2}
photometric	light exposure	H_v	$\dfrac{\text{luminous energy}}{\text{area}}$	$\dfrac{lm \cdot s}{m^2} ; lx \cdot s$

Table 2.8
Summary of irradiation and light exposure parameters

2.3.3
Relationshio between irradiance or illuminance and the reflected radiance or luminance

Originally, the reflected luminance $L_{v,r}$ from a diffusely reflecting plane surface A (see Section 6.3, Reflection of radiation) could be determined very simply from the incident illuminance E_v, since the product of the illuminance E_v in Lux and the reflectivity ρ gave the reflected luminance $L_{v,r}$ in Apostilbs. Hence the reflected luminance or reflected radiance calculation will be carried out with new and old measurement units, for greater ease of understanding. The reflected luminance or radiance value will be needed typically for calculations involving reflection opto-couplers. Here, only a uniformly diffusely relecting plane surface A_r will be considered as the reflecting light or radiation source. The reflected radiance $L_{e,r}$ is firstly dependent on the irradiance E_e, or the reflected luminance $L_{v,r}$ on the illuminance E_v and the reflectivity ρ. The irradiance or illuminance is calculated from the equation (2.40). The reflectivity (see Chapter 6) is the ratio of the reflected radiant flux $\Phi_{e,r}$ to the incident radiant flux $\Phi_{e,o}$ or of the reflected luminous flux $\Phi_{v,r}$ to the incident luminous flux $\Phi_{v,o}$.

$$\rho = \frac{\Phi_r}{\Phi_o} \tag{2.48}$$

The reflected radiance $L_{e,r}$ or the reflected luminance $L_{v,r}$ is calculated in accordance with equation (2.28):

$$L_r = \frac{\Phi_r}{A_r \cdot \cos\varphi \cdot \Omega} \tag{2.49}$$

If the radiant flux $\Phi_{e,o}$ falling on the reflecting surface A_r, or the luminous flux $\Phi_{v,o}$, falls perpendicularly on the surface A_r, the factor $\cos\varphi = 1$ and can therefore be omitted:

$$L_r = \frac{\Phi_r}{A_r \cdot \Omega} \tag{2.50}$$

Because Lambert's law applies, the solid angle Ω is equal to $\pi \cdot \Omega_o$ (see Sections 2.24, 3.2, 6.3 or DIN 5469, Item 1.8). If Ω is replaced by $\pi \cdot \Omega_o$, then equation (2.50)

$$L_r = \frac{\Phi_r}{A_r \cdot \pi \cdot \Omega_o} \tag{2.51}$$

If the reflected radiant flux $\Phi_{e,r}$ or the reflected luminous flux $\Phi_{v,r}$ is replaced by the equations (2.40) and (2.48), then the final simplified form becomes:

$$L_r = \frac{\rho \cdot E \cdot A_r}{A_r \cdot \pi \cdot \Omega_o} = \frac{\rho \cdot E}{\pi \cdot \Omega_o} \tag{2.52}$$

In the obsolescent measurement unit, the Apostilb, the factor πsr is taken into account, so that for the previously-mentioned calculation of the reflected illuminance $L_{v,r}$ in Apostilbs, the following formula applies:

$$\frac{L_{v,r}}{\text{asb}} = E_{v,r} \cdot \rho \tag{2.53}$$

For the calculation of the reflected luminance $L_{v,r}$, the equation (2.52) reads:

$$L_{v,r} = \frac{\rho \cdot E_v}{\pi \cdot \Omega_o} \tag{2.54}$$

For the calculation of the reflected radiance $L_{e,r}$, the equation (2.52) becomes:

$$L_{e,r} = \frac{\rho \cdot E_e}{\pi \cdot \Omega_o} \tag{2.55}$$

2.3.4
Spectral radiation-distribution units

The definitions stated so far for physical or radiometric units such at radiant flux

Φ_e, radiant intensity I_e, radiant emittance M_e, radiance L_e and irradiance E_e do not relate to the spectral distribution of the radiation. The spectral radiation distribution is taken into account by the "spectral density of the radiation-distribution units". These units are related either to a differential range of the wavelength $d\lambda$ or of the frequency $d\nu$ i.e. they are functions of the wavelength λ or of the frequency ν.

Example:
The spectral radiance $L_{e,\lambda}$ is the proportion of the radiance dL_e related to a small wavelength interval $d\lambda$.

$$L_{e,\lambda} = \frac{dL_e}{d\lambda} \qquad (2.56)$$

Through integration of the spectral radiance $L_{e,\lambda}$ over all wavelength intervals $d\lambda$, the radiance L_e is obtained:

$$L_e = \int_0^\infty L_{e,\lambda} \cdot d\lambda \qquad (2.57)$$

By identification with the index λ for the wavelength or ν for the frequency, the symbols for the radiation-distribution units are converted into those for the spectral radiation distribution units (*Table 2.9*).

Provided there is no risk of confusion with other spectral units, the designation "spectral density of a radiation-distribution unit" can be shortened to "spectral unit". For example, the "spectral density of the radiant flux" can be abbreviated to "spectral radiant flux".

Parameter	Formula	Practical unit	
Spectral irradiation	$H_{e,\lambda} = \dfrac{dH_e}{d\lambda}$	$Wsm^{-2} (nm^{-1})$	(2.58)
Spectral radiance	$L_{e,\lambda} = \dfrac{dL_e}{d\lambda}$	$Wm^{-2} sr^{-1} (nm^{-1})$	(2.56)
Spectral radiant power	$\Phi_{e,\lambda} = \dfrac{d\Phi_e}{d\lambda}$	$W (nm^{-1})$	(2.59)
Spectral radiant power	$\Phi_{e,\nu} = \dfrac{d\Phi_e}{d\nu}$	$W (Hz^{-1})$	(2.60)
Spectral radiant intensity	$I_{e,\lambda} = \dfrac{dI_e}{d\lambda}$	$Wsr^{-1} (nm^{-1})$	(2.61)
Spectral radiant emittance	$M_{e,\lambda} = \dfrac{dM_e}{d\lambda}$	$Wm^{-2} (nm^{-1})$	(2.62)
Spectral irradiance	$E_{e,\lambda} = \dfrac{dE_e}{d\lambda}$	$Wm^{-2} (nm^{-1})$	(2.63)

Table 2.9
Summary of spectral units

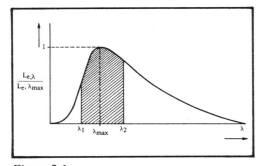

Figure 2.1
Relative spectral radiance distribution of a
Planck radiator as a function of λ

Figure 2.1 shows the relative spectral
radiance distribution of a Planck's radiator
as a function of the wavelength λ. The
relative spectral radiance $L_{e,\lambda,\text{rel}}$ is the
quotient of the spectral radiance $L_{e,\lambda}$ at the
wavelength λ, divided by the maximum
spectral radiance $L_{e,\lambda,\text{max}}$ at the maximum
wavelength λ_{max}:

$$L_{e,\lambda,\text{rel}} = \frac{L_{e,\lambda}}{L_{e,\lambda\text{max}}} \qquad (2.64)$$

The proportional radiance q is the quotient
of the integral of the spectral radiance
$L_{e,\lambda}$ over the wavelength range between
λ and λ_2 divided by the integral of the
spectral radiance $L_{e,\lambda}$ over the whole
wavelength range:

$$q = \frac{\int_{\lambda_1}^{\lambda_2} L_{e,\lambda} \cdot d\lambda}{\int_0^\infty L_{e,\lambda} \cdot d\lambda} \qquad (2.65)$$

This equation is most easily solved either
graphically, with a special slide rule, or
scientific calculator.

The proportional radiance q is a relative
value. In practice, it is often sufficient to
use relative calculations and models of the
spectral distribution of the radiation. In
these it is immaterial, which radiation unit
is taken into account.

For radiation such as that from a temperature
radiator with a known distribution
temperature T_v (see Chapter 4), the
relative spectral radiation distribution $S\lambda$
has been introduced as a relative unit for
each wavelength interval $d\lambda$. $S\lambda$ will also
be called a radiation function. If the
spectral radiation distribution units are
represented generally by $X\lambda$, then the
radiation function $S\lambda$ is the quotient of the
value $X\lambda$ at the wavelength λ, divided by
the maximum value $X\lambda_{\text{max}}$ at the
wavelength λ_{max}.

$$S\lambda = \frac{X\lambda}{\dot{X}\lambda_{\text{max}}} \qquad (2.66)$$

The radiation leaving any radiation source
is exactly described in its spectral distribution
by the radiation function $S\lambda$. As an
example, let us mention the standard light
source (as defined in DIN specs) which is
very important in optoelectronics. The
spectral radiation distribution $S(A)\lambda$ of the
standard light A has been stated in the
wavelength range from 320 nm to 780 nm
in a table in DIN 5033, Part 7.

Since, for standard light A, the value for
$S(A)_{560\,\text{nm}}$ at the wavelength $\lambda = 560$ nm
is made equal to 100, the equation (2.65)
must be completed by a constant factor.

$$S(A)\lambda = \frac{X(A)\lambda}{X(A)\lambda_{\text{max}}} \cdot \text{const.}_A \qquad (2.67)$$

In general, the standard light A is produced
by a temperature radiator with distribution
temperature $T_v = 2856^\circ$K. The values for $S\lambda$
of this temperature radiation can be
determined, apart from the calculation
stated in Chapter 4, with the "Radiation
Calculator" available from GE. At $T_v =
2856^\circ$K and $\lambda = 560$ nm, the value for
$S\lambda \approx 0.35$ is read off. The constant
which applies for the standard light A can
be calculated with the formula (2.68).

$$\text{const.}_A = \frac{S(A)_{560\,\text{nm}}}{S_{560\,\text{nm}}} = \frac{100}{0.35} = 286 \qquad (2.68)$$

31

With the aid of this constant, the value for $S(A)_\lambda$ can be calculated from S_λ.

$$S(A)_\lambda = S_\lambda \cdot \text{const.}_A \qquad (2.69)$$

At a wavelength $\lambda = 660$ nm, the value for $S_\lambda \approx 0.6$ is read off. The value for $S(A)_{660 \text{ nm}}$ amounts to:

$$S(A)_{660\text{nm}} = S_{660\text{nm}} \cdot \text{const.}_A = 0.6 \cdot 286 = \underline{\underline{171.6}} \qquad (2.70)$$

This value agrees approximately with the DIN standard value for $S(A)_{660 \text{ nm}} = 171.96$.

Since the spectral distribution of Planck radiators is described unambiguously by the distribution temperature T_V (see Chapter 4), the radiation function S_λ for these radiators is often to be found as a function of T_V and λ in books of tables.

3
Laws of radiation

3

Laws of radiation

3.1
Solid angle

The solid angle is a sterometric value. *Figure 3.1* shows a spherical sector of a sphere. The value of the solid angle Ω of a spherical sector (*Figure 3.2*) is determined by the ratio of the section of the spherical surface A to the square of the radial distance r. From this:

$$\Omega = \frac{A}{r^2} \cdot \Omega_0 \tag{3.1}$$

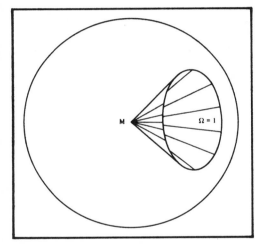

Figure 3.1
Sector of a sphere

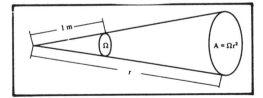

Figure 3.2
Calculation of the solid angle Ω of a spherical sector

The unit of the solid angle is the steradian, abbreviated to sr. The general dimension of the solid angle is, in accordance with

equation (3.1), area per area. In radiation calculations, the possibilities arise, that either the unit sr is not converted with respect to m^2/m^2, or m^2/m^2 is not converted into sr. The correction factor $\Omega_0 = 1$ sr, which is inserted, prevents such calculation errors (see equation (2.19) and section 10.1).

The greatest possible solid angle is formed by the sphere. If the radial distance is related to r = 1, e.g., r = 1 m, then the solid angle is:

$$\Omega = \frac{4\pi\,m^2}{1\,m^2} \cdot \Omega_0 = 4\pi \cdot \Omega_0 = 4\pi\;sr \tag{3.2}$$

Semi-infinite space, or a hemisphere with unit radius r = 1 m, has the solid angle:

$$\Omega = \frac{2\pi\,m^2}{1\,m^2} \cdot \Omega_0 = 2\pi \cdot \Omega_0 = 2\pi sr\cdot\;sr \tag{3.3}$$

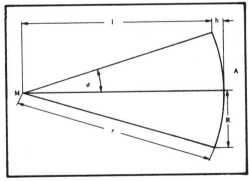

Figure 3.3
Two-dimensional representation of a spherical sector. A = Area cut out of the spherical surface by the spherical sector (spherical cap), h = height of the section, R = radius of the circumscribing circle, φ = half aperture angle, r = radial distance (from the centre to the surface of the sphere) = 1, l = r − h.

Figure 3.3 shows a two-dimensional representation of a spherical sector. Thus the solid angle Ω can be calculated with the half aperture angle φ.

$$\Omega = \frac{A}{r^2} \cdot \Omega_0 = \frac{2r \cdot \pi \cdot h}{r^2} \cdot \Omega_0$$

$$= \frac{2r^2 \cdot \pi (1-\cos\varphi)}{r^2} \cdot \Omega_0$$

$$= 2\pi (1-\cos\varphi) \cdot \Omega_0$$

$$(3.4)$$

With the values R and 1, the solid angle can also be calculated:

$$\Omega = \frac{A}{r^2} \cdot \Omega_0 = \frac{2\pi r (r-l)}{r^2} \cdot \Omega_0$$

$$= 2\pi \left(1 - \frac{1}{\sqrt{R^2 + l^2}}\right) \Omega_0$$

$$= 2\pi \left(1 - \frac{1}{\sqrt{\dfrac{R^2}{l^2} + 1}}\right) \cdot \Omega_0$$

$$(3.5)$$

To clarify this, *Figure 3.4* shows the relationship between the solid angle Ω and the half aperture angle φ. For example:

$$1 \text{ sr} = \varphi = 32.72°$$

3.2
Lambert radiator

The lambert radiator is an imaginary, ideal, exactly diffusely radiating black-body radiator (see Laws of Radiation). These conditions are satisfied by a uniformly heated black spherical radiator.

The Lambert radiator has a uniform radiance in all directions (*Figure 3.5*). Therefore, with this radiation source, the calculations are very simple

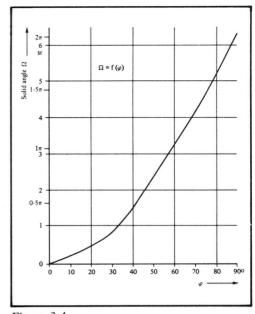

Figure 3.4
Dependence of the solid angle Ω on the half aperture angle φ of a spherical sector

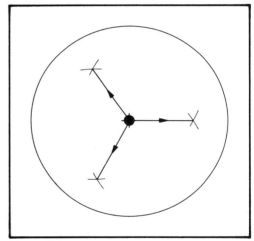

Figure 3.5
Radiation diagram of the Lambert radiator

3.2.1
Fundamental law of photometry

The radiances of the surface elements of a Lambert radiator are equal, even if they are observed at different angles. The areas A_{S1} and A_{S2} are simplified plane surfaces. From *Figure 3.6*:

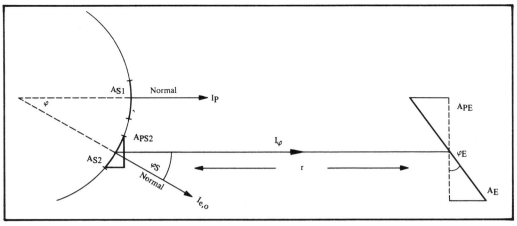

Figure 3.6
Section of the radiator surface of a Lambert radiator

$$L_{e1} = L_{e2} = \frac{I_{e,o}}{A_{S1}} = \frac{I_{e,o}}{A_{S2}} = \frac{I_{e,\varphi}}{A_{PS2}}$$

$$= \frac{I_{e,\varphi}}{A_{S2} \cdot \cos \varphi_S} \qquad (3.6)$$

$I_{e,o}$ = Radiant intensity in the normal direction

$I_{e,\varphi}$ = Radiant intensity in the direction of angle φ

(Suffixes: S = Transmitter, E = Receiver, P = Projected).

If the power radiated at the angle φ_S from A_{S2} falls perpendicularly at a great distance r on the small receiver surface A_{PE}, then the incident radiant power is:

$$\Phi_e = L_e \cdot \cos\varphi_S \cdot A_{S2} \cdot \Omega$$

$$= \frac{L_e \cdot \cos\varphi_S \cdot A_{S2} \cdot A_{PE}}{r^2} \cdot \Omega_o \qquad (3.7)$$

If the receiver surface element A_{PE} is a projection of A_E, the incident radiant power is:

$$\Phi_e = L_e \frac{A_{S2} \cdot \cos \varphi_S \cdot A_E \cdot \cos \varphi_E}{r^2} \cdot \Omega_o \qquad (3.8)$$

The fundamental law of photometry can only be expressed exactly with higher mathematics. It reads:

$$d^2 \Phi_e = L_e \frac{dA_{S2} \cdot \cos \varphi_S \cdot dA_E \cdot \cos \varphi_E}{r^2} \cdot \Omega_o \qquad (3.9)$$

In this form, in addition, it is restricted to vacuum only. The distance r should always be greater than the photometric limiting distance (see Section 3.3).

3.2.2
Lambert's Cosine Law

The proportionality of the radiated and of the incident radiant power to $\cos \varphi$ can be determined by experiment. This is called Lambert's cosine law (*Figure 3.7*).

In this diagram, dA_S is a surface element of a Lambert radiator. The radiant flux per element of solid angle, the radiant intensity, shows a maximum in the normal direction. The radiant flux per unit of solid angle decreases symmetrically on all sides, in proportion with $\cos \varphi$:

$$I_{e,\varphi} = I_{e,o} \cdot \cos \varphi \qquad (3.10)$$

The total radiant power φ_e of the surface element dA_S of a Lambert radiator is the

37

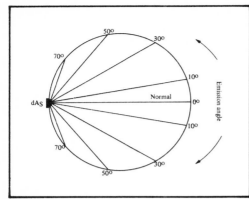

Figure 3.7
Lambert's cosine law

product of $I_{e,o}$ and the integral over all solid angle elements $d\Omega$ in the semi-infinite space.

$$\Phi_e = I_{e,o} \int \cos \varphi \, . \, d\Omega = I_{e,o} \, . \, \pi \, . \, \Omega_o$$
(3.11)

The equation (3.6) stated, that the radiance L_e of a small plane surface A_S (strictly speaking surface element dA_S) of a Lambert radiator is equal for all directions. Thus, from equation (2.25), one obtains, by

integration over semi-infinite space, for the radiant emittance M_e:

$$M_e = L_e \int \cos \varphi \, . \, d\Omega = L_e \, . \, \pi \, . \, \Omega_o$$
(3.12)

3.3
Calculation of radiation with small surface radiators and surface receivers

A small radiator surface can be regarded, in simplified form, as a point radiator, if the distance r to the receiver is very large in comparison with the maximum dimension of the radiator surface (at least a factor of 10). This minimum distance is called the photometric limiting distance. The surface radiator has its whole radiant flux in semi-infinite space. Therefore it is advantageous to work with the radiant emittance M_e, i.e., the total radiant flux of a surface element in semi-infinite space is related to the small finite area of the surface element dA_S.

The following relationships are taken as the basis for the radiation calculations:

Exact Formulae	Simplified Formulae
$$M_e = \frac{d\Phi_e}{dA_S}$$ (2.9)	In the case of homogeneous radiation as with the Lambert radiator: $$M_e = \frac{\Phi_e}{A_S}$$ (2.11)

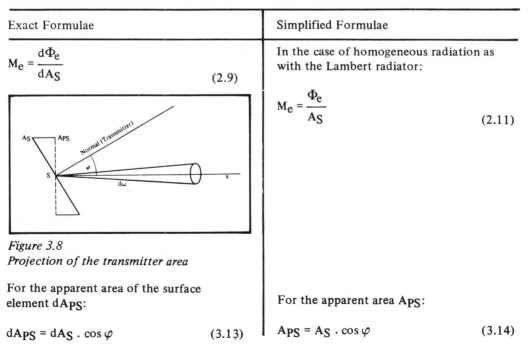

Figure 3.8
Projection of the transmitter area

For the apparent area of the surface element dA_{PS}:

$$dA_{PS} = dA_S \, . \, \cos \varphi \qquad (3.13)$$

For the apparent area A_{PS}:

$$A_{PS} = A_S \, . \, \cos \varphi \qquad (3.14)$$

38

For the radiant intensity in the direction $\vec{SE}$:

$$I_{e,\varphi} = I_{e,o} \cdot \cos\varphi \qquad (3.10)$$

For the radiance of the point S in the direction $\vec{SE}$ (see quation (2.22):

$$L_e = \frac{dI_{e,\varphi}}{dA_{PS}} = \frac{dI_{e,\varphi}}{dA_S \cdot \cos\varphi} \qquad (3.15)$$

$$L_e = \frac{d^2 \Phi_e}{dA_S \cdot \cos\varphi \cdot d\Omega} \qquad (2.23)$$

$$d^2 \Phi_e = L_e \, d\Omega \cdot dA_S \cdot \cos\varphi \qquad (3.16)$$

(see equation 2.26)

$$I_{e,\varphi} = I_{e,o} \cdot \cos\varphi \qquad (3.10)$$

With uniform radiance (see Section 2.2.3 and 3.2.1):

$$L_e = \frac{I_{e,\varphi}}{A_{PS}} = \frac{I_{e,\varphi}}{A_S \cdot \cos\varphi} \qquad (3.19)$$

$$L_e = \frac{\Phi_e}{A_S \cdot \cos\varphi \cdot \Omega} \qquad (2.28)$$

$$\Phi_e = L_e \cdot \Omega \cdot A_S \cdot \cos\varphi \qquad (2.31)$$

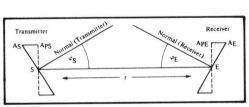

Figure 3.9
Projection of the transmitter and receiver areas.

For the solid angle from the point S to dA_E:

$$d\Omega_S = \frac{dA_{PE}}{r^2} \cdot \Omega_o = \frac{dA_E \cos\varphi_E}{r^2} \cdot \Omega_o \qquad (3.17)$$

For the solid angle from point E to dA_{PS}:

$$d\Omega_E = \frac{dA_{PS}}{r^2} \cdot \Omega_o = \frac{dA_S \cos\varphi_S}{r^2} \cdot \Omega_o \qquad (3.18)$$

In order that the rays meet the receiver surface at about the same angle, the distance r must be great in comparison with the size of the area. This minimum distance is the photometric limiting distance.

$$\Omega_S = \frac{A_{PE}}{r^2} \cdot \Omega_o = \frac{A_E \cos\varphi_E}{r^2} \cdot \Omega_o \qquad (3.20)$$

$$\Omega_E = \frac{A_{PS}}{r^2} \cdot \Omega_o = \frac{A_S \cdot \cos\varphi_S}{r^2} \cdot \Omega_o \qquad (3.21)$$

If $d\Omega$ is replaced in equation (3.16) by the equation (3.17), then the radiant flux Φ_e falling on dA_E, and thus the fundamental photometric law, is obtained (see equation 3.9):

$$d^2\Phi_e = L_e \frac{dA_E \cdot \cos E \cdot dA_S \cdot \cos\varphi_S}{r^2} \cdot \Omega_o$$

(3.22)

The symmetrical construction of equation (3.22) allows dA_E to be considered as the radiator and dA_S as the receiver, if the radiance L_e is the same in both cases.

The expression $\dfrac{dA_S \cdot \cos\varphi_S}{r^2}$ can be replaced by $d\Omega_E$:

$$d^2\Phi_e = L_e \cdot d\Omega_E \cdot dA_E \cdot \cos\varphi_E$$

(3.23)

In equation (3.22), L_e and $dA_S \cdot \cos\varphi_S$ can be replaced by $dI_{e,\varphi}$:

$$d^2\Phi_e = dI_{e,\varphi}\frac{dA_E \cdot \cos\varphi_E}{r^2}\Omega_o$$

(3.24)

In equation (3.24), $d^2\Phi_e$, dA_E and $dI_{e,\varphi}$ can be replaced by E_e and $I_{e,\varphi}$ (see equation 2.38). The photometric inverse square law is obtained:

$$E_e = \frac{I_{e,\varphi} \cdot \cos\varphi_E}{r^2} \cdot \Omega_o$$

(3.25)

$$\Phi_e = L_e \frac{A_E \cdot \cos\varphi_E \cdot A_S \cdot \cos\varphi_S}{r^2} \cdot \Omega_o$$

(3.26)

$$\Phi_e = L_e \cdot \Omega_E \cdot A_E \cdot \cos\varphi_E$$

(3.27)

$$\Phi_e = I_{e,\varphi} \cdot \frac{A_E \cdot \cos\varphi_E}{r^2} \cdot \Omega_o$$

(3.28)

$$E_e = \frac{I_{e,\varphi} \cdot \cos\varphi_E}{r^2} \cdot \Omega_o$$

(3.25)

3.3.1
Inverse square law

This states, that the irradiances at two surfaces, which are struck by the rays from a point source of radiation at equal angles, are inversely proportional to the square of the distance of these areas from the source of radiation (*Figure 3.10*)

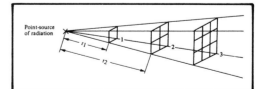

Figure 3.10
Decrease in irradiance with distance (inverse square law)

40

The mathematical expression for this reads:

$$\frac{E_{e,1}}{E_{e,2}} = \frac{\dfrac{I_{e,\varphi} \cdot \cos\varphi_E \cdot \Omega_o}{r_1^2}}{\dfrac{I_{e,\varphi} \cdot \cos\varphi_E \cdot \Omega_o}{r_2^2}} \qquad (3.29)$$

$$\frac{E_{e,1}}{E_{e,2}} = \frac{r_2^2}{r_1^2} \qquad (3.30)$$

4
Laws of Radiation
from a Black Body

4.1
Black and "Non-black" Bodies

4.1.1
Black bodies

Every body which has a higher temperature than $0°$ Kelvin emits radiation due to its own temperature. These radiation sources are called *temperature radiators*.

The laws of radiation are derived from a theoretical ideal radiator, the *black-body* or *black radiator*. This is understood to mean a body which completely absorbs radiation of any wavelength falling upon it at any angle. Furthermore, the black body emits, in every direction and at any wavelength, the maximum possible radiant energy, as compared with other temperature radiators of the same temperature, geometrical shape and dimensions.

In practice, black-body radiators can only be produced for limited temperature ranges. The best-known model of a black body is the *cavity radiator* (*Figure 4.1*): an enclosed cavity, the walls of which are impervious to heat and are at the same temperature, is filled with "black radiation" (see DIN 5496). In this case, the cavity radiation depends only on the wall temperature, so that it is almost equivalent to the radiation of a black body of equal temperature. The cavity emerges through an aperture, while its character as black radiation does not change.

In order to obtain substantially ideal black radiation, first the aperture is kept small in relation to the surface of the cavity and secondly the reflectivity of the cavity walls is reduced with suitable blackening agents. Radiating surfaces differ more or less in their radiation performance from that of a black body, depending on their quality.

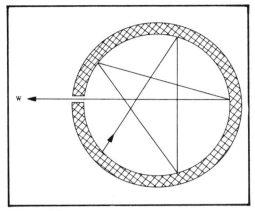

Figure 4.1
Cavity radiator

1 or the identification of temperature radiators, various temperature terms and data are needed: The unit of absolute temperature T is the Kelvin (K).

$$[T] = K \qquad (4.1)$$

The other commonly used unit of temperature t is the degree Celsius ($°C$).

$$[t] = °C \qquad (4.2)$$

The Kelvin is defined as the 273·16th part of the thermodynamic temperature of the triple point of water (see DIN 1301).

$$1 \, K = \frac{1}{273·16} . \, 273·16 \, K \qquad (4.3)$$

The temperature t in degrees Celsius is the difference between any given thermodynamic temperature T in K as compared with the freezing point of water i.e. $T_0 = 273·16$ K.

$$t = T - T_0 = T - 273·16 \, K \qquad (4.4)$$

Absolute zero, T = 0 K, is obtained from equation (4.4) with

$$t = 0 \text{ K} - 273 \cdot 16 \text{ K} = -273 \cdot 16 \text{°C} \qquad (4.5)$$

If the thermodynamic temperature is represented as the sum of T_O and a Celsius temperature, the following equation applies:

$$T = T_O + t \qquad (4.6)$$

The black body, at any temperature T_s and for any wavelength λ, has a spectral radiance which is determined by Planck's law of radiation. For every wavelength λ, any temperature radiator is allocated that temperature T_s of the black body, at which the temperature radiator has the same spectral radiance as the black body.

For the temperature radiator under consideration, at wavelength λ, this temperature is called the *spectral radiation temperature* T_s or *black-body* temperature. For non-black temperature radiators, it is always lower than the true temperature.

The *colour temperature* T_f of a radiator is that temperature of the black body, at which the black body gives the same impression of colour (has approximately the same spectral radiation distribution in the visible range) as the radiator under consideration. A colour temperature can be allocated to temperature radiators, but not in all cases to luminescence radiators.

The *distribution temperature* T_v of a radiator is that temperature of the black body, at which the radiation function of the radiator under consideration, in a wavelength range to be stated, between λ_1 and λ_2, is strictly or approximately proportional to the radiation function of the black body. If the stated spectral range includes the visible range, the distribution temperature corresponds to the colour temperature.

The *ratio temperature* T_r is that temperature of the black body, at which the ratio of the spectral radiances for two different wavelengths λ_1 and λ_2 is equal to that of the radiator under consideration.

4.1.2
Non-black bodies

Grey emitters are described as non-black radiators and their spectral distribution of the radiance is strictly or approximately proportional to the distribution of the spectral radiance of the black body. The emittivity in semi-infinite space (see Section 4.2.4) is equal or constant for all wavelengths of the spectral range under consideration. The colour temperature T_f, the distribution temperature T_v, the ratio temperature T_r and the true temperature T of a grey emitter have the same value in each case (see DIN 5496).

$$T_f = T_v = T_r = T \qquad (4.7)$$

In practice, non-black temperature radiators are only grey emitters for a given spectral range.

Selective radiators are non-black radiators. Their spectral distribution of radiance does *not* correspond to the spectral distribution of radiance of the black body. They have spectral emission bands or lines. Selective radiators are mostly luminescence emitters.

Mixed radiators are also non-black bodies. Their spectral distribution of radiance corresponds to that of a grey emitter, but with additional superimposed emission bands or lines. Mixed radiators emit temperature and luminescence radiation simultaneously.

4.2
Laws of radiation

4.2.1
Planck's law of radiation

Planck's law of radiation permits spectral radiation calculations for the black body. To simplify the radiation conditions, a plane radiating surface A_S of a black body is assumed. The radiance of every surface element dA_S is constant, since the black-body is also a Lambert radiator. The total radiant flux from the plane surface A_S, the radiant emittance $M_{e,s}$, is emitted in semi-infinite space only. The spectral radiant emittance $M_{e,s,\lambda}$ will be calculated as follows:

$$M_{e,s,\lambda} = \frac{2\pi \cdot c_O^2 \cdot h}{\lambda^5 \cdot (e^{\frac{c_O \cdot h}{\lambda \cdot k \cdot T}} - 1)} \qquad (4.8)$$

c_O = Velocity of light
k = Boltzmann's constant
h = Planck's constant

The factor 2 applies for unpolarised radiation, while it is omitted for linearly polarised radiation. With linearly polarised radiation, the vector of the electrical field strength of an electromagnetic wave only oscillates in a certain direction, perpendicular to the direction of propagation. With unpolarised radiation, the direction of oscillation is subject to a continuous irregular change. This radiation can be visualised as the statistical superimposition of two field-strength vectors, perpendicular to each other, $\vec{E_x}$ and $\vec{E_y}$, each with half the radiant power as compared with the unpolarised radiation. Since Lambert's law applies exactly for black-body radiation, the spectral radiance can be calculated from equations (3.12) and (4.8).

$$L_{e,s,\lambda} = \frac{2\pi \cdot c_O^2 \cdot h}{\Omega_o \cdot \pi \cdot \lambda^5 \cdot (e^{\frac{c_O \cdot h}{\lambda \cdot k \cdot T}} - 1)} = \frac{2 \cdot c_O^2 \cdot k}{\Omega_o \cdot \lambda^5 \cdot (e^{\frac{c_O \cdot h}{\lambda \cdot k \cdot T}} - 1)} \qquad (4.9)$$

The spectral radiant intensity can be determined in accordance with equations (2.29) and (4.9), if the size of the emitting area A_S is known:

$$I_{e,s,\lambda} = \frac{2 \cdot c_O^2 \cdot h \cdot A_S \cdot \cos\varphi}{\Omega_o \cdot \lambda^5 \cdot (e^{\frac{c_O \cdot h}{\lambda \cdot K \cdot T}} - 1)} \qquad (4.10)$$

4.2.2
Stefan-Boltzmann Law

The total radiation in semi-infinite space is given by integration of radiation law (4.8) over the whole wavelength range

$$M_{e,s} = \int M_{e,s,\lambda} \cdot d\lambda \qquad (4.11)$$

$$M = \sigma \cdot T^4 \qquad (4.12)$$

$$\sigma = \frac{2\pi^5 \cdot k^4}{15 \, h^3 \, c_O^2} \qquad (4.13)$$

$\sigma = 5 \cdot 67 \cdot 10^{-8} \, Wm^{-4}$ or

$\sigma = 5 \cdot 67 \cdot 10^{-12} \, Wcm^{-2}K^{-4}$

The Stefan-Boltzmann law, corresponding to equation (4.12), states that the radiant emittance of the black body is proportional to the fourth power of its temperature. In this, σ is the Stefan-Boltzmann constant.

In optoelectronics, the radiant emittance $M_{e,s}$ of a black body often relates to the temperature T_1 = 2042 K (previous temperature for the definition of a candela; now 2045 K) and T_2 = 2850 K previous colour temperature of a tungsten lamp for the measurement of the photo-current sensitivity of Si photodetectors; now 2856 K).

For T_1:

$$M_{e,s,1} = 5 \cdot 67 \cdot 10^{-12} Wcm^{-2}K^{-4} \cdot 2045^4 \cdot K^4 \qquad (4.14)$$

$$M_{e,s,1} = 5 \cdot 67 \cdot 10^{-12} \cdot 17 \cdot 5 \cdot 10^{12} Wcm^{-2}$$

$$\underline{M_{e,s,1} = 99 \, Wcm^{-2}}$$
==================

The radiance $L_{e,s,1}$ in any direction in semi-infinite space can be calculated with equation (3.12). For the temperature $T_1 = 2045$ K:

$$L_{e,s,1} = \frac{M_{e,1}}{\pi \cdot \Omega_o} \qquad (4.15)$$

$$L_{e,s,1} = \frac{99\ W}{\pi \cdot 1sr \cdot cm^2} \qquad (4.16)$$

$$\underline{\underline{L_{e,s,1} = 31 \cdot 55\ Wcm^{-2} sr^{-1}}}$$

For $T_2 = 2856$ K, the following is obtained for the radiant emittance:

$$M_{e,s,2} = 5 \cdot 67 \cdot 10^{-12} Wcm^{-2}\ K^{-4} \cdot 2856^4 K^4 \qquad (4.17)$$

$$M_{e,s,2} = 5 \cdot 67 \cdot 10^{-12} \cdot 6 \cdot 65 \cdot 10^{13} Wcm^{-2}$$

$$\underline{\underline{M_{e,s,2} = 377\ Wcm^{-2}}}$$

According to equation (4.15), the radiance $L_{e,s,2}$ in any direction amounts to:

$$L_{e,s,2} = \frac{377\ W}{\pi \cdot 1sr \cdot cm^2} \qquad (4.18)$$

$$\underline{\underline{L_{e,s,2} = 120\ Wcm^{-2} \cdot sr^{-1}}}$$

4.2.3
Wien Displacement Law

It is deduced from the radiation law, that a black body emits its maximum radiant power at a given wavelength λ_{max}, (λ_{max} is temperature-dependent). The spectral radiant emittance at λ_{max} increases very steeply in proportion with the fifth power of the Kelvin temperature (*Figure 4.2*).

$$M_{e,s,\lambda max} = 1 \cdot 309 \cdot 10^{-18} \frac{W}{cm^2 (nm) K^5} \cdot T^5 \qquad (4.19)$$

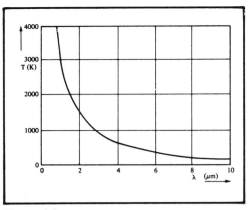

Figure 4.2
The Wien displacement law
$\lambda_{max} \cdot T = 2080\ \mu m\ K$

In the literature, the following dimensions are often found:

$$M_{e,s,\lambda max} = 1 \cdot 309 \cdot 10^{-15} \frac{W}{cm^2 (\mu m) K^5} \cdot T^5$$

As in Section 4.2.2, the values of $M_{e,s,\lambda max}$ for the temperatures $T_1 = 2045$ K are also of interest here, as is the temperature of the sun, with $T_3 = 6000$ K, for comparison.

$M_{e,s,1,\lambda max}$

$$= \frac{1 \cdot 309 \cdot 10^{-15}\ W \cdot 2045^5 \cdot K^5}{cm^2 \cdot \mu m \cdot K^5}$$

$$\underline{\underline{= 46\ W \cdot cm^{-2}\ (\mu m^{-1})}} \qquad (4.20)$$

$M_{e,s,2,\lambda max}$

$$= \frac{1 \cdot 309 \cdot 10^{-15}\ W \cdot 2856^5 \cdot K^5}{cm^2 \cdot \mu m \cdot K^5}$$

$$\underline{\underline{= 248\ W \cdot cm^{-2}\ (\mu m^{-1})}} \qquad (4.21)$$

$M_{e,s,3,\lambda max}$

$$= \frac{1 \cdot 309 \cdot 10^{-15}\ W \cdot 6000^5 \cdot K^5}{cm^2 \cdot \mu m \cdot K^5}$$

$$\underline{\underline{= 10178\ W \cdot cm^{-2}\ \mu m^{-1})}} \qquad (4.22)$$

As a further example, $M_{e,s,\lambda max}$ will be calculated for the temperature $T = 2000$ K

$M_{e,s,\lambda_{max}}$

$$= \frac{1 \cdot 309 \cdot 10^{-15} \, W \cdot 2000^5 \cdot K^5}{cm^2 \cdot (\mu m) \cdot K^5}$$

$$= 41 \cdot 8 \, Wcm^{-2} \, (\mu m^{-1}) \qquad (4.23)$$

This value agrees with that of $4.18 \cdot 10^5$ $Wm^{-2} (\mu m^{-1})$ in *Figure 4.4*.

From equations (4.15) and (4.23), the maximum spectral radiance $L_{e,s,\lambda max}$ at the emission temperature $T = 2000$ K amounts to:

$$L_{e,s,\lambda max} = \frac{M_{e,s,\lambda max}}{\pi \cdot \Omega_o} \qquad (4.24)$$

$$L_{e,s,\lambda max} = \frac{41 \cdot 8 \, W}{\pi \cdot 1sr \cdot cm^2 (\mu m)} \qquad (4.25)$$

$$L_{e,s,\lambda max} = 13 \cdot 6 \, Wcm^{-2} sr^{-1} \, (\mu m^{-1})$$

This value agrees with that of $1 \cdot 36 \cdot 10^5 W.m^{-2}$ $\cdot sr^{-1} (\mu m^{-1})$ in *Figure 4.5*.

It is specified in the radiation law, that a black body shifts its maximum wavelength λ_{max} with the maximum emission to shorter wavelengths with increasing temperature. The equation (4.26) states, that the product of the maximum wavelength λ_{max} and the emission temperature T remains constant:

$$\lambda_{max} \cdot T = \text{constant} \qquad (4.26)$$

$$\lambda_{max} \cdot T = \frac{h \cdot c_o}{K \cdot 4 \cdot 965} = 2880 \cdot 10^{-6} \, mK \qquad (4.27)$$

or

$$\lambda_{max} \cdot T = 2880 \, \mu m \, K \text{ (Wien constant)}$$

By inversion of the formula (4.27), if the temperature is known, the maximum wavelength λ_{max} at maximum emission of the black body can be calculated.

$$\lambda_{max} = \frac{2880 \, \mu m \, K}{T} \qquad (4.28)$$

The following examples indicate the significance of the law (4.28):

An IR vidicon is to take a thermal radiation picture of the human body. The IR vidicon target must therefore have high sensitivity to the maximum energy radiated from the human body at the maximum wavelength λ_{max}. In simplified form, the human body is assumed to be a black-body radiator with a temperature $T = 300$ K. From equation (4.28), its maximum wavelength is

$$\lambda_{max} = \frac{2880 \, \mu m \, K}{300 \, K} = 9 \cdot 6 \, \mu m \qquad (4.29)$$

At this wavelength, the vidicon target must have its maximum sensitivity. On the other hand, a black-body radiator, at a temperature of $2045°$K (definition temperature of the candela) has a maximum wavelength:

$$\lambda_{max} = \frac{2880 \, \mu m \, K}{2045 \, K} = 1 \cdot 4 \, \mu m \qquad (4.30)$$

Furthermore, at a temperature of 2856 K (colour temperature of the scientific tungsten lamp for measurement of the photocurrent sensitivity of Si photodetectors) the black-body radiator has a maximum wavelength of

$$\lambda_{max} = \frac{2880 \, \mu m \, K}{2856 \, K} = 1 \cdot 04 \, \mu m \qquad (4.31)$$

and finally, at a temperature of 6000 K (temperature of the sun), it has a maximum wavelength of

$$\lambda_{max} = \frac{2880 \, \mu m \, K}{6000 \, K} = 480 \, nm \qquad (4.32)$$

49

By conversion of the formula (4.27), if the maximum wavelength λ_{max} is known, the temperature of the black body can be calculated.

$$T = \frac{2880 \, \mu m \, K}{\lambda_{max}} \qquad (4.33)$$

If the maximum wavelength λ_{max} of the black body corresponds with the wavelength of the maximum photopic sensitivity of the eye (daylight vision) $V\lambda_{max} = 555$ nm, then its temperature is

$$T = \frac{2880 \, \mu m \, K}{0 \cdot 555 \, \mu m} = 5189 \, K \qquad (4.34)$$

The Wien displacement law is illustrated in a graph in *Figure 4.2*. In this, the results of equations (4.29) to 4.31) and (4.34) can be read.

4.2.4
Emittivity

The radiation from all bodies in a wavelength range under consideration is dependent firstly on the temperature and secondly, in the case of non-black bodies, on the material composition and surface condition.

The ratio of the emitted radiation of any given temperature radiator to the emitted radiation of a black body of the same temperature is known as the emittivity ϵ. The ratio of the spectral radiance $L_{e,\lambda}$ of a temperature radiator in the normal direction to the surface (optical axis) to the spectral radiance $L_{e,s,\lambda}$ of the black body at equal temperature is defined as the *spectral emittivity normal to the surface* $\epsilon(\lambda)_n$

$$\epsilon(\lambda)_n = \frac{L_{e,\lambda,n}}{L_{e,s,\lambda}} \qquad (4.35)$$

The ratio of the spectral radiance $L_{e,\lambda,\varphi}$ of a temperature radiator in a given direction to

the spectral radiance $L_{e,s,\lambda}$ f the black-body of equal temperature is called the *directional spectral emittivity* $\epsilon(\lambda,\varphi)$:

$$\epsilon(\lambda,\varphi) = \frac{L_{e,\lambda,\varphi}}{L_{e,s,\lambda}} \qquad (4.36)$$

All non-black bodies have a directional spectral emittivity $\epsilon(\lambda,\varphi)$ of <1; in many cases only a few percent. Integration of the directional spectral emittivity over the wavelength range between λ_1 and λ_2 will be defined as the directional band emittivity $\epsilon(\varphi)_b$ and that over the whole wavelength range as the directional total emittivity $\epsilon(\varphi)_t$:

$$\epsilon(\varphi) = \frac{\int L_{e,\lambda,\varphi} \cdot d\lambda}{\int L_{e,s,\lambda} \cdot d\lambda} \qquad (4.37)$$

The ratio of the radiant emittance M_e of a temperature radiator to the radiant emittance $M_{e,s}$ of the black body of equal temperature is called the emittivity in semi-infinite space $\epsilon_\cap$.

$$\epsilon_\cap = \frac{M_e}{M_{e,s}} \qquad (4.38)$$

As is shown by *Figure 4.3*, the spectral radiant emittance $M_{e,\lambda}$ of a non-black body is always less than the spectral radiant emittance $M_{e,s,\lambda}$ of a black body at the same radiator temperature and wavelength.

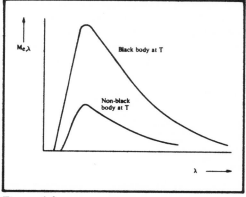

Figure 4.3
Spectral radiant emittance of a black and non-black body as functions of the wavelength λ

4.2.5
Kirchhoff's Law

In the case of a temperature radiator, the directional spectral emittivity, for every temperature and every wavelength, is equal to the spectral absorption for incident radiation from the same direction (see Section 6.1).

$$\epsilon(\lambda,\varphi) = \alpha(\lambda,\varphi) \qquad (4.39)$$

4.2.6
Radiation isotherms

The spectral radiation units of the black body can be represented firstly, as a function of temperature with the wavelength as a variable, as isochromatic lines and secondly as a function of the wavelength with the temperature as a variable as radiation isotherms.

Figure 4.4 shows radiation isotherms of the black body on a linear scale.

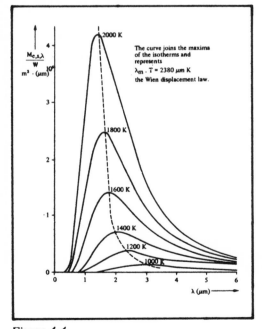

Figure 4.4
Radiation isotherms $M_{e,s,\lambda} = f(\lambda)$ on a linear scale.

The spectral radiant emittance $M_{e,s,\lambda}$ is represented as a function of the wavelength λ. The calculation in equation (4.23) can be verified in this graph.

Figure 4.5, on the other hand, shows radiation isotherms on a log-log scale. The spectral radiance $L_{e,s,\lambda}$ is represented as a function of the wavelength λ. The calculation in equaiton (4.25) can be verified with this graph.

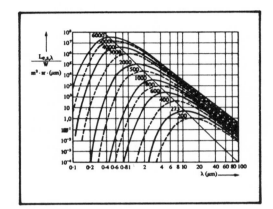

Figure 4.5
Radiation isotherms $L_{e,s,\lambda} = f(\lambda)$ on log-log scale

4.2.7
Reduced law of radiation

In practice, both absolute values of the spectral density of a radiation unit at a wavelength which is of interest, λ, and also relative units of a proportion of the radiation (see equation 2.65) are needed from a Planck radiator with a given temperature T. If the spectral density of the radiation units is represented generally by X_λ, then the spectral density of a radiation unit from a Planck radiator with constant temperature T and at a given wavelength λ can be calculated from the following general equation:

$$X_\lambda = \left(\frac{X_\lambda}{X_{\lambda_{max}}}\right) \cdot X_{\lambda_{max}} \qquad (4.40)$$

51

The ratio of X_λ to $X_{\lambda_{max}}$ in the bracket in equation (4.40) is the radiation function S_λ. From the equations (2.66) and (4.40), one obtains:

$$X_\lambda = S_\lambda \cdot X_{\lambda_{max}} \qquad (4.41)$$

The spectral radiant emittance is usually selected as a spectral radiation unit for X_λ. The spatial distribution of the radiance or radiant intensity from the Planck radiator under consideration will not then be needed.

The values of the spectral radiant emittance for the black body are inserted in equation (4.41):

$$M_{e,s,\lambda} = S_\lambda \cdot M_{e,s,\lambda_{max}} \qquad (4.42)$$

The radiation function S_λ can be read off, with the product of the desired wavelength λ and the temperature T of the black body $(\lambda T)/cm. K$ in the *Table 4.1* of the graph in *Figure 4.6*. From this graph, the radiation function S_λ can also be seen from the quotient of the desired wavelength λ and the maximum wavelength λ_{max} of the black body.

$\dfrac{\lambda \cdot T}{cm \cdot K}$	S_λ	$\dfrac{\lambda \cdot T}{cm \cdot K}$	S_λ	$\dfrac{\lambda \cdot T}{cm \cdot K}$	S_λ
0·0800	0·0015	0·31	0·984	0·70	0·249
0·0825	0·0020	0·32	0·965	0·74	0·216
0·0850	0·0030	0·33	0·945	0·78	0·187
0·0875	0·0042	0·34	0·921	0·82	0·161
0·0900	0·0060	0·35	0·898	0·86	0·139
0·0925	0·0083	0·36	0·875	0·90	0·121
0·0950	0·011	0·37	0·851	0·94	0·106
0·00975	0·014	0·38	0·829	0·98	0·093
0·10	0·017	0·39	0·806	1·0	0·089
0·11	0·038	0·40	0·783	1·1	0·064
0·12	0·074	0·41	0·759	1·2	0·049
0·13	0·128	0·42	0·734	1·3	0·038
0·14	0·191	0·43	0·711	1·4	0·028
0·15	0·265	0·44	0·686	1·5	0·023
0·16	0·349	0·45	0·664	1·6	0·019
0·17	0·430	0·46	0·641	1·7	0·015
0·18	0·521	0·47	0·618	1·8	0·0125
0·19	0·613	0·48	0·595	1·9	0·010
0·20	0·685	0·49	0·572	2·0	0·0085
0·21	0·735	0·50	0·548	2·2	0·0060
0·22	0·790	0·52	0·507	2·4	0·0044
0·23	0·844	0·54	0·473	2·6	0·0032
0·24	0·895	0·56	0·435	2·8	0·0024
0·25	0·931	0·58	0·400	3·0	0·0019
0·26	0·972	0·60	0·367	3·2	0·0015
0·27	0·982	0·62	0·339	3·4	0·0012
0·28	0·996	0·64	0·313	3·6	0·0010
0·29	1·000	0·66	0·289		
0·30	0·997	0·68	0·267		

Table 4.1
Radiation function $S_\lambda = f(\lambda \cdot T)$

The spectral radiant emittance $M_{e,s,\lambda}$ of a black body with a temperature of T = 2045 K, at the wavelength λ = 0·93 μm, is to be determined. At its maximum wavelength λ_{max} = 1·4 μm (see equation 4.30), the spectral raidant emittance $M_{e,s,\lambda max}$ = 46 Wcm^{-2}(μm^{-1}) (see equation 4.20). The product ($\lambda \cdot$ T) gives:

$$\lambda \cdot T = 0\cdot93 \ \mu m \cdot 2045 \ K = 0\cdot19 \ cm \ K \tag{4.43}$$

The quotient of the desired wavelength λ and the maximum wavelength λ_{max} is

$$\frac{\lambda}{\lambda_{max}} = \frac{0\cdot93 \ \mu m}{1\cdot4 \ \mu m} = 0\cdot664 \tag{4.44}$$

The radiation function S_λ can be read off from *Table 4.1* or from *Figure 4.6* with the value from equation (4.43), or from *Figure*

4.6 only with the value from equation (4.44). One obtains:

$$S_\lambda = 0\cdot613 \tag{4.45}$$

The spectral radiant emittance $M_{e,s,\lambda}$, from equation (4.42), is:

$$M_{e,s,\lambda} = 0\cdot613 \cdot 46 \ Wcm^{-2} \ (\mu m^{-1})$$

$$= 28\cdot2 \ Wcm^{-2} \ (\mu m^{-1}) \tag{4.46}$$

If, from a Planck radiator at a given temperature T, the ratio is formed, from equivalent radiation units, of the shorter-wavelength radiation up to a selected wavelength

$$S_\lambda = f \left(\frac{\lambda}{\lambda_{max}}\right) \ \text{and} \ S_\lambda = f \ (\lambda \cdot T)$$

λ, to the total radiation, this ratio is called the proportional radiation q (see also equation 2.65):

$$q = \frac{\int_o^\lambda X_{e,\lambda} \cdot d\lambda}{\int_o^\infty X_{e,\lambda} \cdot d\lambda} \tag{4.47}$$

If the spectral radiant emittance of a black-body radiator is inserted in the equation (4.47), then the following formula is obtained:

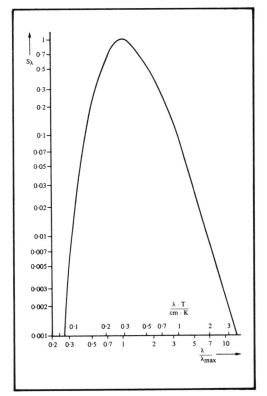

Figure 4.6
Radiation function

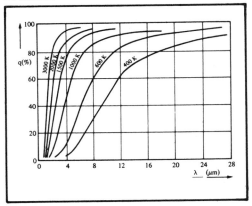

Figure 4.7
Proportional radiation q = f(λ)

$$q = \frac{\int_o^\lambda M_{e,s,\lambda} \cdot d\lambda}{\int_o^\infty M_{e,s,\lambda} \cdot d\lambda} \qquad (4.48)$$

Figure 4.7 shows the proportional radiation q as a function of the selected wavelength λ, with various temperatures T of the black body:

With any given temperature T of the black body, and with the product $(\lambda \cdot T)/cm \cdot K$, which is formed from the selected wavelength λ and the temperature T of the black body, the proprotional radiation q is read off from the *Table 4.2* or from the graph in *Figure 4.8*.

The shorter-wavelength proportional radiation q_1 from a black body with temperature T = 2856 K, up to the wavelength λ_1 = 615 nm, is to be determined. The product $(\lambda_1 \cdot T)$ is:

$$\lambda_1 \cdot T = 0.615 \cdot 10^{-4}\ cm \cdot 2856\ K$$

$$= 0.1756\ cmK \qquad (4.49)$$

$\dfrac{\lambda \cdot T}{cm \cdot K}$	q	$\dfrac{\lambda \cdot T}{cm \cdot K}$	q	$\dfrac{\lambda \cdot T}{cm \cdot K}$	q
0.050	$1.3652 \cdot 10^{-9}$	0.140	$7.9053 \cdot 10^{-3}$	0.460	$5.8057 \cdot 10^{-1}$
0.052	$3.6788 \cdot 10^{-9}$	0.150	$1.3023 \cdot 10^{-2}$	0.480	$6.0880 \cdot 10^{-1}$
0.054	$9.1749 \cdot 10^{-9}$	0.160	$1.9962 \cdot 10^{-2}$	0.500	$6.3494 \cdot 10^{-1}$
0.056	$2.1358 \cdot 10^{-8}$	0.170	$2.8858 \cdot 10^{-2}$	0.520	$6.5912 \cdot 10^{-1}$
0.058	$4.6745 \cdot 10^{-8}$	0.180	$3.9754 \cdot 10^{-2}$	0.540	$6.8146 \cdot 10^{-1}$
0.060	$9.6798 \cdot 10^{-8}$	0.190	$5.2613 \cdot 10^{-2}$	0.560	$7.0209 \cdot 10^{-1}$
0.062	$1.9069 \cdot 10^{-7}$	0.200	$6.7331 \cdot 10^{-2}$	0.580	$7.2116 \cdot 10^{-1}$
0.064	$3.5907 \cdot 10^{-7}$	0.210	$8.3750 \cdot 10^{-2}$	0.600	$7.3877 \cdot 10^{-1}$
0.066	$6.4902 \cdot 10^{-7}$	0.220	$1.0168 \cdot 10^{-1}$	0.620	$7.5505 \cdot 10^{-1}$
0.068	$1.1302 \cdot 10^{-6}$	0.230	$1.2091 \cdot 10^{-1}$	0.660	$7.8402 \cdot 10^{-1}$
0.070	$1.9025 \cdot 10^{-6}$	0.240	$1.4122 \cdot 10^{-1}$	0.700	$8.0885 \cdot 10^{-1}$
0.072	$3.1045 \cdot 10^{-6}$	0.250	$1.6239 \cdot 10^{-1}$	0.740	$8.3020 \cdot 10^{-1}$
0.074	$4.9236 \cdot 10^{-6}$	0.260	$1.8423 \cdot 10^{-1}$	0.780	$8.4861 \cdot 10^{-1}$
0.076	$7.6070 \cdot 10^{-6}$	0.270	$2.0653 \cdot 10^{-1}$	0.820	$8.6455 \cdot 10^{-1}$
0.078	$1.1473 \cdot 10^{-5}$	0.280	$2.2911 \cdot 10^{-1}$	0.860	$8.7840 \cdot 10^{-1}$
0.080	$1.6923 \cdot 10^{-5}$	0.290	$2.5183 \cdot 10^{-1}$	0.900	$8.9048 \cdot 10^{-1}$
0.082	$2.4453 \cdot 10^{-5}$	0.300	$2.7454 \cdot 10^{-1}$	0.940	$9.0105 \cdot 10^{-1}$
0.084	$3.4668 \cdot 10^{-5}$	0.310	$2.9712 \cdot 10^{-1}$	0.980	$9.1033 \cdot 10^{-1}$
0.086	$4.8287 \cdot 10^{-5}$	0.320	$3.1947 \cdot 10^{-1}$	1.00	$9.1455 \cdot 10^{-1}$
0.088	$6.6159 \cdot 10^{-5}$	0.330	$3.4150 \cdot 10^{-1}$	1.10	$9.3217 \cdot 10^{-1}$
0.090	$8.9269 \cdot 10^{-5}$	0.340	$3.6314 \cdot 10^{-1}$	1.20	$9.4532 \cdot 10^{-1}$
0.092	$1.1874 \cdot 10^{-4}$	0.350	$3.8432 \cdot 10^{-1}$	1.30	$9.5331 \cdot 10^{-1}$
0.094	$1.5586 \cdot 10^{-4}$	0.360	$4.0502 \cdot 10^{-1}$	1.40	$9.6304 \cdot 10^{-1}$
0.096	$2.0204 \cdot 10^{-4}$	0.370	$4.2518 \cdot 10^{-1}$	1.50	$9.6909 \cdot 10^{-1}$
0.098	$2.5885 \cdot 10^{-4}$	0.380	$4.4479 \cdot 10^{-1}$	1.60	$9.7390 \cdot 10^{-1}$
0.100	$3.2804 \cdot 10^{-4}$	0.390	$4.6382 \cdot 10^{-1}$	1.70	$9.7777 \cdot 10^{-1}$
0.110	$9.2957 \cdot 10^{-4}$	0.400	$4.8227 \cdot 10^{-1}$	1.80	$9.8091 \cdot 10^{-1}$
0.120	$2.1727 \cdot 10^{-3}$	0.420	$5.1738 \cdot 10^{-1}$	1.90	$9.8349 \cdot 10^{-1}$
0.130	$4.3866 \cdot 10^{-3}$	0.440	$5.5012 \cdot 10^{-1}$	2.00	$9.8563 \cdot 10^{-1}$

Table 4.2
Proportional radiation q = f ($\lambda \cdot$ T)

54

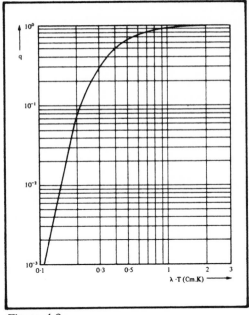

Figure 4.8
Proportional radiation $q = f(\lambda \cdot T)$

With the value from equation (4.49), the proportional radiation q_1 is determined from *Figure 4.8* or by interpolation from *Table 4.2*, or better still from a substantially extended table.

$$q_1 \approx 3 \cdot 45 \cdot 10^{-2} = 3 \cdot 45\% \qquad (4.50)$$

From the same black body, the shorter-wavelength proportional radiation q_2, up to the wavelength $\lambda_2 = 510$ nm is to be determined. The product $(\lambda_2 \cdot T)$ is:

$$\lambda_2 \cdot T = 0 \cdot 51 \cdot 10^{-4} \text{ cm} \cdot 2856 \text{ K}$$

$$= 0 \cdot 1456 \text{ cmK} \qquad (4.51)$$

The proportional radiation q_2 is determined with the value from equation (4.51), as before. We obtain:

$$q_2 \approx 1 \cdot 05 \cdot 10^{-2} = 1 \cdot 05\% \qquad (4.52)$$

If, as in section 2.3.4 and similarly to equation (2.65), the ratio is formed for a Planck radiator with a given temperature T and with equivalent radiation values, from

the radiation in the wavelength range between λ_1 and λ_2 in relation to the total radiation, then this ratio is also called the "proportional radiation q":

$$q = \frac{\int_{\lambda_1}^{\lambda_2} X_{e,\lambda} \cdot d\lambda}{\int_{o}^{\infty} X_{e,\lambda} \cdot d\lambda} \qquad (4.53)$$

If, for easier calculation, the spectral radiant emittance of a black body is again inserted in equation (4.53), the new formula:

$$q = \frac{\int_{\lambda_1}^{\lambda_2} M_{e,s,\lambda} \cdot d\lambda}{\int_{o}^{\infty} M_{e,s,\lambda} \cdot d\lambda} \qquad (4.54)$$

is obtained.

From the previous formula, the proportional radiation q_1 up to the wavelength λ_1 and the proportional radiation q_2 up to the wavelength λ_2 can be determined. If the longer wavelength is selected for λ_1 and the shorter wavelength for λ_2, then the difference between the proportional radiation q_1 and the proportional radiation q_2 gives the desired proportional radiation

$$q = q_1 - q_2 \qquad (4.55)$$

The proportional radiation q from a black body with temperature T = 2856 K is to be determined in the wavelength range between $\lambda_1 = 615$ nm and $\lambda = 510$ nm. The value for $q_1 = 3 \cdot 45\%$ has been determined in equation (4.50) and the value for $q_2 = 1 \cdot 05\%$ in equation (4.52). From equation (4.55), the proportional radiation q is:

$$q = 3 \cdot 45\% - 1 \cdot 05\% = 2 \cdot 4\% \qquad (4.56)$$

Note:
The temperature slide-rule "Temperaturstab Nr.922", from Aristo, is suitable for the calculations of equations (4.16, 4.18, 4.28 to 4.34, 4.47, 4.48, 4.50 and 4.52).

55

The calculations of equations (4.12, 4.14 to 4.23, 4.28 to 4.34, 4.38, 4.45, 4.47, 4.48, 4.50 and 4.52) can be carried out with the "Radiation Calculator" or the Radiation Calculation Slide Rule from General Electric.

5
General and Photometric Evaluation of Radiation

5.1
The human eye as a light receiver

The luminous flux from an observed object passes through the optical system of the eye to the retina, so that a true image of the object is produced there. The optical system of the eye consists of the transparent cornea, the aqueous humour, the iris with the pupil, the lens and the vitreous humour. Since the curvature of the lens of the eye can be varied (variation of the total refraction), the effective focal length of this optical projection system is variable. This capability of the eye to adapt itself to produce focussed images of objects at different distances, is called *accommodation*.

The pupil of the iris adapts the area of its aperture within certain limits to the effective luminance in the field of vision. With high luminances in the field of vision, the pupil has a small aperture area and conversely with low luminances it has a large aperture area. The aperture area of the pupil acts like a variable stop, so that the solid angle for the incident luminous flux on the individual points of the retina is variable in the ratio of about 1:16. The retina of the human eye consists of about 10^8 rod-shaped visual cells and 7.10^6 cone-shaped visual cells. After their sensitivity threshold is exceeded, the cones respond to differences in brightness and the wavelength of the incident light. They pass their stimuli, due to the incoming light, in the form of electrical pulses, each through an optic nerve fibre to the cerebral cortex. Above a luminance in the field of vision of about 10 cd.m^{-2}, predominantly only the cones are excited. The cones are therefore always excited in daylight, so that cone vision is also called "daylight vision".

The cones are located, in a very dense

concentration, mainly in the central part of the retina (yellow spot). Here, the eye has its greatest resolution capability of about one minute of arc. Through continuous oscillating movements, the eye can make use of this high resolution capability over a relatively wide angle of the field of view.

The rods only respond to differences in brightness. They are mainly excited at a luminance, in the field of vision, below about 10^{-3} cd.m^{-2}. The cones can no longer process such low luminance values in the field of vision. Therefore, at night or in darkness, only the rods are still sensitive, so that rod vision is also designated "night vision". The rods pass their excitation, due to light stimuli, in the form of electrical pulses, in inter-connected groups, each through an optic nerve fibre to be cerebral cortex. In the central part of the retina there are hardly any rods, but they are distributed, in relatively lower concentrations as compared with the cones, over the remainder of the surface of the retina. Through the interconnection of the individual rods into groups, each group works on one optic nerve fibre, so that the effective receiving area is increased. This characteristic causes a considerable multiplication of the sensitivity in the case of rod vision.

This capability of the eye to adapt itself to different brightness sensitivities with cone and rod vision and the capability of varying the pupil area according to the luminance in the field of view, is called *adaptation*.

5.2
Optical sensitivities

From considerations of radiation physics, the excitation of the visual cells of the human eye occurs through absorption of the incident visible radiation. The photo-sensitive material in the rods and the three different photosensitive materials in the cones absorb the incident radiation with different spectral sensitivity distributions. Although there are three types of cone, in the case of cone vision, the resultant spectral absorption of all cones is assessed together for the sake of simplicity. With daylight vision, the eye only works in the spectral brightness sensitivity range of the cones. The spectral brightness sensitivity of the eye, adapted for brightness, with a luminance of 10 cd.m^{-2} in the field of view, is defined as the photopic sensitivity $V(\lambda)$ of the eye. Here, the eye with normal colour vision has its highest sensitivity $V(\lambda)_{max}$ at the maximum wavelength λ_{max} = 555 nm. The limits of the photopic sensitivity range of the eye lie around λ_1 = 380 nm and λ_2 = 780 nm.

The ratio of the spectral radiance $L_{e,555nm}$ or $L_{e,\lambda max}$ at the maximum wavelength λ_{max} = 555 nm to the spectral radiance $L_{e,\lambda}$ at another wavelength λ in the photopic sensitivity range of the eye gives, with equal excitation of the cones of the eye, adapted to brightness, the photopic sensitivity $V(\lambda)$ for this wavelength:

$$V(\lambda) = \frac{L_{e,\lambda max}}{L_{e,\lambda}}$$

(5.1)

The sensitivity $v(\lambda)$ of the eye is a relative value. In *Table 5.1*, the $V(\lambda)$ values are listed, in coarse steps, as a function of the wavelength λ. *Figure 5.1* shows the $V(\lambda)$ curve plotted with these $V(\lambda)$ values.

In the case of night-vision, the eye only works in the spectral brightness sensitivity range of the rods. The spectral bright-ness sensitivity of the eye, adapted to darkness, with a luminance of 10^{-3} cd.m^{-2} n the field of view, is defined as the scotopic sensitivity $V'(\lambda)$ of the eye. Here, the eye with normal vision has its highest sensitivity $V'(\lambda)_{max}$ at a maximum

wavelength λ_{max} = 507 nm. The limits of the scotopic sensitivity range of the eye lie around λ_1 = 330 nm and λ_2 = 730 nm. The ratio of the spectral radiance $L_{e,507nm}$ or $L_{e,\lambda max}$ at the maximum wavelength λ_{max} = 507 nm, to the spectral radiance $L_{e,\lambda}$ at another wavelength λ in the scotopic sensitivity range of the eye, which is necessary for equal excitation of the rods of the eye, when adapted for darkness, gives the scotopic sensitivity $V'(\lambda)$ of the eye.

$$V'(\lambda) = \frac{L_{e,\lambda max}'}{L_{e,\lambda}'}$$

(5.2)

The scotopic sensitivity $V'(\lambda)$ of the eye is a relative value. In *Figure 5.1*, the scotopic sensitivity $V'(\lambda)$ is shown by a broken line, as a function of the wavelength λ. All the $V(\lambda)$ and $V'(\lambda)$ values represent the measured values ascertained by the CIE through extensive series of measurements. Because of the change in definition of the candela, in which the temperature of the black body has been redefined from 2042 K to 2045·5 K, the latest $V(\lambda)$ and $V'(\lambda)$ vlaues can be looked up in DIN 5031, Part 3.

5.3
Photometric evaluation of radiation

An absolute evaluation of a radiation from the brightness perceived by the eye is not possible. Sensations of brightness can give different results under different conditions of observation.

Monochromatic radiation in the visible range, entering the eye, causes an excitation of the receiver cells, which is proportional to $V(\lambda) \cdot L_{e,\lambda} \cdot \Delta\lambda$. If several different monochromatic radiations strike the retina simultaneously, then the total excitation of the visual cells concerned is equal, to a sufficient degree of accuracy, to the sum of the individual excitations. Even with a large number of monochromatic radiations,

60

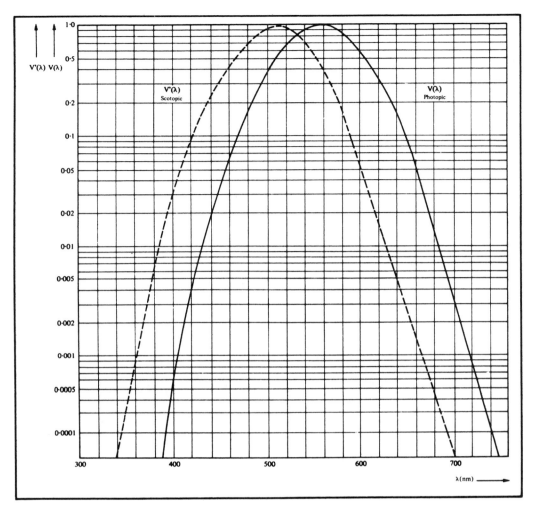

Figure 5.1
Sensitivities V(λ) and V'(λ) of the eye as functions of the wavelength λ

or in the extreme case with a continuous spectrum, this law, which is called the theorem of additivity, is still valid. In this case, the excitation is proportional to $\int V(\lambda) \cdot L_{e,\lambda} \cdot d\lambda$. Additivity is the prerequisite for a physiological evaluation of the radiation by the human eye. If two different monochromatic or mixed radiations give rise to the same impression of brightness under the same observation conditions, they are given the same evaluation factor.

A physical radiant unit, evaluated physiologically by the eye, is defined as a photometric unit. Photometric parameters have their own system of measurement units (see Chapter 2). The scotopic and photopic units have the same numerical values, if the black-body radiation to be evaluated has the solidification point of platinum as its base temperature. The scotopic units are identified, for distinction from the photopic units, by the stroke on the corresponding symbols.

61

Every photometric unit is basically derived from the equivalent physical radiant unit. The value of a particular photometric unit $X_{v,\lambda}$ is calculated for a monochromatic radiation at the wavelength λ_1, by multiplying the equivalent spectral radiation unit X_{e,X_1} either by the evaluation factor of the photopic sensitivity $V(\lambda)_1$ of the eye and the conversion constant C or by the evaluation factor of the scotopic sensitivity $V'(\lambda)$ of the eye and its conversion factor C'. In general, the following relationship applies for monochromatic radiation:

$$X_{v,\lambda} = X_{e,\lambda} \cdot V(\lambda) \cdot C \qquad (5.3)$$

and

$$X'_{v,\lambda} = X_{e,\lambda} \cdot V'(\lambda) \cdot C' \qquad (5.4)$$

If the value of a photometric quantity X_v is to be determined for radiation made up of various wavelengths, the total radiation must be evaluated wavelength by wavelength with the value of the corresponding sensitivities $V(\lambda$ or $V'(\lambda)$ of the eye. The spectral values of the physical radiant unit X_e, multiplied by the corresponding sensitivity $V(\lambda)$ or $V'(\lambda)$ are integrated over the relevant wavelength range. The product of the integral and the corresponding conversion factors C or C' give the value of the required equivalent photometric parameter.

$$X_v = C\int_{380\text{ nm}}^{780\text{ nm}} X_{e,\lambda} \cdot V(\lambda) \cdot d\lambda \qquad (5.5)$$

$$X'_v = C'\int_{330\text{ nm}}^{730\text{ nm}} X_{e,\lambda} \cdot V'(\lambda) \cdot d\lambda \qquad (5.6)$$

The photometric evaluation of radiation is carried out, in accordance with previous definitions, with the radiance and luminance in the field of vision. These values are inserted in the equations (5.3) to (5.6). For the luminance, the following relationship applies in the case of monochromatic radiation:

$$L_{v,\lambda} = L_{e,\lambda} \cdot V(\lambda) \cdot C \qquad (5.7)$$

$$L'_{v,\lambda} = L_{e,\lambda} \cdot V'(\lambda) \cdot C' \qquad (5.8)$$

For wide band radiation, the following applies:

$$L_v = C\int_{380\text{ nm}}^{780\text{ nm}} L_{e,\lambda} \cdot V(\lambda) \cdot d\lambda \qquad (5.9)$$

$$L'_v = C'\int_{330\text{ nm}}^{730\text{ nm}} L_{e,\lambda} \cdot V'(\lambda) \cdot d\lambda \qquad (5.10)$$

5.3.1
Determination of the conversion constants C and C'

Since the sensitivity $V(\lambda)$ or $V'(\lambda)$ of the eye, as a relative value, has no units, the conversion constants C or C' each contain a constant numerical factor and corresponding units, according to the different definitions of the physical radiation and photometric units. The unit of the conversion constant C is obtained by inversion of the equation (5.3):

$$[C] = \frac{[X_{v,\lambda}]}{[X_{e,\lambda}] \cdot [V(\lambda)]} \qquad (5.11)$$

The luminance $L_{v,\lambda}$ and the spectral radiance $L_{e,\lambda}$ are inserted in equation (5.11):

$$[C] = \frac{[L_{v,\lambda}]}{[L_{e,\lambda}] \cdot [V(\lambda)]} \qquad (5.12)$$

After insertion of the units in equation (5.12), the unit of the conversion constant C is obtained.

$$[C] = \frac{\text{lm} \cdot \text{sr}^{-1} \cdot \text{m}^{-2} \, (\mu\text{m}^{-1})}{\text{W} \cdot \text{sr}^{-1} \cdot \text{m}^{-2} \, (\mu\text{m}^{-1})} = \frac{\text{lm}}{\text{W}} \qquad (5.13)$$

The unit of the conversion constant C' is obtained by inverting the equation (5.4):

$$[C'] = \frac{[X'_{v,\lambda}]}{[X_{e,\lambda}] \cdot [V'(\lambda)]} \qquad (5.14)$$

$$[C'] = \frac{[L'_{v,\lambda}]}{[L_{e,\lambda}] \cdot [V'(\lambda)]} \qquad (5.15)$$

$$[C'] = \frac{lm \cdot sr^{-1} \cdot m^{-2} \; (\mu m^{-1})}{W \cdot sr^{-1} \cdot m^{-2} \cdot (\mu m^{-1})} = \frac{lm}{W} \qquad (5.16)$$

The numerical value of the conversion constant C must be calculated in accordance with the definition of the light unit in DIN 5031, by inversion of the formula (5.9) for wide band radiation:

$$C = \frac{L_v}{\int\limits_{380 \; nm}^{780 \; nm} L_{e,s,\lambda} \cdot V(\lambda) \cdot d\lambda} \qquad (5.17)$$

The candela, abbreviated cd, applies as the light unit for photopic vision. The luminous intensity 1 cd is defined as $1/60$ cm^2 or $1/600\,000$ m^2 of the surface of a black-body radiator at the temperature of platinum solidifying under a pressure of $101\,325$ N/m^2 (T = 2045 K) in the normal direction. According to equation (2.27), with a radiator surface $A_S = 1$ cm^2 the luminance L_v is:

$$L_v = \frac{I_v}{A_S \cdot \cos\varphi} = \frac{60 \; cd}{1 \; cm^2 \cdot 1} = 60 \; \frac{cd}{cm^2}$$

$$= 60 \; \frac{lm}{cm^2 \cdot sr} \qquad (5.18)$$

The integral in the denominator of the equation (5.17) is solved, by multiplying the absolute value of the total radiation described above by the relative photopically evaluated proportional radiation q (see equation 2.65 and 4.47).

$$L_{e,s,2045 \; K} \cdot q = \int\limits_{380 \; nm}^{780 \; nm} L_{e,s,\lambda} \cdot V(\lambda) \cdot d\lambda \qquad (5.19)$$

The absolute value of $L_{e,s,2045 \cdot K}$ was calculated in Section 4.2.2. From equation (4.16) it is:

$$L_{e,s,2045 \; K} = 31 \cdot 55 \; Wcm^{-2} \; sr^{-1}$$

The proportional radiation q to be evaluated photopically is determined graphically with the aid of Figures 5.2 and 5.3. Here, firstly the radiation function $S_{\lambda,2045 \; K}$ of a black body at the temperature 2045 K and secondly the photopic sensitivity V(λ) are plotted as functions of the wavelength λ. The calculation of the radiation function S_λ is stated in Section 4.2.7. The product of the ordinates V(λ) and S_λ is given by the curve V(λ)·S_λ. The ratio of the area A_1 measured by planimetry, under the curve V(λ)·S_λ to be measured area A_2 under the curve S_λ gives the proportional radiation q.

$$q = \frac{A_1}{A_2} \qquad (5.20)$$

For the better measurement of the area A_1 under the curve V(λ)·S_λ, this section is shown enlarged in *Figure 5.3.*

The values inserted in equation (5.20) give the proportional radiation

$$q = \frac{18 \cdot 2 \; mm^2}{6440 \; mm^2} = 2 \cdot 82 \cdot 10^{-3} \qquad (5.21)$$

If the values from equation (4.16) and (5.21) are inserted in equation (5.19), then the photopically evaluated radiance is obtained:

$$\int\limits_{380 \; nm}^{780 \; nm} L_{e,s,\lambda} \cdot V(\lambda) \cdot d\lambda$$

$$= 31 \cdot 55 \; Wcm^{-2} sr^{-1} \cdot 2 \cdot 82 \cdot 10^{-3}$$

$$= 89 \cdot 10^{-3} \; Wcm^{-2} sr^{-1} \qquad (5.22)$$

The required numerical value of the conversion constant C is obtained by insertion

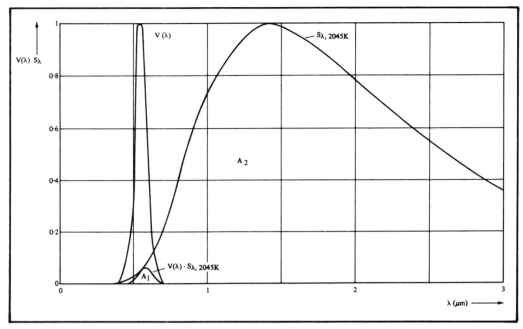

Figure 5.2
Photopic evaluation of the black-body radiation of T = 2045 K

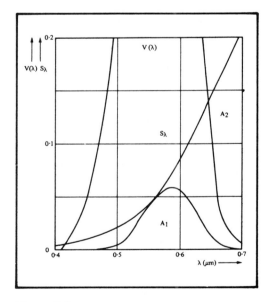

Figure 5.3
Photopic evaluation of the black-body radiation of T = 2045 K. Enlarged section from Figure 5.2, for planimetric measurement of the area A₁

of the values from the equations (5.18) and (5.22) in the equation (5.17):

$$C = \frac{60 \text{ lm cm}^{-2} \text{sr}^{-1}}{89 \cdot 10^{-3} \text{Wcm}^{-2} \text{sr}^{-1}} \approx 674 \text{ lmW}^{-1}$$

(5.23)

By the same calculation method, the ratio of the luminance (see equation 5.18) to the radiance of the black body radiator at the solidification temperature of platinum, T = 2045 K, evaluated in accordance with the scotopic sensitivity V'(λ), gives the conversion factor

$$C' = 1725 \text{ lmW}^{-1}$$

(5.24)

5.3.2
Photometric radiation equivalent

To simplify the conversion of physical radiation units into photometric units, the product of the corresponding conversion constants C or C' with the relevant sensitivity $V(\lambda)$ or $V'(\lambda)$ is formed. The product $C \cdot V(\lambda)$ is defined as the photometric radiation equivalent $K(\lambda)$, and $C'V'(\lambda)$ as the photometric radiation equivalent $K'(\lambda)$.

$$K(\lambda) = C \cdot V(\lambda) \qquad (5.25)$$

$$K'(\lambda) = C' \cdot V'(\lambda) \qquad (5.26)$$

The photometric radiation equivalent $K(\lambda)$ or $K'(\lambda)$ represents the relevant conversion factor between radiant and photometric units with monochromatic radiation. From the equations (5.3) and (5.25), the general simplified formula for the calculation of the photometric radiation equivalent $K(\lambda)$ for monochromatic radiation of wavelength λ is obtained:

$$K(\lambda) = \frac{X_{v,\lambda}}{X_{e,\lambda}} \qquad (5.27)$$

From equations (5.4) and (5.26), the simplified formula is obtained for the calculation of the photometric radiation equivalent $K'(\lambda$:

$$K'(\lambda) = \frac{X'_{v,\lambda}}{X_{e,\lambda}} \qquad (5.28)$$

If equivalent photometric and radiometric values are inserted in equations (5.27) and (5.28), then, as in equation (5.13), by reciprocal cancellation of their solid-angle and area units, for example, the ratio of the luminous flux $\Phi_{v,\lambda}$ to the radiant flux $\Phi_{e,\lambda}$ is obtained. The photometric radiation equivalent for monochromatic radiation $K(\lambda)$ is therefore defined as "the ratio of the luminous flux $\Phi_{v,\lambda}$ to the radiant flux $\Phi_{v,\lambda}$ for the monochromatic radiation of wavelength λ". The simplified equation reads:

$$K(\lambda) = \frac{\Phi_{v,\lambda}}{\Phi_{e,\lambda}} \qquad (5.29)$$

The simplified equation for the calculation of the photometric radiation equivalent $K'(\lambda)$ reads, correspondingly:

$$K'(\lambda) = \frac{\Phi'_{v,\lambda}}{\Phi_{e,\lambda}} \qquad (5.30)$$

If the maximum value for $V(\lambda)$ at the maximum wavelength $\lambda_{max} = 555$ nm, of $V(\lambda)_{max} = 1$, is inserted in equation (5.25), the conversion constant C represents the maximum value of the photometric radiation equivalent $K(\lambda)_{max}$.

$$K(\lambda)_{max} = C \qquad (5.31)$$

If the conversion constant C is replaced by the equation (5.17), the basic equation is obtained for the calculation of the maximum value of the photometric radiation equivalent $K(\lambda)_{max}$:

$$K(\lambda)_{max} = \frac{L_v}{\int\limits_{380\,nm}^{780\,nm} L_{e,s,\lambda} \cdot V(\lambda) \cdot d\lambda} \qquad (5.32)$$

If the corresponding values from equation (5.23) are inserted in the equation (5.32), the maximum value of the photometric radiation equivalent is defined as "the ratio of the luminance, fixed at $L_v = 60$ lm cm^{-2}sr^{-1} of the black body at the solidification temperature of platinum $T = 2045$ K, to the radiance of the black body at the same temperature, evaluated in accordance with the photopic sensitivity $V(\lambda)$ of the eye".

$$K(\lambda)_{max} = \frac{60\ \text{lm} \cdot \text{cm}^{-2} \cdot \text{sr}^{-1}}{89.10^{-3}\ \text{Wcm}^{-2}\text{sr}^{-1}} \approx 674\,\text{lmW}^{-1} \qquad (5.33)$$

According to DIN 5031, 1970 Edition, this value has been laid down at 673 lm/W.

$$K(\lambda)_{max} = 673\ \text{lmW}^{-1} \qquad (5.34)$$

The reciprocal of the maximum value of

the photometric radiation equivalent $K(\lambda)_{max}$ or $K(\lambda)_{555\,nm}$ is defined in the literature as the "mechanical light equivalent M".

$$M = \frac{1}{K(\lambda)_{max}} \qquad (5.35)$$

From equation (5.33), the value for $K(\lambda)_{max}$ is inserted in equation (5.35). The previous value of the mechanical light equivalent M is:

$$M = \frac{1\ W}{680\ lm} = \underline{\underline{1 \cdot 47\ mW\ lm^{-1}}} \qquad (5.36)$$

The maximum value of the photometric radiation equivalent $K(\lambda)_{max}$ permits the photometric radiation equivalent $K(\lambda)$ to be calculated at any given wavelength λ. For this, the conversion factor C from equation (5.25) is replaced by $K(\lambda)_{max}$ in accordance with equation (5.31):

$$K(\lambda) = K(\lambda)_{max} \cdot V(\lambda) \qquad (5.37)$$

For $K(\lambda)_{max}$, the value 673 lmW^{-1} is inserted:

$$K(\lambda) = V(\lambda) \cdot 673\ lmW^{-1} \qquad (5.38)$$

In *Table 5.1*, values of the photometric radiation equivalent $K(\lambda)$ are listed according to the wavelength λ (prior to 1970: $K(\lambda)_{max} = 680\ lmW^{-1}$).

By the same calculation method, a value of the photometric radiation equivalent $K'(\lambda)$ can be determined at the wavelength λ. The maximum value of the photometric radiation equivalent $K'(\lambda)_{max}$ is:

$$K'(\lambda)_{max} = 1725\ lmW^{-1} \qquad (5.39)$$

In *Figure 5.4*, the photometric radiation equivalents $K(\lambda)$ and $K'(\lambda)$ are shown as functions of the wavelength λ.

The following calculations now only relate to the photometric radiation equivalent $K(\lambda)$ of the photopic sensitivity $V(\lambda)$ of the eye.

Wave-length $\frac{\lambda}{nm}$	Spectral brightness sensitivity of the eye $V(\lambda);$ ($L_v = 10\ cd.m^{-2}$)	Photometric radiation equivalent $\dfrac{K(\lambda)}{\frac{lm}{W}}$
380	0·00004	0·0272
400	0·0004	0·272
410	0·0012	0·816
420	0·0040	2·72
430	0·0116	7·89
440	0·023	15·64
450	0·038	25·84
460	0·060	40·8
470	0·091	61·88
480	0·139	94·52
490	0·208	141·44
500	0·323	219·64
510	0·503	342·04
520	0·710	482·8
530	0·862	586·16
540	0·954	648·72
550	0·995	676·6
555	1	680
560	0·995	676·6
570	0·952	647·36
580	0·870	591·6
590	0·757	514·76
600	0·631	429·08
610	0·503	342·04
620	0·381	259·08
630	0·265	180·2
640	0·175	119
650	0·107	72·76
660	0·061	41·48
670	0·032	21·76
680	0·017	11·56
690	0·0082	5·576
700	0·0041	2·788
710	0·0021	1·428
720	0·00105	0·714
730	0·00052	0·3536
740	0·00025	0·170
750	0·00012	0·0816
760	0·00006	0·0408

Table 5.1

The photopic sensitivity of the eye, $V(\lambda)$, and the photometric radiation equivalent $K(\lambda)$ as functions of the wavelength λ

66

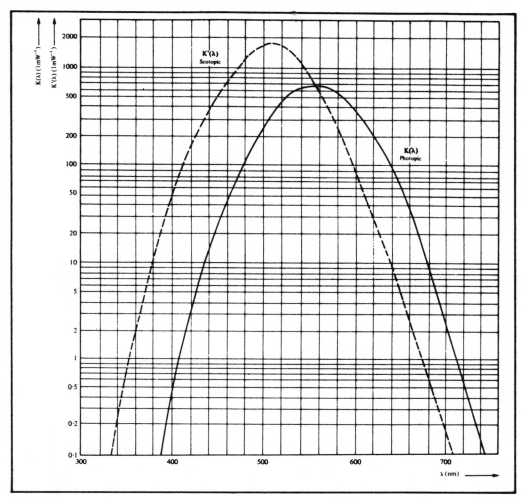

Figure 5.4
Photometric radiantion equivalent K(λ) and K'(λ) as functions of the wavelenth λ

5.3.3
Photometric radiation equivalent of the photopic sensitivity of the eye

With the aid of the equation (5.5), for a photopically evaluated mixed radiation, the radiation equivalent K of the total radiation will be formed, by putting the photopically evaluated radiation in relation to the total radiation. For this, the conversion constant C is replaced, in accordance with equation (5.31), by the maximum value of the photometric radiation equivalent $K(\lambda)_{max}$:

$$K = \frac{K(\lambda)_{max} \cdot \int_{380\,nm}^{780\,nm} X_{e,\lambda} \cdot V(\lambda) \cdot d\lambda}{\int_{0}^{\infty} X_{e,\lambda} \cdot d\lambda} \qquad (5.40)$$

If the same radiometric values, e.g., the radiant flux, are inserted for $X_{e,\lambda}$ in equation (5.40), then after calculation, the radiation equivalent K is obtained as the ratio of the photopically evaluated radiant flux to the total radiant flux:

67

$$K = \frac{K(\lambda)_{max} \cdot \int\limits_{380\ nm}^{780\ nm} \Phi_{e,\lambda} \cdot V(\lambda) \cdot d\lambda}{\int\limits_{o}^{\infty} \Phi_{e,\lambda} \cdot d\lambda} \qquad (5.41)$$

The ratio of radiation evaluated by the eye to the total radiation is the visual efficiency V

$$V = \frac{\int\limits_{380\ nm}^{780\ nm} \Phi_{e,\lambda} \cdot V(\lambda) \cdot d\lambda}{\int\limits_{o}^{\infty} \Phi_{e,\lambda} \cdot d\lambda} \qquad (5.42)$$

Equation (5.42), inserted in (5.41), gives the following relationship for K:

$$K = K(\lambda)_{max} \cdot V \qquad (5.43)$$

The photopically evaluated radiant power in the numerator of equation (5.41) corresponds, as shown in equation (5.5), to the photometric luminous power Φ_v. The denominator corresponds to the total radiant flux Φ_e. The photometric radiation equivalent K of a mixed radiation is therefore defined as the quotient of the luminous power Φ_v and the total radiant flux Φ_e.

$$K = \frac{\Phi_v}{\Phi_e} \qquad (5.44)$$

Corresponding to the equations (5.27) and (5.40), the general equation must read:

$$K = \frac{X_v}{X_e} \qquad (5.45)$$

When the visual efficiency V is known, the photometrically evaluated radiant flux is obtained from the equations (5.43) and (5.44):

$$\Phi_v = \Phi_e \cdot V \cdot K(\lambda)_{max} \qquad (5.46)$$

For any given radiometric parameters, in general:

$$X_v = X_e \cdot V \cdot K(\lambda)_{max} \qquad (5.47)$$

As shown by equations (2.34) and (2.35), the ratio Φ_v to Φ_e also corresponds to the ratio of the luminous efficiency η_v of a radiation source to the radiant efficiency η_e of the same radiation source

$$K = \frac{\eta_v}{\eta_e} \qquad (5.48)$$

In addition to this, *Figure 5.5* shows the variation of the radiation equivalent K and the luminous efficiency η_v of the black body as a function of the absolute temperature.

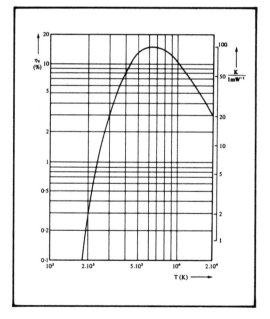

Figure 5.5
Radiation equivalent K and luminous efficiency η_v of the black body as a function of the absolute temperature T

68

5.4
Calculation of the photometric radiation equivalent K for different radiations

5.4.1
Calculation of the photometric radiation equivalent K for a Planck radiator at the temperature specified on the definition of the candela

The photometric radiation equivalent K for the radiation of a black body at the temperature T = 2045 K, which applies for the definition of luminous intensity, will be calculated in accordance with equation (5.45). In this equation, the values of the luminance and radiance for the radiation source described can be inserted from the equations (4.16) and (5.18).

We obtain:

$$K = \frac{L_v}{L_e} \tag{5.49}$$

$$K = \frac{60 \text{ lm} . \text{cm}^{-2} . \text{sr}^{-1}}{31 \cdot 55 \text{ W} . \text{cm}^{-2} . \text{sr}^{-1}} = 1 \cdot 91 \text{ lmW}^{-1} \tag{5.50}$$

The result can be verified by means of *Figure 5.5* with K = f(T).

5.4.2
Calculation of the photometric radiation equivalent K for the standard light A

The standard light A has already been mentioned in Section 2.34. It is used for the measurement of the photocurrent sensitivity of germanium and silicon radiation detectors. If the user has a tungsten lamp calibrated to the distribution temperature of T_v = 2856 K, the illuminance, measured with relatively wide tolerances by a Luxmeter, can be converted by means of the radiation equivalent into the irradiance. The photometric radiation equivalent of the standard light A can be calculated from equation (5.49):

$$K = \frac{L_v}{L_{e,2856 \text{ K}}} \tag{5.49}$$

From equation (4.18), the value of the radiance $L_{e,2856 \text{ K}}$ of a black body at the temperature T = 2856 K or of a Planck radiator with distribution temperature T_v = 2856 K is:

$$L_{e,2856 \text{ K}} = 120 \text{ Wsr}^{-1} \text{cm}^{-2}$$

The luminance L_v is calculated by inversion of equation (5.32); this gives:

$$L_v = K(\lambda)_{max} . \int_{380 \text{ nm}}^{780 \text{ nm}} L_{e,s,\lambda} . V(\lambda) . d\lambda \tag{5.51}$$

The integral is solved in accordance with equation (5.19):

$$L_{e,2856 \text{ K}} . q = \int_{380 \text{ nm}}^{780 \text{ nm}} L_{e,s,\lambda} . V(\lambda) . d\lambda \tag{5.52}$$

The proportional radiation q, to be evaluated photopically, will be determined graphically with the aid of *Figures 5.6* and *5.7*. The product of the ordinates of the photopic sensitivity $V(\lambda)$ of the eye and the radiation function $S_{\lambda,2856 \text{ K}}$ gives the curve $V(\lambda) \cdot S_{\lambda,2856 \text{ K}}$. The ratio of the measured area A_1 under the curve $V(\lambda) \cdot S_{\lambda,2856 \text{ K}}$ to the measured area A_2 under the curve $S_{\lambda,2856 \text{ K}}$ gives the proportional radiation q, as was stated in equation (5.20):

$$q = \frac{A_1}{A_2}$$

For a better planimetric measurement, the area A_1 under the curve $V(\lambda) . S_{\lambda,2856 \text{ K}}$ is shown enlarged in Figure 5.7.

$$q = \frac{376 \text{ mm}^2}{15036 \text{ mm}^2} = 0 \cdot 025 \tag{5.53}$$

If the values from equations (4.18) and

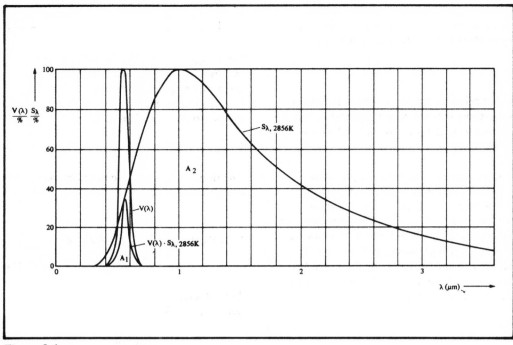

Figure 5.6
Photopic evaluation of the black-body radiation at T = 2856 K

(5.53) are inserted in equation (5.52), the photopically evaluated radiance is obtained:

$$= 3 \text{ Wcm}^{-2}\text{sr}^{-1} \qquad (5.54)$$

From equation (5.51), the luminance is:

$$L_v = 680 \text{ lmW}^{-1} \cdot 3 \text{ Wcm}^{-2}\text{sr}^{-1}$$

$$= 2040 \text{ lm cm}^{-2}\text{sr}^{-1} \qquad (5.55)$$

The required photopic radiation equivalent K of the standard light A can be calculated by inserted of the values equations (4.18) and (5.55) in the formula (5.49).

$$K = \frac{2040 \text{ lm cm}^{-2}\text{sr}^{-1}}{120 \text{ Wcm}^{-2}\text{sr}^{-1}} = 17 \text{ lmW}^{-1} \qquad (5.56)$$

The result can be verified with *Figure 5,5*, which shows the curve K = f(T).

5.4.3
Calculation of the photometric radiation equivalent K of the luminescence radiation from light-emitting diodes (L.E.D.S)

The photometric measuring instruments which have been common up to the present are not very suitable for the measurement of the selective radiation from light-emitting diodes. The spectral sensitivity of their photodetectors is corrected with a special filter to the overall photopic sensitivity of the eye. This correction shows relatively large tolerances in a few wavelength ranges. The selective radiation from light-emitting diodes can therefore only be measured with a photodetector which is exactly photopically corrected to their relatively small wavelength range. The light-emitting diodes of various colours must be measured in each case with the appropriately corrected photodetector or sensor.

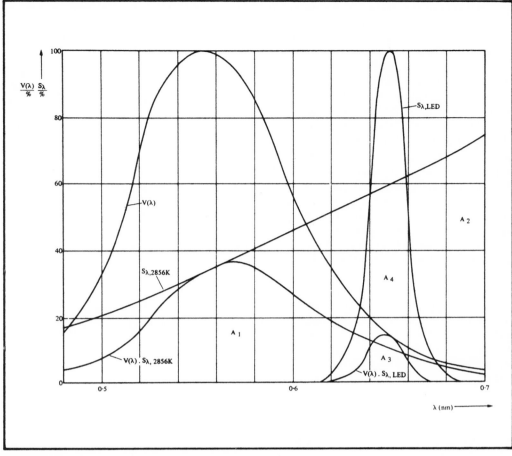

Figure 5.7
Photopic evaluation of the black-body radiation at T = 2856 K. Enlarged section from
Figure 5.6 for planimetric measurement of the area A₁ (on left of diagram)
Photopic evaluation of the radiation from a red-emitting GaAsP diode, e.g., for the
diodes TIL 209A and TIL 220 (on right of diagram)

At the start of the development of light-emitting diodes, there were still no suitable photometric measuring instruments available. Therefore, the measured radiant power from a light-emitting diode was previously stated in data sheets.

With the following calculation of the photometric radiation equivalent for luminescence diodes with selective light emission, the user can determine the photometric luminous power with an uncorrected radiant power meter, provided that the radiation function $S_{\lambda, LED}$ of an

LED luminescence radiation is determined exactly with a monochromator. If, on the other hand, the relative spectral radiation distribution, stated in the data sheets, is guaranteed within close tolerances by the semiconductor manufacturer, the luminescent radiant power Φ_e can be converted directly from its radiation equivalent K into the photometric luminous power Φ_v. With LED's giving a green light output, the spectral distribution can be assessed superficially by subjective colour comparison. For LED's with a red output, however, the subjective colour

71

comparison can give totally false assessments of the photometric luminous power. As a result of poor manufacture, LED's intended to have a red light output can also radiate in the near infra-red range. Such defects can easily be detected with an infra-red converter. The radiant power determined with an unweighted radiant power meter may only be used, if the infra-red part of this radiation is also taken into account. Texas Instruments guarantee the minimum, typical or maximum wavelength stated in their data sheets with the maximum emission of the corresponding light-emitting diodes or display units. In addition, the typical relative spectral radiation distribution, the radiation function S_λ. is stated as a function of the wavelength in graphs.

As an example, the typical radiation function $S_{\lambda,LED}$ of the red-emitting diodes TIL 209 and TIL 220 is plotted in *Figure 5.7*, with the photopic sensitivity of the eye $V(\lambda)$, as a function of the wavelength λ. The product of the ordinates $V(\lambda)$ and $S_{\lambda,LED}$ gives the curve $V(\lambda) \cdot S_{\lambda,LED}$. The ratio of the area A_3 under the curve $V(\lambda) \cdot S_{\lambda,LED}$ to the area A_4 under the curve $S_{\lambda,LED}$ gives the effective visible proportional radiation q (see equation 5.20).

$$q = \frac{A_3}{A_4} \tag{5.57}$$

$$q = \frac{160 \text{ mm}^2}{1350 \text{ mm}^2} = 0\cdot1185 \tag{5.58}$$

The photopically evaluated radiant power is calculated similarly to equation (5.19):

$$\Phi_{e,LED} \cdot q = \int_{380 \text{ nm}}^{780 \text{ nm}} \Phi_{e,\lambda} \cdot V(\lambda) \cdot d\lambda \tag{5.59}$$

For example, the minimum guaranteed radiant power $\Phi_{e,min,LED}$ is taken from the data sheet of the luminescent diode TIL 220 under "Radiant Power Output" $P_{o,min}$. It is 25 μW. The value from

equation (5.58) and the data-sheet value are inserted in equation (5.59). The photopically evaluated radiant power is then:

$$\int_{380 \text{ nm}}^{780 \text{ nm}} \Phi_{e,\lambda} \cdot V(\lambda) \cdot d\lambda = 25 \cdot 10^{-6} \text{ W} \cdot 0\cdot1185$$

$$= 2\cdot96 \ \mu W \tag{5.60}$$

The required typical radiation equivalent K_{typ} of the luminescence radiation from the red-emitting GaAsP diodes TIL 209 and TIL 220 can be calculated with equation (5.41) or the inverted equation (5.46) as follows:

$$K_{typ} = \frac{K(\lambda)_{max} \int_{380 \text{ nm}}^{780 \text{ nm}} \Phi_{e,\lambda} \cdot V(\lambda) \cdot d\lambda}{\int_{o}^{\infty} \Phi_{e,\lambda} \cdot d\lambda} \tag{5.41}$$

$$K_{typ} = \frac{680 \text{ lmW}^{-1} \cdot 2,96 \ \mu W}{25 \ \mu W} = 80\cdot5 \text{ lmW}^{-1} \tag{5.61}$$

A simplified calculation of the photometric radiation equivalent K_{typ} for the selective luminescence radiation from light-emitting diodes is possible, by reading the $K(\lambda)$ value from *Table 5.1* for the typical wavelength with the maximum emission of the corresponding diode. The typical wavelength λ_p with maximum emission of the light-emitting diodes TIL 209 and TIL 220 is:

$$\lambda_p = 650 \text{ nm} \tag{5.62}$$

For this wavelength, the value of the photometric radiation equivalent is read from *Table 5.1*

$$K(\lambda)_{typ} = 72\cdot76 \text{ lmW}^{-1} \tag{5.63}$$

This value does not agree exactly with that from equation 5.61, since on the one hand a graphical integration is always subject to a certain inaccuracy and also the function $V(\lambda)$ is curved.

A further very simple calculation with reasonable accuracy is as follows: The arithmetic mean is formed from the radiation equivalents of the shorter-wavelength half-power point H_1 and the longer-wavelength half-power point H_2 (see Section 10.4):

$$K_{typ} = \frac{K(\lambda)_{H1} + K(\lambda)_{H2}}{2} \qquad (5.64)$$

The red luminescence radiation from the GaAsP diodes TIL 209 and TIL 220 has its typical wavelength λ_p for maximum emission at $\lambda_p = 650$ nm, as is already shown in equation (5.62). Its typical bandwidth between the half-power points H_1 and H_2 is 20 nm. The wavelengths of the typical half-power points therefore lie around

$$\lambda_{H1} = 640 \text{ nm} \qquad (5.65)$$

and

$$\lambda_{H2} = 660 \text{ nm} \qquad (5.66)$$

The corresponding values of the photo-metric radiation equivalent for these wavelengths are read off from *Table 5.1* and are inserted in equation (5.64):

$$K(\lambda)_{H1} = 119 \text{ lmW}^{-1} \qquad (5.67)$$

$$K(\lambda)_{H2} = 41 \cdot 48 \text{ lmW}^{-1} \qquad (5.68)$$

$$K_{typ} = \frac{119 \text{ lmW}^{-1} + 41 \cdot 48 \text{ lmW}^{-1}}{2}$$

$$= 80 \cdot 24 \text{ lmW}^{-1} \qquad (5.69)$$

This result is near to the graphical solution using equation (5.61).

A somewhat more accurate method is to calculate with "Kepler's rule". For this,

the radiation equivalents are needed at λ_p, λ_{H1} and λ_{H2}:

$$K_{typ} = \frac{K(\lambda)_{H1} + 4 \cdot K(\lambda)_p + K(\lambda)_{H2}}{6} \qquad (5.70)$$

Applying this to the light-emitting diodes TIL 209 and TIL 220, with the aid of *Table 5.1:*, one obtains:

$$K_{typ}$$
$$= \frac{119 \text{ lmW}^{-1} + 4 \cdot 72 \cdot 76 \text{ lmW}^{-1} + 41 \cdot 48 \text{ lmW}^{-1}}{6}$$

$$= 75 \cdot 25 \text{ lmW}^{-1} \qquad (5.71)$$

This value is more accurate than that previously determined.

The green luminescence radiation from the GaP diodes TIL 211 and TIL 222 has its typical wavelength λ_p for maximum emission at

$$\lambda_p = 565 \text{ nm} \qquad (5.72)$$

Their typical bandwidth between the half-power points H_1 and H_2 is 35 nm. The wavelengths of the typical half-power points thus lie at

$$\lambda_{H1} = 547 \cdot 5 \text{ nm} \qquad (5.73)$$

and

$$\lambda_{H2} = 582 \cdot 5 \text{ nm} \qquad (5.74)$$

The values of the photopic sensitivity $V(\lambda)$ of the eye can be taken from DIN 5031, Part 3. We obtain:

$$V(\lambda)_{565 \text{ nm}} = 0 \cdot 9786 \qquad (5.75)$$

$$V(\lambda)_{547 \text{ nm}} = 0 \cdot 9874 \qquad (5.76)$$

$$V(\lambda)_{582 \text{ nm}} = 0 \cdot 8494 \qquad (5.77)$$

From equation (5.37), the radiation equivalents $K(\lambda)$ for these wavelengths are:

$$K(\lambda) = K(\lambda)_{max} \cdot V(\lambda) \qquad (5.37)$$

$$K(\lambda)_p = 680 \text{ lmW}^{-1} \cdot 0 \cdot 9786$$
$$= 665 \cdot 6 \text{ lmW}^{-1} \qquad 5.78)$$

$$K(\lambda_{H1} = 680 \text{ lmW}^{-1} \cdot 0 \cdot 9874$$
$$= 672 \text{ lmW}^{-1} \qquad (5.79)$$

$$K(\lambda)_{H2} = 680 \text{ lmW}^{-1} \cdot 0 \cdot 8494$$
$$= 577 \text{ lmW}^{-1} \qquad (5.80)$$

Because of the sharper curvature of the $V(\lambda)$ function in the green spectral range, Kepler's rule, as in equation (5.71), is advisable for the calculation:

$$K_{typ}$$
$$= \frac{672 \text{ lmW}^{-1} + 4. \ 665 \cdot 5 \text{ lmW}^{-1} + 577 \text{ lmW}^{-1}}{6}$$

$$= 652 \text{ lmW}^{-1} \qquad (5.81)$$

With the examples shown, the user can himself determine the corresponding radiation equivalents for other selective luminescence radiations from light-emitting diodes.

5.5
Conversion of radiometric units into photometric, photopic units

5.5.1
Conversion of radiometric units into photometric units for Planck radiation

The calculation of the total radiant power of a Planck radiator for the distributor temperatures which are of interest has been described in Chapter 4. Planck radiation is used mainly for the measurement of the photocurrent sensitivity of germanium and silicon photodetectors. The measured or calculated photocurrent of a photodetector does not, however, depend only on the incident irradiance, but also on the weighting of the radiation given by the spectral sensitivity of the photodetector. Irradiance measurements of Planck radiation are carried out exactly with wide-band radiant power meters. But very often only photometric and selective radiant power meters are available. The evaluation of radiation by selective radiant power meters and by silicon photodetectors requires knowledge of the spectral sensitivity values, which have not yet been dealt with (see Chapter 9).

The photometric evaluation of radiation is carried out with the radiation equivalent, which is described in detail in Sections 5.3 and 5.4. The photometric radiation equivalent K permits the simple conversion of radiometric units into photometric units. The general conversion formula for total radiation is obtained by inversion of equation (5.45):

$$X_v = K \cdot X_e \qquad (5.82)$$

If the irradiance E_e is inserted for X_e in equation (5.82), the illuminance E_v is obtained for X_v.

$$E_v = K \cdot E_e \qquad (5.83)$$

The irradiance values used for the measurement of the photocurrent sensitivity of germanium and silicon photodetectors lie between 1 mW/cm^2 and 20 mW/cm^2. Very sensitive phototransistors are measured with the irradiance values 1 mW/cm^2, 2 mW/cm^2 and 5 mW/cm^2, while less sensitive or older phototransistors and photodiodes are measured with the irradiance values 9 mW/cm^2 and 20 mW/cm^2. From equation (5.50), the photometric radiation equivalent at the temperature for the definition of the specified light unit of Planck radiation is:

$$K = 1 \cdot 91 \text{ lmW}^{-1}$$

For this Planck radiation, the following illuminance values are obtained in accordance with equation (5.83):

74

for $E_e = 1$ mWcm^{-2}:

$E_v = 1·91$ 1 mW^{-1} . 1 mWcm$^{-2} = 19·1$ lx

$$(5.84)$$

for $E_e = 2$ mWcm^{-2}:

$E_v = 1·91$ lmW^{-1} . 2 mW.cm$^{-2} = 38·2$ lx

$$(5.85)$$

for $E_e = 5$ mWcm^{-2}:

$E_v = 1·91$ lmW^{-1} . 5 mWcm$^{-2} = 95·5$ lx

$$(5.86)$$

for $E_e = 9$ mWcm^{-2}:

$E_v = 1·91$ lmW^{-1} . 9mWcm$^{-2} = 171·9$ lx

$$(5.87)$$

for $E_e = 20$ mWcm^{-2}:

$E_v = 1·91$ lmW^{-1} . 20 mWcm$^{-2} = 382$ lx

$$(5.88)$$

The photometric radiation equivalent of the standard light A which is specified for the measurement of the photocurrent sensitivity for germanium and silicon photodetectors, from equation (5.56), is:

$K = 17$ lmW^{-1}

For Planck radiation related to the standard light A, the following illuminance values are obtained in accordance with equation (5.83):

for $E_e = 1$ mWcm^{-2}:

$E_v = 17$ lmW^{-1} . 1 mWcm$^{-2} = 170$ lx

$$(5.89)$$

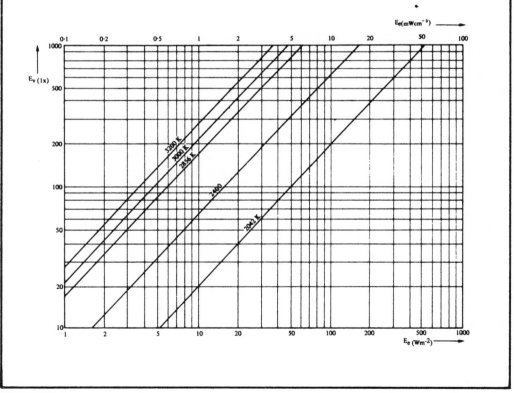

Figure 5.8
Conversion of the irradiance into illuminance; with the distribution temperatures of Planck radiators as variables

for $E_e = 2$ mWcm^{-2}:

$E_v = 17$ lmW^{-1} . 2 mWcm^{-2} = 340 lx

$$(5.90)$$

for $E_e = 5$ mWcm^{-2}:

$E_v = 17$ lmW^{-1} . 5 mWcm^{-2} = 850 lx

$$(5.91)$$

for $E_e = 9$ mWcm^{-2}:

$E_v = 17$ lmW^{-1} . 9 mWcm^{-2} = 1530 lx

$$(5.92)$$

for $E_e = 20$ mWcm^{-2}:

$E_v = 17$ lmW^{-1} . 20 mWcm^{-2} = 3400 lx

$$(5.93)$$

Comparison of the examples (5.84) to (5.88) with the corresponding examples (5.89) to (5.93) will clearly show, to the practical worker, that a deviation from the specified distributor temperature for the standard light A causes large measurement errors. The conversion of the irradiance to the illuminance of Planck radiation can be seen from the graph in *Figure 5.8*. The distribution temperature is the variable. Corresponding to the above conversion, the irradiance can be determined more simply by measuring the illuminance with an accurately calibrated Luxmeter. In doing this, care is to be taken, that no reflected or ambient radiation falls on the photodetector of the Luxmeter.

5.5.2
Conversion of radiometric units into photometric units for the luminescence radiation of light-emitting diodes

The radiometric units of the luminescence radiation of a light-emitting diode will be converted in accordance with equation (5.82) into the equivalent photometric units:

$$X_v = K_{typ} \cdot X_e \qquad (5.94)$$

The calculation of typical radiation equivalent K_{typ} for luminescence radiation was described in Section 5.4.3. From equation (5.71), the typical radiation equivalent K_{typ} for the luminescence radiation of the red-emitting GaAsP diode TIL 220 is:

$$K_{typ} = 72 \cdot 25 \text{ lmW}^{-1}$$

The minimum guaranteed radiant power $\Phi_{e,min}$ has a value of 25 μW (see Section 5.4.3 or data sheet of the TIL 220). If this radiant power and the radiation equivalent from equation (5.71) are inserted in equation (5.94), the minimum guaranteed luminous power Φ_v is obtained for the typical spectral radiation distribution of the TIL 220:

$$\Phi_{v,min} = \Phi_{e,min} \cdot K_{typ} \qquad (5.95)$$

$$\Phi_{v,min} = 25 \text{ } \mu W \text{ . } 72 \cdot 25 \text{ lmW}^{-1} = 1 \cdot 88 \text{ min} \qquad (5.96)$$

This minimum luminous power still gives no information on the photometric data of a semiconductor radiation source. These depend mainly on the construction and also on the radiation distribution pattern of the element. In the early days of light-emitting diodes, the radiance or luminance perpendicular to the wafer surface, i.e., in the normal direction, was stated in the data sheets. For four different reasons, however, this practice has been abandoned and a number of well-known American firms agreed, for future light-emitting diodes and display units, to state the luminous intensity or radiant intensity in the normal direction, with the units cd or mcd and Wsr^{-1} or mWsr^{-1}. The reasons for this measure were as follows:

1
The statements of luminance in the initial period had been erroneously related to the whole wafer area. As a result of the

geometrical shape of the element and the arrangement of bonds pads the whole area does not light up.

2

Similarly, in the case of light-emitting diodes with either clear or coloured lenses or plastic cases the corresponding, considerably larger effective luminous area had not been taken into account, but here too the data was related to the total wafer area.

3

In addition, the different units of luminance confused the users (see Section 2.2.4)

4

The result was that some users adopted the practice of assessing the light-emitting diodes by subjective visual observation.

For the calculation and measurement of the luminous and radiant intensity of light-emitting and radiation-emitting diodes, reference is made to Chapter 10, Section 10.1.

The light-emitting diode Type TIL 210 has, to a first approximation, a radiant intensity distribution in accordance with Lambert's law. By rearrangement of the equation (3.11), the luminous intensity can be calculated in simplified form.

$$I_{v,min} = \frac{\Phi_{v,min}}{\pi \cdot \Omega_0} \qquad (5.97)$$

If the value from equaton (5.96) is inserted in equation (5.97), the value of the minimum luminous intensity in the normal direction for this diode is obtained:

$$I_{v,min} = \frac{1 \cdot 88 \text{ mlm}}{\pi \cdot \Omega_0} = \frac{0 \cdot 6 \text{ mlm}}{sr} = 0 \cdot 6 \text{ mcd} \qquad (5.98)$$

The necessary radiant powers of a red-emitting GaAsP diode and a green-emitting

GaP diode, for an equal impression of brightness on the eye under equal conditions of observation, are very different, assuming that both light-emitting diodes have the same luminous intensity distribution. To compare the green-emitting GaP diode with the red-emitting GaAsP diode, the luminous power of the red-emitting GaAsP diode from equation (5.96) and the typical radiation equivalent for the green luminescence radiation of the GaP diode will be inserted in the rearranged equation (5.95). With $\Phi_v = 1 \cdot 88 \cdot 10^{-3}$ lm and $K_{typ} = 652$ lmW^{-1}, we thus obtain:

$$\Phi_e = \frac{\Phi_v}{K_{typ}} \qquad (5.99)$$

$$\Phi_e = \frac{1 \cdot 88 \cdot 10^{-3} \text{ lm}}{652 \text{ lmW}^{-1}} = \underline{2 \cdot 88 \ \mu W} \qquad (5.100)$$

Under the assumptions described above for equal brightness excitation of the eye, from equation (5.100), only $2 \cdot 88 \ \mu W$ is needed for the green-emitting GaP diode, while the red-emitting GaAsP diode has to deliver $25 \ \mu W$.

5.6
Actinic value

5.6.1
The actinic value of radiation spectrum for photodetectors and for the human eye

The spectral sensitivity of the eye has been described in Sections 5.3 and 5.4. There, it has also been shown, how various radiation spectra can be evaluated with the aid of the curves $V(\lambda)$ and $K(\lambda)$. Just like the eye, every semiconductor photodetector has a sensitivity curve, which is characteristic of the material and its doping and with which similar calculations can be carried out. Further details on the spectral sensitivity of photodetectors are contained in Part II.

The absolute sensitivity of a photo-detector is designated by s. The general definition of the sensitivity of the eye or of an electronic photodetector is described by the following equation.

$$s = \frac{Y}{X} \qquad (5.101)$$

The general value Y denotes the output value of the photodetector, while the general value X represents an input value to the photodetector, evaluated by radiometric, photometric or other means.

In relation to the eye:

1

The relative sensitivity s_{rel} corresponds to the visual efficiency V, see equation (5.42):

$$s_{rel} = V = \frac{\int\limits_{380\ nm}^{780\ nm} \Phi_{e,\lambda} \cdot V(\lambda) \cdot d\lambda}{\int\limits_{0}^{\infty} \Phi_{e,\lambda} \cdot d\lambda} \qquad (5.102)$$

2

The absolute spectral sensitivity $s(\lambda)$ corresponds to the photometric radiation equivalent $K(\lambda)$:

$$S(\lambda) = K(\lambda) = K(\lambda)_{max} \cdot V(\lambda) \qquad (5.103)$$

3

The relative spectral sensitivity $s(\lambda)_{rel}$ corresponds to the photopic sensitivity $V(\lambda)$ of the eye:

$$s(\lambda)_{rel} = \frac{s(\lambda)}{s(\lambda)_{max}} = V(\lambda) \qquad (5.104)$$

4

The absolute sensitivity s corresponds to the photometric radiation equivalent K:

$$s = K = K(\lambda)_{max} \cdot V \qquad (5.105)$$

The values of these sensitivity quantities depend on the wavelength λ of a mono-chromatic radiation or on the total spectral distribution of the radiation and on the spectral sensitivity distribution of the eye. Similar, corresponding relationships apply for any given photodetector.

The sensitivity for a radiation spectrum Z will be denoted by s(Z) and that for a radiation spectrum N by s(N). Here, the radiation Z is any radiation to be investigated in more detail, while the radiation N represents a selected reference radiation. The corresponding input values are denoted by X(Z) or X(N) and the corresponding output values by Y(Z) or Y(N). The sensitivity with radiation Z or radiation N is determined by

$$s(Z) = \frac{Y(Z)}{X(Z)} \qquad (5.106)$$

$$s(N) = \frac{Y(N)}{X(N)} \qquad (5.107)$$

An unweighted radiometric input value X_e is obtained by integration of the corresponding spectral values $X_{e,\lambda}$ over the wavelength range which is of interest. The radiometric input values $X(Z)_e$ and $X(N)_e$ are given by

$$X(Z)_e = \int X(Z)_{e,\lambda} \cdot d\lambda \qquad (5.108)$$

$$X(N)_e = \int X(N)_{e,\lambda} \cdot d\lambda \qquad (5.109)$$

The input values $X(Z)_e$ and $X(N)_e$ can be represented as relative units, related to a maximum value, if, in accordance with equations (5.108), (5.109) and (2.66), the integration is carried out over the wavelength range which is of interest with the radiation function $S(Z)\lambda$ or $S(N)\lambda$ of the particular radiation Z or N:

$$X(Z)_{rel} = \int S(Z)\lambda \cdot d\lambda \qquad (5.110)$$

$$X(N)_{rel} = \int S(N)\lambda \cdot d\lambda \qquad (5.111)$$

The output values Y(Z) and Y(N) are determined by integration of the

78

corresponding spectral input values, in this case $X(Z)_{e,\lambda}$ and $X(N)_{e,\lambda}$, and the spectral sensitivity $s(\lambda)$ of the photodetector over the wavelength range which is of interest:

$$Y(Z) = \int X(Z)_{e,\lambda} \cdot s(\lambda) \cdot d\lambda \qquad (5.112)$$

$$Y(N) = \int X(N)_{e,\lambda} \cdot s(\lambda) \cdot d\lambda \qquad (5.113)$$

The output units $Y(Z)$ and $Y(N)$ can also be stated in relative terms, if, in accordance with equations (5.112), (5.113) and (2.66), the integration is carried out over the wavelength range which is of interest with the relevant radiation function $S(Z)_\lambda$ or $S(N)_\lambda$ and the relative spectral sensitivity $s(\lambda)_{rel}$ of the photodetector:

$$Y(Z)_{rel} = \int S(Z)_\lambda \cdot s(\lambda)_{rel} \cdot d\lambda \qquad (5.114)$$

$$Y(N)_{rel} = \int S(N)_\lambda \cdot s(\lambda)_{rel} \cdot d\lambda \qquad (5.115)$$

In optoelectronics, the actinic value $a(Z)$ of a radiation spectrum Z is the ratio of its effect on a photodetector, related to the effect of a reference radiation spectrum N. Here, the effect of the radiation corresponds to the sensitivity $s(Z)$ or $s(N)$ of the photodetector.

$$a(Z) = \frac{s(Z)}{s(N)} \qquad (5.116)$$

The equations (5.112 to 5.116) presuppose, that the output parameters $Y(Z)$ and $Y(N)$ are proportional to the corresponding input parameters $X(Z)$ and $X(N)$. If there is no proportional relationship between the input and the corresponding output parameter then in many cases the actinic value can be calculated by equating the output parameters:

$$Y(Z) = Y(N) \qquad (5.117)$$

For this case, the extended form of the equation (5.116) is obtained by insertion of equations (5.106) and (5.107) in equation (5.116):

$$a(Z) = \frac{s(Z)}{s(N)} = \frac{X(N)}{X(Z)} \qquad (5.118)$$

The general extended form of equation (5.116) reads:

$$a(Z) = \frac{s(Z)}{s(N)} = \frac{Y(Z) \cdot X(N)}{X(Z) \cdot Y(N)} \qquad (5.119)$$

The actinic value, which is calculated for two radiation spectra with radiometric input parameters for a given photodetector, is given the definition $a_e(Z)$.

To calculate the actinic value of a radiation spectrum for a given linear photodetector the absolute values can be inserted in equation (5.119) in accordance with equations (5.108), (5.109), (5.112) and (5.113), or the relative values in accordance with equations (5.110), (5.111), (5.114) and (5.115):

$$a_e(Z)$$
$$= \frac{\int X(Z)_{e,\lambda} \cdot s(\lambda) \cdot d\lambda \cdot \int X(N)_{e,\lambda} \cdot d\lambda}{\int X(Z)_{e,\lambda} \cdot d\lambda \cdot \int X(N)_{e,\lambda} \cdot s(\lambda) \cdot d\lambda} \qquad (5.120)$$

$$a_e(Z)$$
$$= \frac{\int S(Z)_\lambda \cdot s(\lambda)_{rel} \cdot d\lambda \cdot \int S(N)_\lambda \cdot d\lambda}{\int S(Z)_\lambda \cdot d\lambda \cdot \int S(N)_\lambda \cdot s(\lambda)_{rel} \cdot d\lambda} \qquad (5.121)$$

The actinic value, which is calculated for two monochromatic radiation spectra, from the radiometric input values for a given photodetector, is designated $a_e(Z)_\lambda$. For monochromatic radiations, the equation (5.120) can be simplified, so that the following formula is obtained from equation (5.119) with spectral sensitivity values:

$$a_e(Z)_\lambda = \frac{s(Z)_\lambda}{s(N)_\lambda} \qquad (5.122)$$

The equations (5.120) and (5.121) can be simplified correspondingly to equations (5.103) and (5.104). The actinic value of a radiation spectrum of the eye is therefore calculated by the ratio of the radiation equivalents for the corresponding radiations sources.

For two radiation spectra Z and N of any given spectral composition, the actinic value $a_e(Z)$ of the radiation Z for the eye can be determined with the following formula:

$$a_e(Z) = \frac{K(Z)}{K(N)} \qquad (5.123)$$

The actinic value $a_e(Z)\lambda$ of a monochromatic radiation Z for the eye can also be calculated with the radiation equivalent $K(\lambda)$ for the corresponding monochromatic radiations:

$$a_e(Z)\lambda = \frac{K(Z)\lambda}{K(N)\lambda} \qquad (5.124)$$

The actinic value, which is calculated from photometric input values for an electronic photodetector, is given the definition $a_v(Z)\lambda$ for two monochromatic radiations and the definition $a_v(Z)$ for two chromatic radiation spectra.

If the input values X(Z) and X(N) are related to the absolute photopic sensitivity $K(\lambda)$ of the eye, then the modified form of equation (5.120) reads:

$a_v(Z)$

$$= \frac{\int X(Z)_{e,\lambda} \cdot s(\lambda) \cdot d\lambda \cdot \int X(N)_{e,\lambda} \cdot K(\lambda) \cdot d\lambda}{\int X(Z)_{e,\lambda} \cdot K(\lambda) \cdot d\lambda \cdot \int X(N)_{e,\lambda} \cdot s(\lambda) \cdot d\lambda} \qquad (5.125)$$

If the input values X(Z) and X(N) are related to the relative spectral sensitivity $V(\lambda)$ of the eye, then the modified form of equation (5.121) reads:

$a_v(Z)$

$$= \frac{\int S(Z)\lambda \cdot s(\lambda)_{rel} \cdot d\lambda \cdot \int S(N)\lambda \cdot V(\lambda) \cdot d\lambda}{\int S(Z)\lambda \cdot V(\lambda) \cdot d\lambda \cdot \int S(N)\lambda \cdot s(\lambda)_{rel} \cdot d\lambda} \qquad (5.126)$$

5.6.2
Actinic value of Plank radiation and luminescence radiation with the eye as a photodetector

The actinic value of a radiation spectrum for the eye can be calculated from the radiations spectra defined in Sections 5.3 to 5.5 with the equations (5.123) and (5.124). To distinguish the various individual radiation spectra,

1
Planck radiation with the temperature $T_v = 2045$ K, which defines the unit of luminous intensity, the candela, is designated by "PL";

2
Planck radiation with the temperature $T_v = 2856$ K, which defines the standard light A, is designated by "PA";

3
The luminescence radiation of the red-emitting GaAsP diode TIL 220 is designated by "LR" and

4
The luminescence radiation of the green-emitting GaP diode TIL 211 is designated by "LG".

The radiation equivalents of these radiations are:

$$K(PL) = 1 \cdot 91 \; lmW^{-1} \qquad (5.50)$$

$$K(PA) = 17 \; lmW^{-1} \qquad (5.56)$$

$$K(LR) = 72 \cdot 25 \; lmW^{-1} \qquad (5.71)$$

$$K(LG) = 652 \; lmW^{-1} \qquad (5.81)$$

If the radiation PL serves as the reference radiation, then, from equation (5.123), the actinic value of the designated radiation on the eye is:

$$a_e(PA) = \frac{K(PA)}{K(PL)} = \frac{17 \; lmW^{-1}}{1 \cdot 91 \; lmW^{-1}} = 8 \cdot 9 \qquad (5.127)$$

$$a_e(LR) = \frac{K(LR)}{K(PL)} = \frac{75 \cdot 25 \text{ lmW}^{-1}}{1 \cdot 91 \text{ lmW}^{-1}} = 39 \cdot 398 \tag{5.128}$$

$$a_e(LG) = \frac{K(LG)}{K(PL)} = \frac{652 \text{ lmW}^{-1}}{1 \cdot 91 \text{ lmW}^{-1}} = 341 \cdot 36 \tag{5.129}$$

If the radiation PA serves as the reference radiation, then the actinic value of the designated radiation on the eye is:

$$a_e(PL) = \frac{K(PL)}{K(PA)} = \frac{1 \cdot 91 \text{ lmW}^{-1}}{17 \text{ lmW}^{-1}} = 0 \cdot 1124 \tag{5.130}$$

$$a_e(LR) = \frac{K(LR)}{K(PA)} = \frac{75 \cdot 25 \text{ lmW}^{-1}}{17 \text{ lmW}^{-1}} = 4 \cdot 4265 \tag{5.131}$$

$$a_e(LG) = \frac{K(LG)}{K(PA)} = \frac{652 \text{ lmW}^{-1}}{17 \text{ lmW}^{-1}} = 38 \cdot 353 \tag{5.132}$$

If the radiation LR serves as the reference radiation, then the actinic value of the designated radiation on the eye is:

$$a_e(PL) = \frac{K(PL)}{K(LR)} = \frac{1 \cdot 91 \text{ lmW}^{-1}}{75 \cdot 25 \text{ lmW}^{-1}} = 0 \cdot 0254 \tag{5.133}$$

$$a_e(PA) = \frac{K(PA)}{K(LR)} = \frac{17 \text{ lmW}^{-1}}{72 \cdot 25 \text{ lmW}^{-1}} = 0 \cdot 226 \tag{5.134}$$

$$a_e(LG) = \frac{K(LG)}{K(LR)} = \frac{652 \text{ lmW}^{-1}}{75 \cdot 25 \text{ lmW}^{-1}} = 8 \cdot 6644 \tag{5.135}$$

If the radiation LG serves as the reference radiation, then the actinic value of the designated radiation on the eye is:

$$a_e(PL) = \frac{K(PL)}{K(LG)} = \frac{1 \cdot 91 \text{ lmW}^{-1}}{652 \text{ lmW}^{-1}} = 0 \cdot 00293 \tag{5.136}$$

$$a_e(PA) = \frac{K(PA)}{K(LG)} = \frac{17 \text{ lmW}^{-1}}{652 \text{ lmW}^{-1}} = 0 \cdot 02607 \tag{5.137}$$

$$a_e(LR) = \frac{K(LR)}{K(LG)} = \frac{75 \cdot 25 \text{ lmW}^{-1}}{652 \text{ lmW}^{-1}} = 0 \cdot 1154 \tag{5.138}$$

If the actinic value is described for the example (5.135), then the luminescence radiation of a green-emitting GaP diode is actinically more effective on the eye by the factor $a_e(LG) = 8 \cdot 6644$, as compared with the luminescence radiation of a red-emitting GaAsP diode TIL 220, which is selected as the reference radiation.

If the actinic value is described for the example (5.138), then the luminescence radiation of a red-emitting GaAsP diode TIL 220 is actinically less effective on the eye by the factor $a_e(LR) = 0 \cdot 1154$, as compred with the luminescence radiation of a green-emitting GaP diode, which is selected as the reference radiation.

For the examples (5.135) and (5.138), the following relationshios apply:

$$a_e(LG) = \frac{1}{a_e(LR)} = \frac{1}{0 \cdot 1154} = 8 \cdot 6644 \tag{5.139}$$

$$a_e(LR) \quad \frac{1}{a_e(LG)} = \frac{1}{8 \cdot 6644} = 0 \cdot 1154 \tag{5.140}$$

6
Interaction between optical radiation and matter

Interaction between optical
radiation and matter

6.1
Absorption, transmission and reflection factors

The conversion of radiation into either a different spectral distribution or into another form of energy, e.g., heat, is called *absorption*. The transparency of a medium to a radiation is the *transmission*. The throwing back of radiation in all possible directions, that is, also to the radiations source, is called *reflection*. The absorption, transmission and reflection of a medium are relative to the incident radiant flux. The absorption factor α is the ratio of the absorbed radiant flux Φ_a to the incident radiant flux Φ_0. The transmission factor τ is the ratio of the radiant flux passed through Φ_{tr} to the incident radiant flux Φ_0.

For this, the following relationships result:

$$\alpha = \frac{\Phi_a}{\Phi_0} \tag{6.1}$$

$$\tau = \frac{\Phi_{tr}}{\Phi_0} \tag{6.2}$$

$$\rho = \frac{\Phi_r}{\Phi_0} \tag{6.3}$$

The following relationship exists between the absorption factor, transmission factor and reflectivity:

$$\alpha + \tau + \rho = 1 \tag{6.4}$$

These values depend on the wavelength of the incident radiation.

6.2
Spectral transmission factor $\tau(\lambda)$ and spectral absorption factor $\alpha(\lambda)$

In either the transmission factor τ or the absorption factor α is considered for a monochromatic radiation, one speaks of the spectral transmission factor $\tau(\lambda)$ or the spectral absorption factor $\alpha(\lambda)$:

$$\tau(\lambda) = \frac{\Phi_{\lambda,tr}}{\Phi_{\lambda,o}} \tag{6.5}$$

$$\alpha(\lambda) = \frac{\Phi_{\lambda,a}}{\Phi_{\lambda,o}} \tag{6.6}$$

The spectral transmission and absorption factors are functions of the wavelength (or the frequency). The adjective "spectral" has a different meaning here, as compared with a spectral radiometric parameter, since the latter is described, precisely, as the spectral density of a particular radiometric parameter. The spectral transmission and absorption factors have the bracketed suffix (λ) attached to their symbols. The following relationships exist between the values τ, α and the spectral values $\tau(\lambda)$, $\alpha(\lambda$

$$\tau = \frac{\Phi_{tr}}{\Phi_0} = \frac{\int \Phi_{\lambda o} \cdot \tau(\lambda \cdot d\lambda}{\int \Phi_{\lambda,o} \cdot d\lambda} \tag{6.7}$$

$$\alpha = \frac{\Phi_a}{\Phi_0} = \frac{\int \Phi_{\lambda,o} \cdot \alpha(\lambda) \cdot d\lambda}{\int \Phi_{\lambda,o} \cdot d\lambda} \tag{6.8}$$

The limits of the integration need to be stated in each case.

A vacuum is completely transparent to radiation. Any solid, liquid or gaseous matter, on the other hand, attenuates radiations. Substances which are transparent to light can also show similar transparency in the near IR range. Metals are only transparent to incident radiations up to a thickness of the order of magnitude of the wavelength. The transmission spectrum can be measured with a spectro-photometer. With a calibrated radiation source, firstly the direct radiant flux and secondly the radiant flux through the medium to be investigated are ascertained at every wavelength. With different film thicknesses of the medium, and also with inhomogeneous substances, the spectral transmission or absorption distribution varies. Zones with high absorption (low transmission) are called *spectral bands*. With spectro-photo-meters of high resolution, they can be broken down into spectral lines.

6.3
Scatter

The deflection of rays from the original direction of propagation in the same medium is called scatter. It occurs through reflection of an incident beam of radiation, while the irradiated bulk elements radiate, as secondary radiation sources, in all possible directions. Part of the stray radiation is also thrown back to the radiation source. Pure scatter, that is, without absorption, can exist in transparent gases. Scatter generally increases the absorption, since the path through the medium is lengthened. On the other hand, scatter always reduces the transmitted radiant flux.

Depending on the size of the scattering particles, a distinction is made between three kinds of scatter:

1
The particle size is very small in proportion to the wavelength (molecular scatter). The scatter is very selec.ive.

2
The particle size is hardly any smaller than the wavelength. The scatter is still selective.

3
The particles are very much larger than the wavelength. The scatter is not selective.

IR radiation is therefore transmitted better in certain media (e.g., mist, clouds) than light would be, since its wavelength is longer. The IR transmission of thin pulverised layers depends on both the particle size, and on the wavelength. Lamp-black completely absorbs visible sunlight, while medium to long-wavelength IR radiation is partially transmitted.

6.4
Reflection of radiation

In the reflection of radiation, a distinction is made between:

1
Directional reflection,

2
Diffuse reflection and

3
Mixed directional and diffuse reflection.

In *Figure 6.1*, a directional reflection, from a plane surface which obeys the optical laws, is illustrated. *Figure 6.2* shows an ideally diffuse reflection. Diffuse reflection occurs at rough reflecting surfaces. Finally, in *Figure 6.3*, a mixture of directional and diffuse reflection is shown.

Diffuse reflections are difficult to measure. If the measurement is to be carried out in one direction, the procedure of the IBK measurement conditions will be followed: the angle of incidence is 45°. The reflected beam is measured in the normal direction.

The law of reflection described the path of the rays at a plane mirror with "Angle of incidence α_e = angle of reflection α_r." In *Figure 6.4*, the plane mirror is rotated through the angle Φ; the reflected beam is deflected by twice this angle. In practice, a

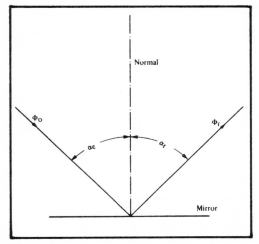

Figure 6.1
Directional (mirror) reflection

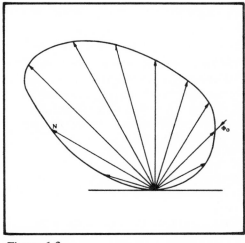

Figure 6.3
Mixed reflection

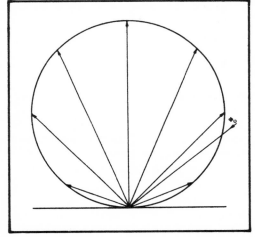

Figure 6.2
Ideal diffuse reflection

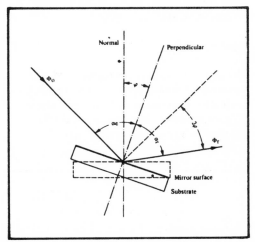

Figure 6.4
The law of reflection

beam, or bundle of rays, is always involved, necessitating careful adjustment of the plane mirror. The plane mirror must have constant reflectivity over the cross-section of the beam and a completely flat mirror surface.

The flatness of optical mirrors is stated in fractions of the wavelength of the yellow sodium or the green mercury line. Metal mirrors are used for uniform reflections over a large wavelength range. The losses are dependent on the wavelength. With new

mirrors they amount to a few percent and with older mirrors up to 50%. Hence for monochromatic radiation sources, interference reflectors (ρ = 99%) are generally used.

6.4.1
Spectral reflectivity $\rho(\lambda)$

If the reflectivity ρ s considered for a monochromatic radiation, it is designated as the spectral reflectivity $\rho(\lambda)$.

87

$$\rho(\lambda) = \frac{\Phi_{\lambda,r}}{\Phi_{\lambda,o}} \qquad (6.9)$$

The spectral reflectivity is a function of the wavelength (or the frequency). The meaning of the adjective "spectral" is explained in Section 6.2. The following relationship exists between the reflectivity ρ and the spectral reflectivity $\rho(\lambda)$:

$$\rho = \frac{\int \Phi_{\lambda,o} \cdot \rho(\lambda) \cdot d\lambda}{\int \Phi_{\lambda,o} \cdot d\lambda} \qquad (6.10)$$

The limits of the integration are to be stated in each case.

Similarly to equation (6.4), the following relationship exists between the spectral values $\alpha(\lambda)$, $\tau(\lambda)$, and $\rho(\lambda)$:

$$\alpha(\lambda) + \tau(\lambda) + \rho(\lambda) = 1 \qquad (6.11)$$

Selective reflections are represented in spectral reflection curves. With metals, the reflectivity increases with the wavelength. For IR radiation, both aluminium and aluminium paint have a very high reflectivity (see *Figure 6.5*).

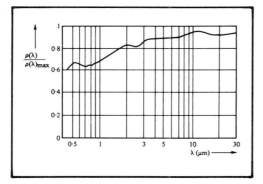

Figure 6.5
Selective reflection from aluminium

Non-metals show reflection bands. They are characteristic of each substance. A spectral band with high reflectivity is described as a metallic reflection. Quartz, for example, shows metallic reflections in several bands (*Figure 6.6*). Transparent

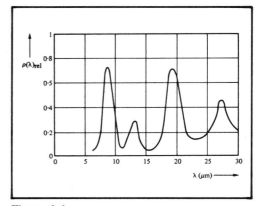

Figure 6.6
Selective reflection of quartz

substances show glass-like reflections. When the incident radiation is perpendicular or at an acute angle to the normal, the reflectivity is low, but with a larger angle to the normal (grazing the surface) the reflectivity obtained is very high.

In the IR range ($\lambda > 1\mu$m), a few substances with very low reflectivities can be classified as black, e.g., asbestos, cotton, rubber, wood, silk.

The reflection capability of thin granulated layers on a solid background depends to a great extent on the base material.

6.5
Basic laws of absorption, attenuation and scatter

The attenuation of radiation is mainly caused by absorption and scatter. If a monochromatic, parallel radiation with the spectral density of the radiant flux Φ_λ falls perpendicularly on a bulk element dV of an attenuating medium, then the attenuation of the radiation, if no reflection occurs, is proportional to the film thickness dx, as long as dΦ_λ is sufficiently small in comparison with Φ_λ. The following applies:

$$\frac{\Phi_{\lambda,x} - \Phi_{\lambda,(x+dx)}}{\Phi_{\lambda,x}} = \frac{d\Phi_{\lambda,x}}{\Phi_{\lambda,x}} = -\beta(\lambda) \cdot dx \qquad (6.12)$$

In this:

$\Phi_{\lambda,(x+dx)}$ = Transmitted spectral radiant flux,

$\Phi_{\lambda,x}$ = Incident spectral radiant flux,

$d\Phi_{\lambda,x}$ = Loss of spectral radiant flux

$\beta(\lambda)$ = Spectral attenuation coefficient

The proportionality factor or attenuation coefficient $\beta(\lambda)$ depends on the properties of the media (or medium) present in the bulk element dV and on the wavelength of the radiation.

If the bulk elements within a particular material do not cause any interactions between one another (e.g. change in aggregation) and if the absorption and the scatter are independent of one another, the attenuation is expressed by

$$\frac{d\Phi_{\lambda,x}}{\Phi_{\lambda,x}} = -[a(\lambda) + \sigma(\lambda)] \cdot dx \qquad (6.13)$$

where $a(\lambda)$ = spectral absorption coefficient in m^{-1} at a wavelength λ,

and $\sigma(\lambda)$ = spectral scatter coefficient in m^{-1} at a wavelength λ.

The absorption coefficient (only if the reduction in intensity is due to true absorption and not to scatter) is the reciprocal value of the depth of penetration d_E. It defines the distance covered in a medium, in which the original incident radiation is attenuated to 1/e (37%). Under the above assumption, the absorption of radiation by an absorbent particle is independent of the particle concentration. At a given wavelength λ the absorption coefficient is proportional to the number of absorbent particles or the particle concentration n_a (Beer's law). This reads as follows:

$$a(\lambda) = a'(\lambda) \cdot n_a \qquad (6.14)$$

where $a'(\lambda)$ = spectral absorption coefficient, related to the concentration.

Under the same conditions, the following applies for the spectral scatter coefficient $\sigma(\lambda)$:

$$\sigma(\lambda) = \sigma'(\lambda) \cdot n_s \qquad (6.15)$$

where n_s = Concentration of scattering particles

and $\sigma'(\lambda)$ = spectral scatter coefficient, related to the concentration.

The coefficients $a'(\lambda)$ and $\sigma'(\lambda)$ have the same units. Thus, for the attenuation in a layer thickness dx, one can insert:

$$\frac{d\Phi_{\lambda,x}}{\Phi_{\lambda,x}} = -[a'(\lambda) \cdot n_a + \sigma'(\lambda) \cdot n_s] \, dx \qquad (6.16)$$

With a finite layer thickness S, the integration produces the Lambert-Beer Law, in which $\Phi_{\lambda,0}$ denotes the incident and $\Phi_{\lambda,tr}$ the transmitted spectral radiant flux.

$$\int_{}^{\Phi_{\lambda,tr}} \frac{d\Phi_{\lambda,x}}{\Phi_{\lambda,x}} = -[a'(\lambda) \cdot n_a + \sigma'(\lambda) \cdot n_s] \int_{x=0}^{x=S} dx \qquad (6.17)$$

$$\ln \frac{\Phi_{\lambda,tr}}{\Phi_{\lambda,0}} = -[a'(\lambda) \cdot n_a + \sigma'(\lambda) \cdot n_s] \cdot S \qquad (6.18)$$

$$\Phi_{\lambda,tr} = \Phi_{\lambda,0} \cdot e^{-[a'(\lambda) \cdot n_a + \sigma'(\lambda, \ n_s] \cdot S} \qquad (6.19)$$

The spectral transmission factor $\tau(\lambda)$ is:

$$\tau(\lambda) = \frac{\Phi_{\lambda,tr}}{\Phi_{\lambda,0}} \qquad (6.20)$$

$$\tau(\lambda) = e^{-[a'(\lambda) \cdot n_a + \sigma'(\lambda) \cdot n_s] \cdot S} \qquad (6.21)$$

89

If the radiation loss is solely due to absorption without scatter, then after integration as above, Lambert's law is obtained:

$$\Phi_{\lambda,tr} = \Phi_{\lambda,o} \cdot e^{-[a'(\lambda \cdot n_a] \cdot S}$$

(6.22)

The spectral transmission factor affected by absorption only, $\tau(\lambda)_a$, is:

$$\tau(\lambda)_a = e^{-[a'(\lambda) \cdot n_a] \cdot S}$$

(6.23)

With constant exponents, the transmission of a material depends only on the quantity of material traversed by the radiation. The assumptions previously made, for example that a' is independent of n_a and σ is independent of N_s, are not always valid. If the absorbent particles interact with one another, the total absorption will be determined not only by the number of absorbent particles, but also by the change in concentration. Under certain circumstances, such changes in concentration can convert the type of absorbent molecule into another. In the case of changes in concentration, Beer's law is no longer applicable.

The spectral transmission factor and the spectral absorption factor are often stated as the pure spectral transmission factor $\tau(\lambda)_R$ and the pure spectral absorption factor $\alpha(\lambda)_R$. In these, the reflection losses are not taken into account. The pure spectral transmission factor $\tau(\lambda)_R$ is defined as the ratio of the transmitted radiant flux $\Phi_{\lambda,tr}$ to the incoming radiant flux $\Phi_{\lambda,i}$:

$$\tau(\lambda)_R = \frac{\Phi_{\lambda,tr}}{\Phi_{\lambda,i}}$$

(6.24)

The pure spectral absorption factor $\alpha(\lambda)_R$ is defined as the ratio of he absorbed radiant flux $\Phi_{\lambda,a}$ to the incoming radiant flux $\Phi_{\lambda,i}$:

$$\alpha(\lambda)_R = \frac{\Phi_{\lambda,a}}{\Phi_{\lambda,i}}$$

(6.25)

6.6
Absorption and transmission spectra

The spectral parameters which depend on the media and substances can be determined for a given layer thickness. The manufacturers of optical components use graphs on their data sheets, to define the characteristics parameter for a given layer thickness as a function of wavelength.

Usually, absoroption or transmission spectra are used. Assuming that no reflection occurs, the absorption scale runs in the opposite direction to the transmission scale. In each case they add up to 100%, since $\tau + \alpha = 1$.

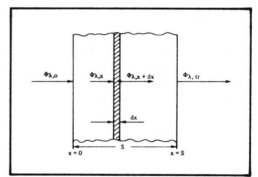

Figure 6.7
The derivation of the absorption law for a differential layer dx

The layer thickness affects the absorption and transmission characteristics. The statement concerning the layer-thickness is omitted if the spectrum or the absorption coefficient is available. Very complicated conditions can arise in attenuation calculations for example in the case of the earth's atmosphere. Here, the absorption, the scatter and the refractive index depend, not only on the wavelength, but also on further variables such as temperature, water vapour content and atmospheric pollution. In most cases, the attenuation or the transmission in the earth's atmosphere is determined for a given small wavelength range, from graphs as a function of distance (*Figure 6.8*). In this graph,

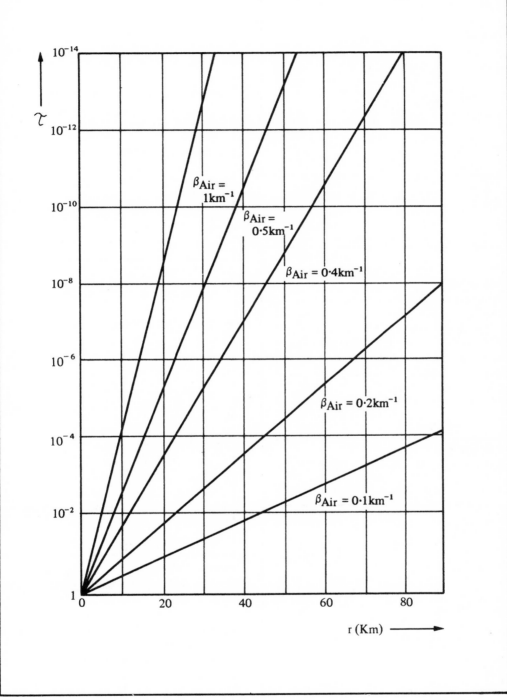

Figure 6.8
Transmission τ in the atmosphere as a function of the distance r, with various attenuation coefficients β as variables

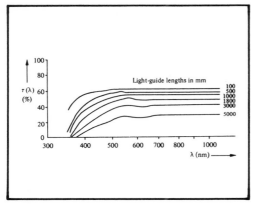

Figure 6.9
Spectral transmission of light-guides

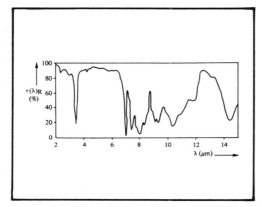

Figure 6.10
Spectral transmission of PVC film

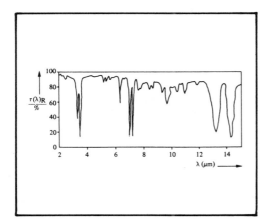

Figure 6.11
Spectral transmission of polystyrene

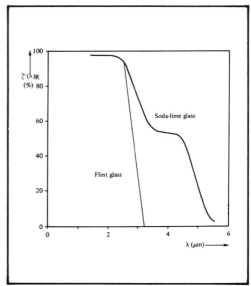

Figure 6.12
Spectral transmission of thin glass films

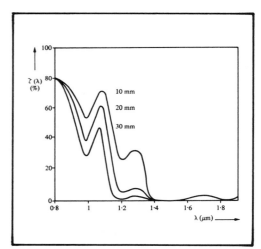

Figure 6.13
Spectral transmission of thin films of water

different attenuation coefficients are used as variables. The attenuation coefficients which occur must be known in the particular application

Figure 6.9 to *6.15* show a number of transmission spectra, as examples. In practice, numerous spectra are needed for

92

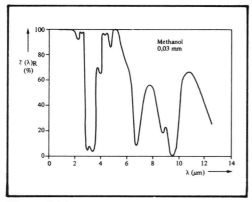

Figure 6.14
Spectral transmission of methanol

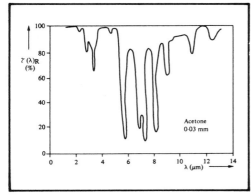

Figure 6.15
Spectral transmission of acetone

the widely varying possible applications. Several physical institutes, in various countries, have extensive documentation of transmission spectra for a large number of substances.

Following these examples, *Figure 6.16* shows the spectral absorption for various water films and *Figure 6.17* the absorption coefficient a = f(λ) for the filter No. 1173 formerly produced by Texas Instruments. Furthermore, *Figures 6.18* and *6.19* give the spectral transmission of this filter.

In addition, *Figures 6.20* and *6.21* show the spectral transmission of various interference filters from the Schott

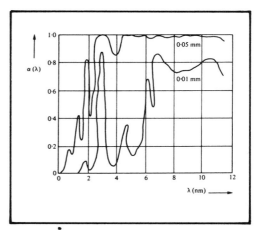

Figure 6.16
Spectral absorption for various water films

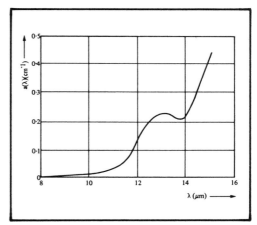

Figure 6.17
Absorption coefficient $\alpha = f(\lambda)$ for the filter No. 1173 formerly from TI

Company, Mainz, West Germany; also in *Figures 6.22* and *6.23* the curves are given for various IR-transmitting black glasses, also from Schott. With these, the pure transmission factor $\tau(\lambda)_R$ is plotted as the ordinate, the reflection losses are not taken into account.

$$\tau(\lambda)_R = \frac{\text{Emergent}}{\text{Incoming}} \text{ radiant flux}$$

In such cases, the reflectivity is usually stated separately.

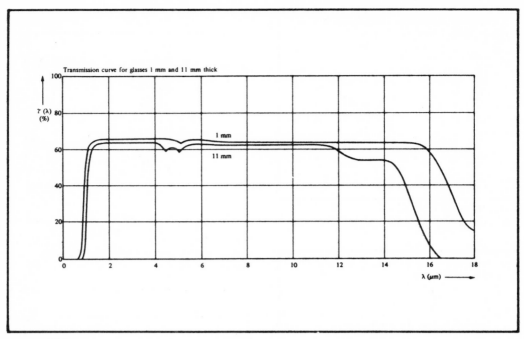

Figure 6.18
Spectral transmission of the filter No. 1173 formerly from TI

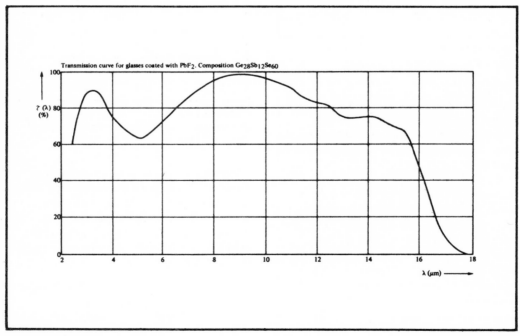

Figure 6.19
Spectral transmission of the filter No. 1173 with the composition stated above

94

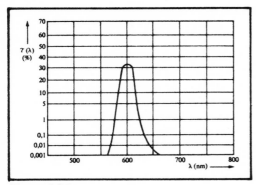

Figure 6.20
Spectral tranmission of a double-band
filter (Schott)

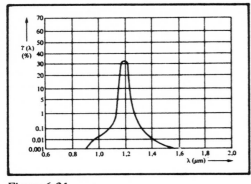

Figure 6.21
Spectral transmission of an IR-band filter
(Schott)

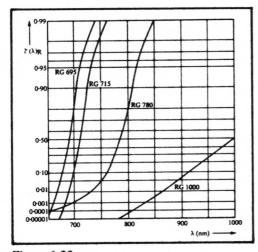

Figure 6.22
Spectral transmission of various IR-
transmitting black glasses from Schott

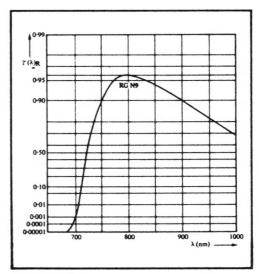

Figure 6.23
Spectral transmission of the IR-transmitting
glass RGN9 from Schott

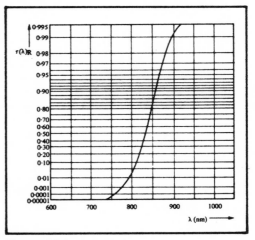

Figure 6.24
Spectral transmission of the filter RG 830
from Schott

The characteristics of another very
interesting filter from Schott, is shown by
Figure 6.24, this has a good transmission
for the 900 nm radiation from GaAs diodes.
The spectral sensitivity of silicon photodetectors can thus be matched optimally to
the spectral emission of GaAs diodes.
Interfering stray short-wave radiation is thus
substantially suppressed.

6.7
Refraction

If a light ray falls at an angle on the
boundary surface of the transparent
media air and water, it is deflected from its
original direction at the boundary surface.
The refracted ray forms the angle of
refraction with the perpendicular to the
surface. (*Figure 6.25*). The wider the
angle with which the incident ray strikes

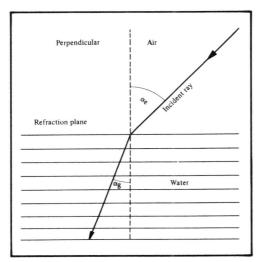
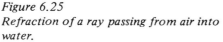

Figure 6.25
Refraction of a ray passing from air into
water.

the refracting plane, the greater is its
deflection. If it strikes the boundary
surface perpendicularly, there is no
deflection. The refractive index, the
ratio of the sine of the angle of incidence,
$\sin\alpha_e$, to the sine of the angle of
refraction, $\sin\alpha_g$, is constant. Thus, with

Index e = incident
Index g = refracted
$$\frac{\sin\alpha_e}{\sin\alpha_g} = \text{const.}$$
(6.25)

The reflective index of a transparent
substance with respect to vacuum is called
the "absolute refractive index" and is
denoted by the index n. Here, a refractive
index of 1 is allocated to vacuum. In

technical calculations, a value of $n = 1$
can also be used for air.

With various materials, the refractive
index n is dependent on the wavelength.
As a result, the angle of refraction is a
function of the wavelength. The separation
of a mixed radiation which thus occurs is
called *dispersion.*

In practice, refractive indices are normally
shown for optical components graphically
as a function of λ. *Figure 6.26* shows, as

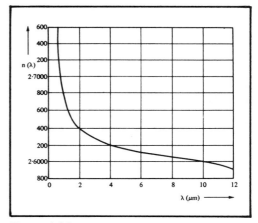

Figure 6.26
The refractive index $n = f(\lambda)$ for the IR
filter No. 1173

an example, the refractive index n as a
function of the wavelength λ for the IR
filter No. 1173 formerly from Texas
Instruments.

The law of refraction describes the path
of a ray when passing from one medium into
another with a different refractive index n.
At the same time as the change in direction,
a change in the propagation speed of the
ray, proportional to the reflective index,
also takes place:

$$n = \frac{\sin\alpha_e}{\sin\alpha_g} = \frac{c_e}{c_g}$$
(6.26)

c_e = Propagation speed in medium e

c_g = Propagation speed in medium g

96

By rearrangement of this equation, constant ratios are obtained:

$$\frac{\sin\alpha_e}{c_e} = \frac{\sin\alpha_g}{c_g} = \text{constant} \qquad (6.27)$$

If these ratios are multiplied by the speed of light in vacuum, c_0/c gives, in each case, the refractive index for the corresponding medium in relation to vacuum as the medium for the incident ray (c_0 = speed of light in vacuum).

$$n_e \cdot \sin\alpha_e = n_g \cdot \sin\alpha_g = \text{constant} \qquad (6.28)$$

The product of the refractive index and the since of the corresponding is constant in refraction (Law of refraction or Snell's Law). The law of refraction gives the ratio of the refractive indices:

$$n_{eg} = \frac{n_g}{n_e} \qquad (6.29)$$

Ray paths are always interchangeable. Therefore it is immaterial, whether the ray starts from the medium e or from the medium g. However the ratio of the refractive indices must be selected in the right direction.

The refraction of a ray from one medium to another medium which is optically denser ($n_e < n_g$), see *Fig. 6.27*, takes place towards the normal to the surface. The refraction of a ray from an optically dense medium into a less dense medium ($n_e > n_g$), see *Fig. 6.28*, takes place away from the normal. In case of reflection at the denser medium, the reflected component of a wave is given a phase shift of $\lambda/2$, but not in case of reflection at the less dense medium.

The transition from an optically denser medium into a less dense medium ($n_e > n_g$) shows strong reflections with increasing angle of incidence. Above a limiting angle (critical angle), total reflection occurs. The limiting angle α_G is obtained as:

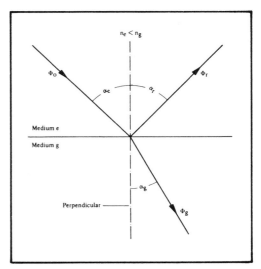

Figure 6.27
Refraction from a less dense to an optically denser medium

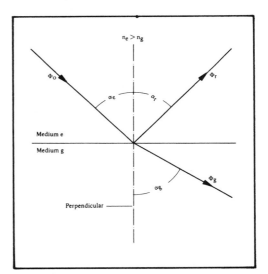

Figure 6.28
Refraction from an optically denser medium to a less dense medium

$$\sin\alpha_G = \frac{n_g}{n_e} \qquad (6.30)$$

Example:
The critical angle for the transition from glass to air is:

97

$$\sin\alpha_G = \frac{n_{Air}}{n_{Glass}} = \frac{1}{1 \cdot 5} = 0 \cdot 667 \Rightarrow \alpha_G = 42° \tag{6.31}$$

For the transition from GaAs to air, the critical angle α_G is:

$$\sin\alpha_G = \frac{n_{Air}}{n_{GaAs}} = \frac{1}{3 \cdot 6} = 0 \cdot 278 \Rightarrow \alpha_G = 16 \cdot 1° \tag{6.32}$$

As is shown by the last example, the small exit angle, above which total reflection occurs, has a very unfavourable effect on the radiant efficiency of GaAs diodes. Total reflection however can be utilised to useful effect for example in light guides.

In general, a reflection occurs whenever a ray is refracted. If the absorption of the second medium is very slight or non-existent, then for *Figures 6.27* and *6.28*:

$$\Phi_g = \Phi_0 - \Phi_r \tag{6.33}$$

$$\Phi_g = \Phi_0 - \Phi_0 \cdot \rho \tag{6.34}$$

$$\Phi_g = \Phi_0 \cdot (1 - \rho) \tag{6.35}$$

Here:

Φ_0 = incident radiation power,

Φ_g = refracted radiant power,

Φ_r = reflected radiant power.

The reflectivity depends on the refractive indices of the two media. If the absorption of the second medium is very slight or non-existent, then one obtains, for the reflectivity with perpendicular incident radiation:

$$\rho = \left(\frac{n_g - n_e}{n_g + n_e}\right)^2 \tag{6.36}$$

From equation (6.3), the reflected radiant power is:

$$\Phi_r = \Phi_0 \cdot \left(\frac{n_g - n_e}{n_g + n_e}\right)^2 \tag{6.37}$$

In the case of the passage of a ray through air-glass-air, the reflectivity is:

$$\rho = \left(\frac{n_2 - n_1}{n_2 + n_1}\right)^2 \cdot 2 = \left(\frac{1 \cdot 5 - 1}{1 \cdot 5 + 1}\right)^2 \cdot 2 = 8\% \tag{6.38}$$

In factor 2 occurs because part of the radiation radiation is reflected twice. In total, 8% of the incident radiant power is reflected. In the IR range, the refractive indices lie between 1·2 and about 4, so that reflectivities between 1 and 36% occur.

The reflection losses and with them the unwanted reflections can be effectively suppressed by application of an anti-reflection layer ("bloom") to both sides of optical glasses. The anti-reflection effect takes place through interference. Of course, dut to the wavelength-dependence of the interference phenomena, it can only be made particularly effective ($\tau > 99\%$) for a limited wavelength range. Also it is dependent on the angle of incidence.

In practice, the radiant power distribution in flat parallel plates is often of interest. When using filters, this power distribution through refraction and reflection must also be taken into account. With a completely absorption-free substance, the radiant power is divided by multiple reflection (see *Figure 6.29*).

On entry into the medium of the plate, the radiant power Φ_0 is attenuated by the factor $(1 - \rho)$ and again, on entering the original medium, by the factor $(1 - \rho)$, that is, in total, by $(1 - \rho)^2$. *Figure 6.29* gives information on the power distribution of the incident radiation Φ_0.

In practice, for the refracted radiant power Φ_g

$$\Phi_g = \Phi_0 \cdot (1 - \rho)^2 \tag{6.39}$$

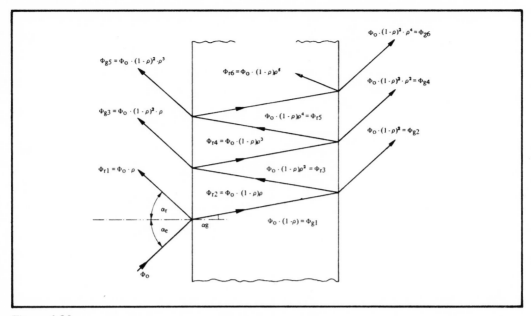

Figure 6.29
Reflection and refraction at a plane, parallel-faced plate.

7
Radiation sources

In semiconductor optoelectronics, three kinds of radiation source are mainly of interest: filament lamps, semiconductor light-emitting diodes and daylight (natural temperature radiators). Other radiation sources, such as discharge lamps, are seldom used in electronic systems except for specific military applications. In special industrial applications, the radiation from incandescent metals or burning gases is analysed.

The ambient and/or background radiation from natural and artificial radiation sources can severely affect an electro-optical system, if the spectral radiation distribution of the interfering radiation source falls within the spectral sensitivity range of the photo-detector.

Radiation sources are classified into groups according to the type of radiation or radiation phenomenon utilised and

taking into account the many basic types which have been developed, as is shown in *Figure 7.1*. Here the most important radiation sources are systematically summarised.

7.1
Natural Radiation Sources

7.1.1
The Sun

The sun has approximately the spectral radiation distribution of a black body with the distribution temperature $T_V = 6000$ K. The solar radiation undergoes absorption in the atmosphere in some wavelength ranges.

On the surface of the earth, the solar spectrum contains characteristic H_2O and

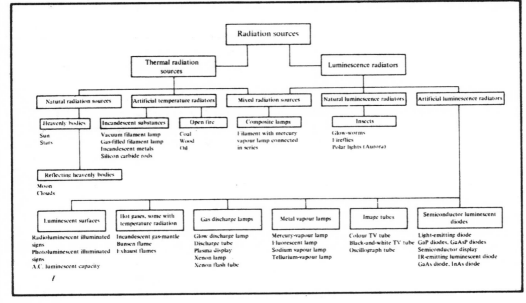

Figure 7.1
Systematic summary of the most important radiation sources.

CO_2 absorption bands (*Figure 7.2*).

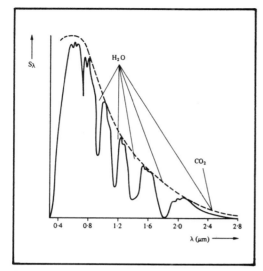

Figure 7.2
The solar spectrum with absorption bands

The broken line shows the solar spectrum without absorption by the atmosphere. In summer, around midday, with solar radiation, an illuminance up to 130 000 lm/m^2 can be measured on the surface of the earth. The illuminance of daylight in winter is about 10 000 lm/m^2.

7.1.2
The Moon

The light radiated from the moon to the earth has an illuminance of about 0.1 lm/m^2.

7.1.3
Clouds

Clouds reflect the radiation from the sun, and also from the earth, very effectively. Although, their radiance is small, there are normally large areas of clouds and the total radiant power is relatively large.

7.2
Artificial Temperature Radiators

7.2.1
Open Fire

In heating installations, fires are often monitored by optoelectronic means. In this case the 12 Hz flicker frequency of the flames is mostly used. The spectral radiation distribution depends on the burning substance and its temperature. In some wavelength ranges, the flames radiate like "grey emitters", since they show the spectral radiation distribution of a Planck radiator. *Figure 7.3* shows,

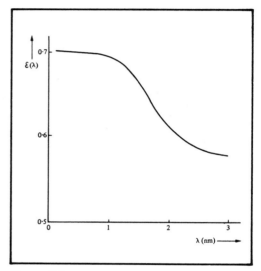

Figure 7.3
Spectral emissivity of coal at 2000 K

as an example, the spectral emissivity of coal at a temperature T = 2000 K. For the visible range, the spectral emissivity is uniformly $E(\lambda) = 0.7$.

7.2.2
Filament Lamps

Filament lamps consist of a glass bulb, which is either evacuated or filled with gas. The radiation is normally produced by an incandescent tungsten filament. Filament lamps are temperature radiators which means they have a continuous spectral radiation distribution (see Chapter 4). The working temperatures lie between 2200 K and 3000 K, the melting point of tungsten being about 3600 K.

The radiant power, luminous power and also the life are dependent, to a great extent, on the voltage. *Figure 7.4* shows the variation in the life time and the

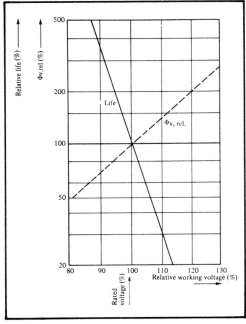

Figure 7.4
Variation in the luminous power and the life of a filament lamp with deviation from the rated voltage

luminous power of a filament lamp with deviation from the rated voltage.

Filament lamps have a positive temperature coefficient and can therefore be used, for example, in Wien bridge circuits, as PTC resistors. The limiting frequency of small filament lamps lies between 15 Hz and 100 Hz. As an example, the frequency response of a scale lamp is illustrated in *Figure 7.5*.

7.3
Luminescence radiators

7.3.1
Hot gases

Hot gases are generally selective radiators. They have a few emission bands. Through the combustion process, CO, CO_2 and H_2O emission bands also occur (CO = 4·8 μm; CO_2 = 2·7 μm, 4·4 μm and 15 μm; H_2O = 2·8 μm and 6·7 μm). Carbon components in the flame cause a continuous spectral emission distribution similar to that of the black body. *Figure 7.6* shows, as an example, the spectral radiation distribution of a low-pressure Bunsen flame. The continuous spectrum very clearly has emission bands superimposed on it.

7.3.2
Discharges through gases

Under normal conditions, gases are good insulators, as long as any applied electrical field strength remains efficiently small. This changes, however, if the field strength exceeds a certain level, depending on the type of gas and the gas pressure. Gas discharges, associated with luminescence phenomena, then take place.

Because of the effect of short-wave and radioactive radiations, which occur practically everywhere, and of thermal excitation, a tiny fraction of the atoms or molecules in every gas is always ionised. These ionised particles (free electrons, positively and negatively charged ions) move under the influence of an electrical field and produce a very weak current, which can only be detected with the most sensitive measuring instruments. Since no luminescence phenomena are yet to be observed at this stage, the term "dark discharge" is used.

The glow or arc discharges which occur with increased field strength can appear in the most widely varying forms. A distinction is made, for example, between low-pressure and high-pressure discharges.

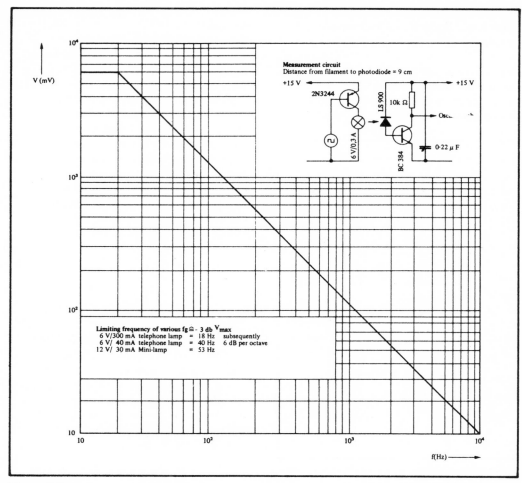

Limiting frequency of various fg ≙ - 3 db V_{max}
6 V/300 mA telephone lamp = 18 Hz subsequently
6 V/ 40 mA telephone lamp = 40 Hz 6 dB per octave
12 V/ 30 mA Mini-lamp = 53 Hz

Figure 7.5
Frequency response of a 6 V/0·3 A filament lamp

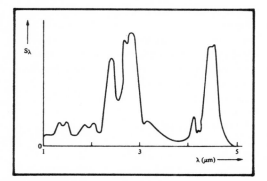

Figure 7.6
Spectral radiation distribution of a low-pressure Bunsen flame

In principle, a glow discharge can be produced at any pressure, but the luminescence phenomena occurring below a pressure of about 100 mbar are the most striking.

In a suitably constructed discharge tube, before a visible flow discharge commences through an increase in the field strength, the current first rises, because the original free electrons, being accelerated more and more, ionise neutral particles by giving up their kinetic energy and thus multiply the number of available charge carriers. If the primary cause of the spontaneously

106

produced charge carriers were now to be suppressed, the whole flow of current would come to a standstill, since now, too, no more additional charge carriers are formed. This transitional range is called the "Townsend discharge" and, like the dark discharge, is not self-supporting, since an external means of ionisation is necessary to maintain the current.

With further increase in the field strength, the positive ions are given so much energy of motion, that finally, on collision with the cathode, they liberate secondary electrons and thus greatly increase the current. Luminescence phenomena, such as are illustrated in *Figure 7.7*, then commence. In this, the negative glow and the positive column are of particular technical interest. *Figure 7.8* shows the potential, field-strength and space-charge distribution of a glow discharge in schematic form. The gas, which is ionised to a certain extent during discharges, is called a *plasma*.

Glow and arc discharges can be maintained, even without external ionisation; they are therefore called self-supporting discharges.

With further increase in the applied voltage, an arc discharge is finally obtained with very heavy currents, while the voltage on the electrodes collapses to a fairly low value. The cause of the high current lies in the fact that, through the intensive ion bombardment on the cathode, the latter is heated up, so that a plentiful thermal emission takes place.

With a variation of the arc discharge, which occurs in tubes with a liquid mercury cathode, it is not a matter of thermal emission but of field emission. Here, the arc constricts itself very tightly just in front of the cathode, while field strengths up to 10^8 V/m are produced.

Figure 7.9 shows a typical I-V discharge characteristic. The glow discharge is a cold discharge, since generally no significant heat is produced. The spectrum of the luminescence radiation, occurring through recombination, is substantially dependent on the pressure and the type of gas. At low pressures, discrete line spectra are obtained, and at higher pressures, through the increasing effect of the particles on one another, these merge into more or less continuous band spectra. The spatial distribution of the luminescent layers is also pressure-dependent. The lower the pressure, the more the areas at the cathode end spread at the expense of those at the anode end. Changes in the electrode spacing only affect the length of the positive column. If the anode is moved into the cathode drop area, the glow discharge is extinguished.

Arc discharges take place both at low and higher pressures. The spectra, which depend on the type of gas and the pressure, are often accompanied by a noticeable temperature radiation. At very high pressures, continuous spectra are obtained over a wide wavelength range.

7.3.3
Discharge lamps

The discharge lamp has a cathode in the form of a metal cylinder or in the shape of a beehive. A rod-shaped anode is located inside the cathode cylinder. Discharge lamps are filled either with neon or helium gas or a corresponding gas mixture. Only the negative cathode glow light is utilised. The colour of the light is reddish or orange. The current is limited by a series resistance. The limiting frequency lies between 10 kHz and 100 kHz, depending on the type.

7.3.4
Discharge tubes

Discharge tubes are gas-filled glass tubes. The unheated electrodes are fitted at each end of the tube. A stray-field transformer

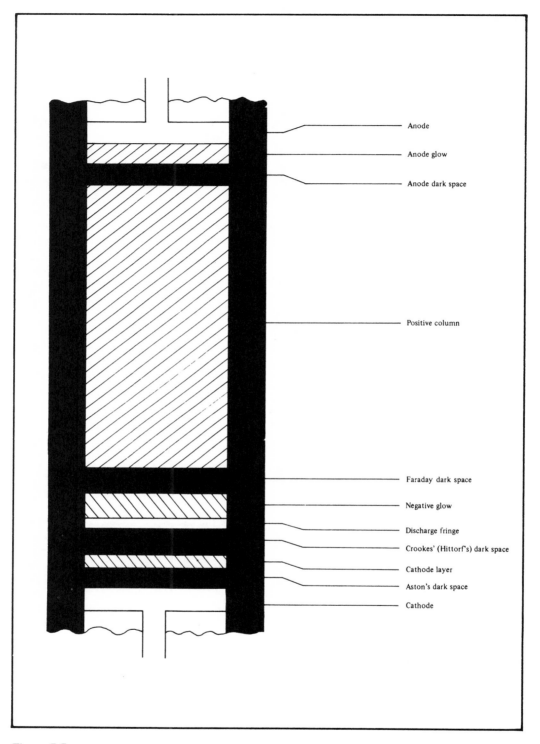

Anode

Anode glow

Anode dark space

Positive column

Faraday dark space

Negative glow

Discharge fringe

Crookes' (Hittorf's) dark space

Cathode layer

Aston's dark space

Cathode

Figure 7.7
The individual regions of a low-pressure glow discharge

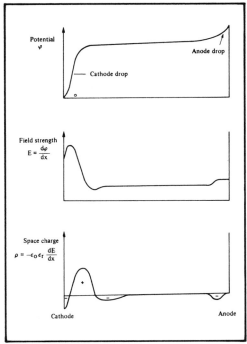

Figure 7.8
Potential distribution, field strength and space charge in a low-pressure glow discharge

supplies the necessary high voltage for ignition and limits the working current through the action of the stray field. With the exception of the xenon tube, discharge tubes produce a discrete line spectrum. *Table 7.1* shows the light colour of discharge tubes with different gas fillings.

Ligh colour	Gas filling
Red	Neon
Yellow	Helium
Blue	Neon and mercury
White	Carbon dioxide
Green	Neon and mercury with brown coloured glasses

Table 7.1
Light colour of discharge tubes with difference gas fillings

7.3.5
Xenon lamps

Xenon lamps consist of quartz bulbs with a xenon gas filling. They are fired with high voltage. In the visible range, xenon lamps have a continuous radiation distribution with a light superimposition of spectral bands. It is therefore possible to allocate a colour temperature to the visible spectrum of the xenon lamp. It is about $T_f = 6000$ K. *Figure 7.10* shows the spectral radiation distribution of a xenon lamp. In the near IR range, two pronounced spectral bands can be seen. Xenon lamps can therefore also be used in IR transmission systems.

7.3.6
Metal Vapour Lamps

Metal vapour lamps are constructed as glass tubes with a metal vapour filling. Ballast units are necessary for ignition and for current limitation in the operating condition. Metal vapour lamps produce a discrete line spectrum.

7.3.6.1
Sodium Vapour Lamps

These produce almost monochromatic yellow light with a wavelength $\lambda = 589$ nm.

7.3.6.2
Low-pressure Mercury-vapour Lamps

These lamps radiate mainly in the UV range.

7.3.6.3
High-pressure and Very-high-pressure Mercury-vapour Lamps

The spectral radiation distribution of high-pressure and very-high-pressure mercury vapour lamps lies in the UV and, with the

109

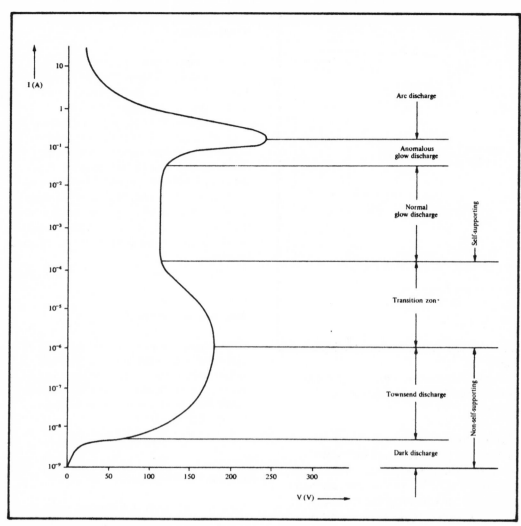

Figure 7.9
Typical I-V characteristic of a gas discharge

colours blue and green, in the visible range (see *Figure 7.11*).

7.3.6.4
Fluorescent lamps

These are low-pressure mercury-vapour lamps, in which the four characteristic Hg-vapour spectral lines are initially used. The invisible UV radiation excites the coating of flurescent material, which is wash-coated onto the interior wall of the glass tube, to fluorescent radiation. A continuous spectrum in the visible range is thus produced. All colours can be achieved by suitable selection of the fluorescent materials. *Figures 7.12* and *7.13* show the spectral radiation distribution $S_\lambda = f(\lambda)$ of a white-light and a warm-white fluorescent tube.

7.4
Mixed light lamps

Mixed-light lamps are composite lamps.

110

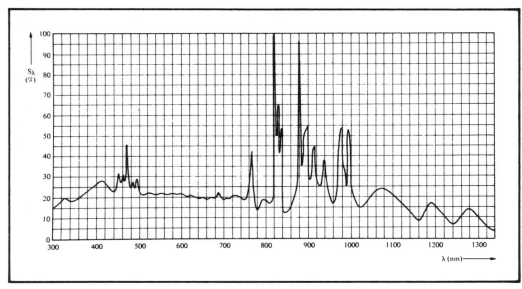

Figure 7.10
Spectral radiation distribution of a high-pressure xenon lamp

They contain, in one glass tube, an incandescent filament and a high-pressure mercury-vapour lamp. The reddish filament lamp light and the greenish-blue mercury vapour light are combined in one lamp. These combined luminescence and temperature radiators are also obtainable with a fluorescent coating on the interior of the glass tube.

7.5
Flash discharge tubes in photography

In amateur flash units, xenon flash lamps or tubes are used as a flash radiation source. *Figure 7.14* shows the construction and the technical data of a standard flash discharge tube. The colour temperature of the xenon-filled flash discharge tubes lies

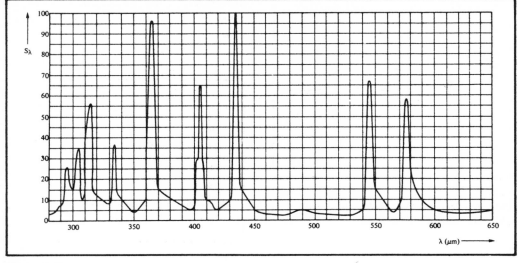

Figure 7.11
Spectral radiation distribution of a very-high-pressure mercury-vapour lamp

111

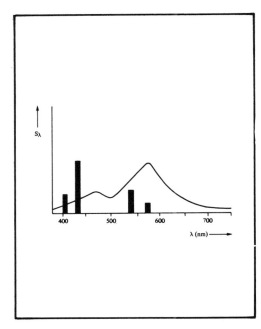

Figure 7.12
Spectral radiation distribution of a white
fluorescent lamp

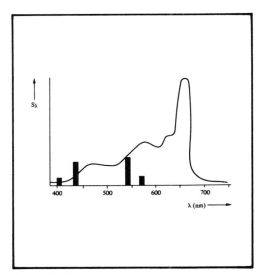

Figure 7.13
Spectral radiation distribution of a warm-
white fluorescent lamp

between 5000 and 7000 K. Short
discharge times with high peak currents

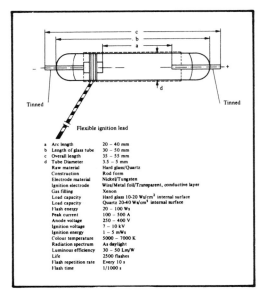

Figure 7.14
Standard flash discharge tube for
photographic applications

produce higher colour temperatures. With
hard-glass tubes, the UV component, up
to about 350 nm, is absorbed by the glass.
The spectral radiation distribution
approximately corresponds to the spectral
distribution of daylight with a superimposed
line spectrum. The spectral radiation
distribution of the xenon flash discharge
tube and that of daylight are illustrated in
Figure 7.15.

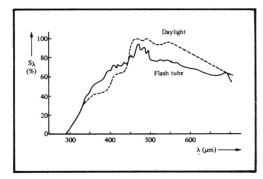

Figure 7.15
Spectral radiation distribution of the xenon
flash discharge tube and of daylight

7.6
Luminescence Diodes

The radiation of a luminescence diode (light-emitting diode) is produced by the recombination of the charge carriers injected in the junction region. Injection luminescence has been dealt with in Section 1.2.3. Wafer geometries will be described in Section 11.1.

7.6.1
Silicon-doped GaAs Diodes

In the mass production of wafers for use as luminescence diodes, two technologies have become principally established. The first method is liquid epitaxy, the principle of which is illustrated, for Si-diped GaAs diodes, in *Figure 7.16*. Here an-N-type GaAs single-crystal slice, doped with silicon, is sunk into a gallium melt, saturated with GaAs at about 900°C and doped with

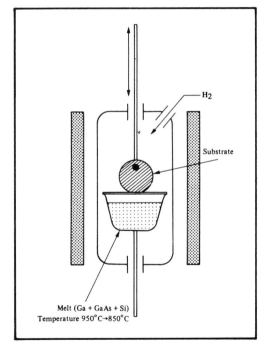

Figure 7.16
Principle of liquid epitaxy (solution growth) in the vertical reactor for Si-doped GaAs diodes

silicon. While the melt cools down, a deposit forms on the substrate slice.

As a doping substance, silicon behaves here in an amphoteric manner, that is, as a quadrivalent element it can be fitted into the GaAs lattice in two possible ways. Silicon atoms either replace gallium atoms and thus form donors or they replace arsenic atoms and act as acceptors.

Above a temperature of 900°C, silicon occupies more donor places than acceptor levels. Below a temperature of 900°C, on the other hand, it occupies more acceptor levels than donor levels. Through a continuous growth process with an initial temperature of about 950°C and

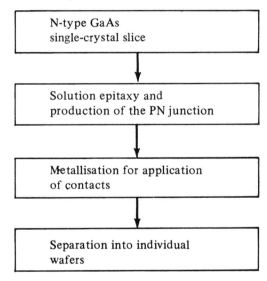

Figure 7.17
Simplified production scheme for Si-doped GaAs wafers

subsequent reduction of the temperature to about 850°C, a PN junction, formed. *Figure 7.17* shows a greatly simplified production scheme for silicon-doped GaAs wafers.

7.6.2
GaP Diodes

GaP diodes are produced in a double

epitaxy process. An N-type GaP single crystal, drawn by the Czochralski process and doped with tellurium, serves as the substrate. In the first liquid epitaxy process to produce red-emitting wafers, a tellurium doped-n-type GaP layer is grown. For clarification, *Figure 7.18* shows the principle of liquid epitaxy for red-emitting GaP wafers.) In the second stage of the liquid epitaxy process, the PN junction is formed by growing a p-layer doped with zinc and oxygen, on top of the tellurium doped n-layer.

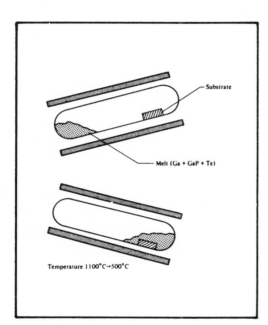

Figure 7.18
Principle of liquid epitaxy in the horizontal reactor for the production of red-emitting GaP diodes

7.6.3
Zinc-doped GaAs and GaAsP Diodes

This second process has been successfully adopted for the mass production of zinc-doped GaAs and GaAsP wafers. In this case, vapour phase epitaxy is combined with planar technology GaAsP cannot be drawn from a melt with the correct

proportion of arsenic and phosphorus to produce a useful diode and this problem is overcome in the vapour phase epitaxy process. Here a single-crystal layer of GaAsP or GaAs is allowed to grow on single-crystal GaAs slices. *Figure 7.19* shows the principle of vapour phase epitaxy used for the production of GaAsP wafers. The PN junction is produced in the subsequent planar process. Finally, in *Figure 7.20*, a

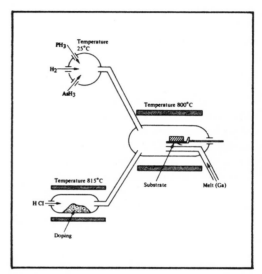

Figure 7.19
Principle of gas epitaxy for GaAsP wafers

simplified production scheme for zinc-doped GaAs and GaAsP wafers is shown.

7.6.4
Consideration of quantum yield

The highest quantum yields are at present obtained with the luminescence diodes produced by the liquid epitaxy process. In the liquid epitaxy process, as compared with vapour phase epitaxy, the slice surface which is composed of disordered diffusion atoms is first dissolved, so that an epitaxial layer of high perfection can be built up out of the saturated gallium melt.

114

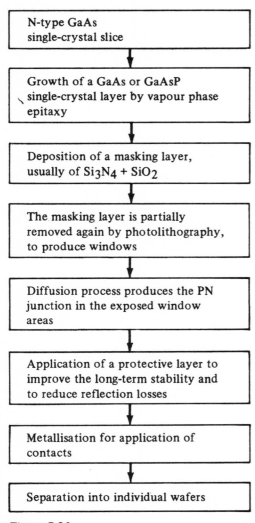

Figure 7.20
Simplified production scheme for zinc-doped GaAs and GaAsP wafers

8
Photodetectors

117

Photodetectors

Photodetectors are classified into various groups according to the photoeffect utilised. The most important kinds are summarised in *Figure 8.1*. The thermal and pneumatic photodetectors, which are mainly used for radiation measurements, are described as far as necessary in Chapter 12 (radiation measurements). The photoelectronic detectors are distinguished according to whether the external or internal photoeffect is used. *Figure 8.2* shows the spectral distribution of the most important semi-conductor radiation sources and photodetectors.

8.1
Photodetectors with external photoeffect

The external photoeffect has already been described in Section 1.3.1. Photodetectors with external photoeffect are divided into non-amplifying and amplifying photocells. In principle, they consist of a photo-cathode and an anode in a vacuum or gas-filled tube.

The photosensitive cathode surface is made from alkali metals or their compounds, such as silicon-antimony, potassium, rubidium, caesium, caesium-antomony or caesium oxide or multi-alkali compounds. The relative spectral sensitivity distribution of alkali metals is shown in *Figure 1.8*.

There is a linear relationship between the photocurrent I_p of a vacuum photocell and the incident irradiance E_e.

In gas-filled photocells, the photoelectrons, which are accelerated as a result of the electric field between the cathode and the anode, produce secondary electrons on collision with gas atoms. These increase the electron current to the anode, while the positive gas ions diffuse towards the cathode. The ion bombardment on the photocathode causes a greater decrease in sensitivity with time then with vacuum photocells. Argon is mostly used as the filling gas. The operating voltage lies around 60 to 75% of the breakover voltage. The limiting frequency of the gas

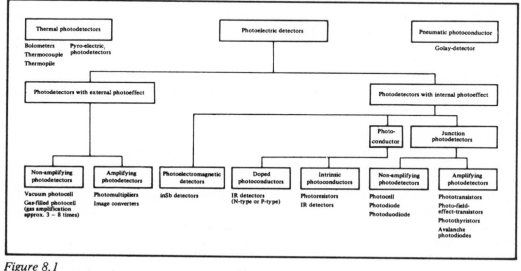

Figure 8.1
Systematic summary of the most important photodetectors

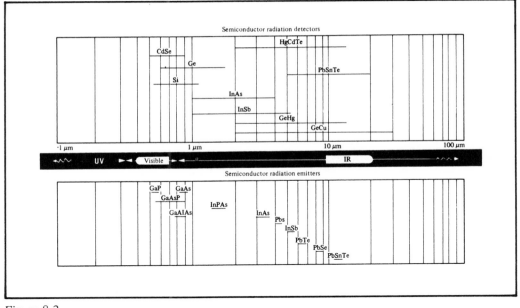

Figure 8.2
Spectral distribution of the most important semiconductor radiation sources and photodetector.

photocell is low in comparison with the vacuum photocell and is usually 5 to 10 kHz (f_g).

Because of the formation of secondary electrons, gas photocells have a small gain of M = 3 to 8. The characterisitc I_p = $f(E_e)$ is not linear and, in the higher voltage range, can be compared with the output characteristics of triode valves. Gas photocells are being increasingly superseded by semiconductor photodetectors.

The combination of a vacuum photocell with a secondary electron multiplier produces a photomultiplier. On impact with a dynode (auxiliary anode), a photoelectron produces several secondary electrons. In this process, no harmful free ions are produced. The amplification of the photocurrent takes place in several stages. It amounts to gains of several orders of magnitude, depending on construction and type.

8.2
Photodetectors with internal photoeffect

The internal photoeffect has been described in Sections 1.3.2 and 1.3.3. The photodetectors with internal photoeffect include photoelectromagnetic detectors, photoconductors and barrier-layer photodetectors. The photoelectromagnetic detectors have so far achieved hardly any practical significance.

8.2.1
Photoconductors or photoresistors

The photoconductor utilises the change in conductivity of a semiconductor under an incident radiation. Photoconductors are either pure intrinsic semiconductors or N-type or P-type doped semiconductors. A distinction is therefore made between intrinsic photoconductors and doped photo-semiconductors.

120

Intrinsic photoconductors are produced both for the visible and for the very near IR range from cadmium chalcogenides such as cadmium sulphide (CdS), cadmium selenide (CdSe) and cadmium telluride (CdTe). Intrinsic photoconductors for the near IR range consist of lead salts such as lead sulphide (PbS), lead selenide (PbSe) and lead telluride. For the middle and far IR range, intrinsic photoconductors consisting of indium arsenide (InAs), indium antimonide (InSb), tellurium (Te) and mercury cadmium telluride (HgCdTe) are used as IR detectors.

The doped photo-semiconductors are doped with appropriate substances, in order to achieve a desired spectral sensitivity in the middle and far IR range up to about 40 μm. So far they are mainly made from germanium or germanium-silicon compounds with appropriate doping. The most important doped photosemi-conductors are:
Gold-doped germanium (Ge:Au);
Gold-, antimony-doped germanium (Ge:Au,Sb);
Zinc-doped germanium (Ge:Zn);
Zinc-, antomony-doped germanium (Ge:Zn,Sb);
Copper-doped germanium (Ge:Cu);
Cadmium-doped germanium (Ge:Cd);
Gold-doped germanium-silicon alloy (Ge-Si:Au);
Zinc-, antimony-doped germanium-silicon alloy (Ge-Si:Zn, Sb).

Generally, the sensitivity of photoresistors are dependent on their "history", so that they show undesired fatigue phenomena, as compared with junction photodetectors. Because of the long life of the charge carriers, the frequency response of sensitive photoresistors is relatively low.

Depending on their design and spectral sensitivity, IR detectors are operated at low temperatures in Dewar flasks or cryostats. The most important sensitivity values of IR detectors are dealt with in more detail in Section 9.9.

8.3
Junction photodetectors

These photodetectors are classified into non-amplifying and amplifying photo-detectors. The non-amplifying devices include photodiodes, photoduodiodes, Schottky barrier photodiodes and photo-cells. The amplifying photodetectors include phototransistors, photo-field-effect-transistors, photothyristors and avalanche photodiodes. The parameters of non-amplifying and amplifying junction photo-detectors are dealt with in detail in Chapter 9.

9
Parameters of IR Detectors and Junction Photodetectors

**Parameters of IR Detectors and
Semiconductor Junction Photodetectors**

9.1

Quantum yield, Q, of junction photodetectors

For the measurement and calculation of quantum yield Q of junction radiation detectors, non-amplifying photodetectors are advantageously operated as photocells. Photodiodes and photocells are physically identical electrical components. They only differ in the electrical mode of operation. Photodiodes are operated with an external voltage source in the reverse direction, while photocells (solar cells) themselves generate a voltage, the photo-voltage.

Junction photodetectors with internal amplification such as avalanche photodiodes or phototransistors, with an external base connection, can also be operated as photocells. Phototransistors obtain their base control current from the photocurrent generated in the collector-base photodiode (see Chapter 1). Their current amplification factor M depends on the magnitude of the photocurrent. Therefore, the quantum yield Q of a phototransistor is conveniently determined from the collector-base junction, operated as a photocell. Similar considerations apply for field effect phototransisitors and photothyristors, since these also come into the category of junction photodetectors with internal amplification.

The photocurrent I_p of a (non-amplifying) junction photocell is proportional to the incident radiant power Φ_e . The quantum yield of a junction photocell can therefore be defined as the ratio of the number of electrons n_E flowing in the external circuit per unit of time to the number of photons n_{Ph} falling on the photodetector. Thus:

$$Q = \frac{n_E}{t} : \frac{n_{Ph}}{t} = \frac{n_E}{n_{Ph}} \qquad (9.1)$$

Here, the number of photons per unit time for monochromatic radiation falling on the photodetector is the ratio of the incident radiant power (see equation 2.4) to the energy of one photon (see equation 1.1)

$$n_{Ph} = \frac{\Phi_{e,\lambda} \cdot t}{h \cdot \nu} \qquad (9.2)$$

The number of electrons per unit time n_E is the ratio of the product of the electrical photocurrent I_p and the time t to the electronic charge $e = 1 \cdot 6 \cdot 10^{-19}$ As. of one electron.

$$n_E = \frac{I_p \cdot t}{e} \qquad (9.3)$$

If the equations (9.2), (9.3), (1.2) and (1.3) are inserted in equation (9.1), the quantum yield for monochromatic radiation $Q(\lambda)$ is obtained:

$$Q(\lambda) = \frac{I_{p,\lambda} \cdot t \cdot h \cdot \nu}{e \cdot \Phi_{e,\lambda} \cdot t} = \frac{I_{p,\lambda} \cdot h \cdot c_0}{\Phi_{e,\lambda} \cdot \lambda \cdot e} \qquad (9.4)$$

The constants h and c_0/e (see Chapter 1) can be replaced by a common constant factor:

$$\frac{h \cdot c_0}{e} = \frac{6 \cdot 62 \cdot 10^{-34} Ws^2 \cdot 3 \cdot 10^8 ms^{-1}}{1 \cdot 6 \cdot 10^{-19} As}$$

$$= 1 \cdot 24 \cdot 10^{-6} WmA^{-1} \qquad (9.5)$$

The quotient of the photocurrent $I_{p,\lambda}$ and the incident monochromatic radiant power $\Phi_{e,\lambda}$ can be replaced by the absolute spectral photosensitivity for photocells (see Sections 9.2 and 9.3):

$$\frac{I_{p,\lambda}}{\Phi_{e,\lambda}} = s(\lambda) \qquad (9.6)$$

If the equations (9.5) and (9.6) are inserted in equation (9.4), then, with the absolute spectral sensitivity $s(\lambda)$ for monochromatic radiation of wavelength λ, the quantum yield $Q(\lambda)$ for the same monochromatic radiation of wavelength λ can be calculated:

$$Q(\lambda) = \frac{s(\lambda)}{\lambda} \cdot 1 \cdot 24 \cdot 10^{-6} \, \text{WmA}^{-1} \tag{9.7}$$

If the quantum yield for monochromatic radiation $Q(\lambda)$ of a junction photocell is equal to unity, then by conversion of equation (9.7), the theoretical absolute spectral sensitivity $s(\lambda)_{th}$ can be determined (see Section 9.2):

$$s(\lambda)_{th} = \frac{\lambda}{1 \cdot 24 \cdot 10^{-6} \, \text{WmA}^{-1}} \tag{9.8}$$

If the absolute spectral sensitivity $s(\lambda)$ of a junction photocell is known, the quantum yield for monochromatic radiation $Q(\lambda)$ can also be calculated by forming the quotient of the absolute spectral sensitivity $s(\lambda)$ at the radiation wavelength λ divided by the theoretical absolute spectral sensitivity $s(\lambda)_{th}$ at the same radiation wavelength λ:

$$Q(\lambda) = \frac{s(\lambda)}{s(\lambda)_{th}} \tag{9.9}$$

The quantum yield for monochromatic radiation $Q(\lambda)$ is also stated as a relative value. The relative quantum yield for monochromatic radiation $Q(\lambda)_{rel}$ is the ratio of the maximum absolute spectral quantum yield $Q(\lambda)_{max}$ for monochromatic radiation of the maximum wavelength λ_{max} to the absolute spectral quantum yield for monochromatic radiation $Q(\lambda)$ at any given radiation wavelength λ:

$$Q(\lambda) = \frac{Q(\lambda)_{max}}{Q(\lambda)} \tag{9.10}$$

The spectral quantum yield for silicon photodiodes is dependent mainly on the temperature, in the longer-wavelength limiting sensitivity range. *Figure 9.1* shows

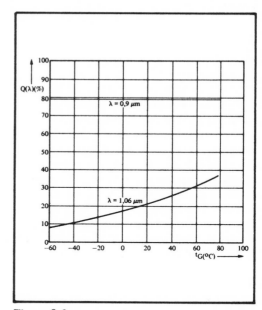

Figure 9.1
Spectral quantum yield $Q(\lambda)$ as a function of the case temperature t_G for the Si photodiode TIXL 80

the spectral quantum yield $Q(\lambda)$ as a function of the case temperature t_G for the large-area Si photodiode TIXL 80. The variable in each case is the wavelength λ of the incident monochromatic radiation.

9.1.1
Quantum yield and photocurrent amplification of avalanche photodiodes

For junction photodetectors with internal amplification, the quantum yield $Q(\lambda)$ for monochromatic radiation can be determined if the internal gain M is known. The quantum yield for monochromatic radiation, $Q(\lambda)$, is measured for avalanche photodiodes, with the amplification factor M set by the applied reverse voltage, for the given wavelength of the laser radiation used. In the data sheet, a typical absolute value for $Q(\lambda)$ and also the function $Q(\lambda)_{rel} = f(\lambda)$ are usually stated. *Figure 9.2*

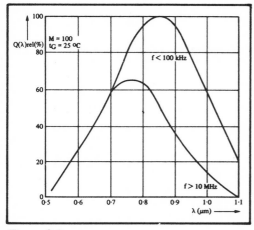

Figure 9.2
Relative quantum yield $Q(\lambda)_{rel}$ as a function of the radiation wavelength λ for the silicon avalanche photodiode TIXL 56.

shows the relative quantum yield $Q(\lambda)$ as a function of the radiation wavelength λ for the silicon avalanche photodiode TIXL 56. Here, the case temperature (t_G) is 25°C and the photocurrent amplification M = 100 The variable is the modulation frequency of the incident radiation.

The maximum quantum yield for avalanche photodiodes is reached with the highest practicable amplification M. To achieve this and to make full use of the avalanche effect, the working point is generally set a few tenths of a volt below the breakdown voltage V_{BR}. If the working point is in the breakdown region, the useful signal power is considerably less than the total noise power produced through the increased current noise. The applied reverse voltage V_R must therefore be adjusted exactly, in order to compensate for the tolerances stated in the data sheet for the breakdown voltage V_{BR}, typically between 140 and 200 V for silicon types and between 30 and 60 V or 85 and 150 V for germanium. Among the individual devices within a production batch, however, a distribution of only a few volts is normally seen for breakdown voltage.

The temperature coefficient $TK_{V_{BR}}$ of the

breakdown voltage V_{BR} is calculated in accordance with the following equations:

$$TK_{V_{BR}} = \frac{V_{BR}(125^\circ C) - V_{BR}(-55^\circ C)}{125^\circ C - (-55^\circ C)}$$

$$TK_{V_{BR},rel} = \frac{V_{BR}(125^\circ C) - V_{BR}(-55^\circ C)}{V_{BR}(25^\circ C)} \cdot \frac{100\%}{180^\circ C}$$

The temperature limits are to be taken from the data sheet in each case. The typical relative temperature coefficient $TK_{V_{BR},rel}$ for Si avalanche photodiodes is about $0 \cdot 11\%/^\circ$C and for Ge devices about $0 \cdot 15\%/^\circ$C.

The photocurrent amplification M for avalanche photodiodes is measured with a fixed avalanche noise response threshold. Here, the amplification M_T is to be determined for the applied reverse voltage V_R, at which the noise deviates from the theoretical characteristic *Figure 9.3* shows the incident radiant power Φ_e and the noise power P_N as

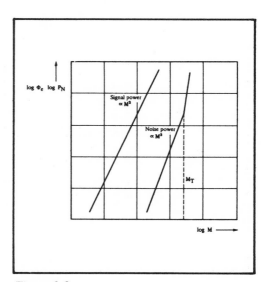

Figure 9.3
Φ_e and P_N as functions of M for the avalanche photodiode TIXL 56.

127

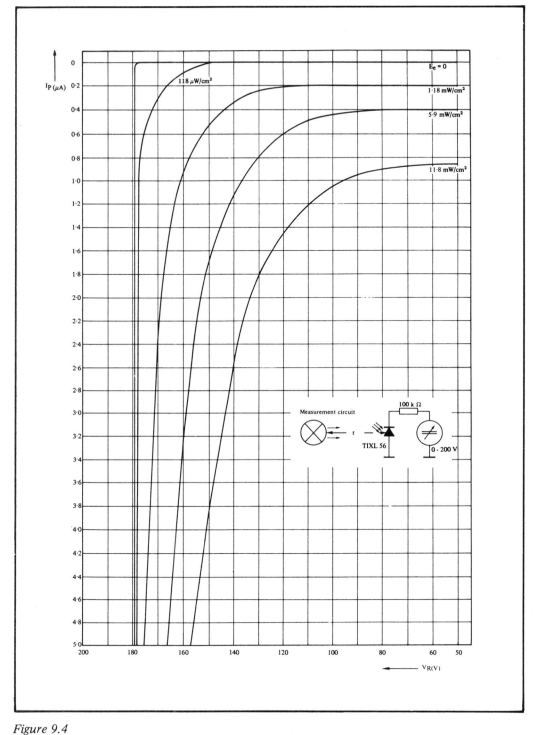

Figure 9.4
$I_p = f(V_R)$ *characteristics for the silicon avalanche photodiode TIXL 56 as a function the irradiance E_e.*

128

functions of the photocurrent amplification M. The incident radiant power Φ_e is determined for the measurement of the photocurrent amplification M, by measuring, with an applied reverse voltage V_R = 100 V for Si avalanche photodiodes, a photocurrent I_p = 1 nA. *Figure 9.4* shows, with the irradiance E_e as a variable, the increasingly rounded form of the function I_p = $f(V_R)$ with increasing irradiance E_e. As the slope of the function I_p = $f(V_R)$ decreases, so too does the photocurrent amplification M.

The typical current amplification for silicon avalanche photodiodes is $M_{typ} > 100$ and for germanium avalanche photodiodes $M_{typ} > 40$. The photocurrent amplification M can be seen as a function of the applied reverse voltage V_R for the silicon avalanche photodiodes TIXL 56 and TIXL 59 in *Figure 9.5* and for the germanium avalanche photodiodes TIXL 57 and TIXL 68 in *Figure 9.6*.

The product of amplification and bandwidth (M . Δf) for a photodetector measures the modulation frequency of the incident radiation, for which the photodetector still has an amplification of M = 1. Thus:

$$f_1 = (M \cdot \Delta f) \qquad (9.11)$$

The (M . Δf) product of a photodetector is stated for the working point, at which it has its maximum amplification M.

For avalanche photodiodes, the amplification M is measured with a modulated laser radiation and the value of the amplification-bandwidth product (M . Δf) is calculated in accordance with equation (9.11). The modulation frequency of the laser radiation is normally about 1 GHz. The wavelength of the laser radiation has been selected at λ = 632·8 nm (HeNe laser) for the measurement of Si avalanche photodiodes and at λ = 1·15 μm for the measurement of Ge avalanche photodiodes. The product of amplification and bandwidth

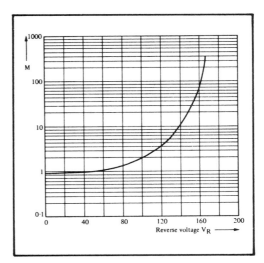

Figure 9.5
M = f(V_R) characteristic for the silicon avalanche photodiodes TIXL 56 and TIXL 59. (The case temperature is t_G = 25°C)

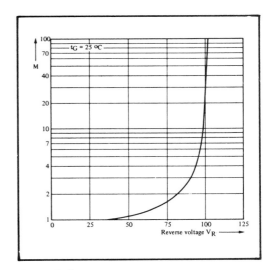

Figure 9.6
M = f(V_R) characteristic for the germanium avalanche photodiodes TIXL 57 and TIXL 68.

(M . Δf) is typically 80 GHz for Si avalanche photodiodes and typically 50 GHz for Ge avalanche photodiodes.

129

9.2
Spectral sensitivity of junction photodetectors

The spectral sensitivity of a junction photodetector depends on the spectral absorption of radiation falling on the junction. The absorbed radiation causes electrons to leave the valency band and climb to the conduction band. The fact, that the electrical field of the space-charge zone separates the pairs of charge carriers produced, was described in Section 1.3.3. The electrons diffuse into the N-region, the holes into the P-region. At the external connections of a junction photocell, in no-load operation the maximum photovoltage and, in short-circuited operation the maximum photocurrent, can be measured. If an incident photon has just the minimum energy to raise an electron from the valency band into the conduction band, this kind of absorption is called "basic lattice" or "intrinsic" absorption. Also, this minimum energy value determines the limiting wavelength, on the long-wave side, for the absorbed radiation. For shorter radiation wavelengths, the absorption generally increases rapidly in semiconductors. The steep transition between relatively low and relatively high absorption is usually defined as the absorption edge, or for semiconductors as the basic lattice absorption edge. With photosemiconductors, there follows, beyond the short-wave side of the absorption edge, the relatively narrow wavelength range of maximum absorption.

If the limiting wavelength on the longwave side of the absorbed radiation, or the wavelength associated with the absorption edge, is defined as that wavelength with the half-value of the maximum spectral photosensitivity, then the width of the "forbidden band" can be determined in simplified form in accordance with equations (1.3) and (1.5):

$$\lambda_G \approx \frac{h \cdot c_o}{\Delta w} = \frac{1 \cdot 24}{\Delta w} \qquad (9.12)$$

In contrast to pure semiconductors, doped semiconductors also absorb a somewhat longer wavelength radiation. To raise an electron from the valency band to the energy level of the foreign atoms, a photon needs less excitation energy, as compared with the intrinsic absorption mode.

Corresponding to the smaller energy difference between the valency band and the dopant level, this absorption follows the basic lattice absorption, on the long-wave side. This kind of absorption is called divergent or "extrinsic" absorption.

Beyond the long-wave side of the absorption edge, in semiconductors, there follows an increasing transmissivity. This is associated with a decrease in absorption and in radiation sensitivity. *Figure 9.7* shows, as an example, the spectral

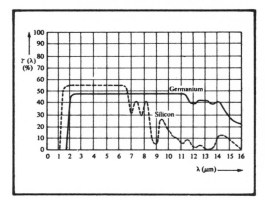

Figure 9.7
Spectral transmission τ(λ), as a function of the radiation wavelength λ, of silicon and germanium; the layer thickness (S) is 10 mm in each case.

transmission of silicon and germanium. These materials can also be used as very effective filters against visible and very near infra-red radiation.

For non-amplifying junction photodetectors, the maximum attainable absolute spectral sensitivity for every wavelength is described by the relationship calculated from the theoretical spectral sensitivity $s(\lambda)_{th}$. The

130

theoretical spectral sensitivity $s(\lambda)_{th}$ corresponds to the absolute spectral sensitivity $s(\lambda)$ for a semiconductor photodetector of quantum yield $Q(\lambda) = 1$. For this, equation (9.8), already stated in Section 9.1, applies:

$$s(\lambda)_{th} = \frac{\lambda}{1 \cdot 24 \cdot 10^{-6} WA^{-1} m}$$

According to this equation, the theoretical spectral sensitivity $s(\lambda)_{th}$ is a linearly increasing function, dependent on the radiation wavelength. At the imaginary shortest wavelength $\lambda_o = 0$, the theoretical spectral sensitivity $s(\lambda)_{th} = 0$. From a certain longer radiation wavelength, characteristic of the particular semiconductor material, λ_T, the semiconductor becomes transparent to this radiation. At this wavelength λ_T, the theoretical spectral sensitivity $s(\lambda)_{th,T}$ is also = 0. The absorption of the longest wavelength radiation which is still possible, i.e., before the photosemiconductor shows transmissivity to radiation, determines the theoretical maximum attainable spectral sensitivity $s(\lambda)_{th,max}$ in accordance with equation (9.8). Here, the wavelength is defined as $\lambda_{th,max}$. For non-amplifying semiconductor photodetectors, all spectral sensitivity curves consequently lie in the triangle defined by:

$s(\lambda)_{th,o} = 0$ at $\lambda_o = 0$

$s(\lambda)_{th,T} = 0$ at λ_T

$s(\lambda)_{th,max} = $ max. at $\lambda_{th,max}$.

The longest wavelength which is still possible for absorbed radiation, $\lambda_{th,max}$ is approximately:

$\lambda_{th,max} = 1 \cdot 12 \ \mu m$ for silicon and
$\lambda_{th,max} = 1 \cdot 8 \ \mu m$ for germanium.

With this data and the equation (9.8), the theoretical spectral sensitivity triangle can be constructed as shown

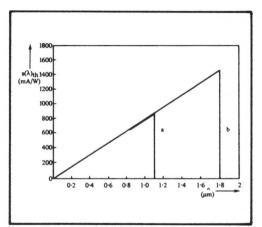

Figure 9.8
Theoretical spectral sensitivity triangle as envelope curve for all spectral sensitivity curves, a) for silicon photodetectors, b) for germanium photodetectors

in Figure 9.8 as an envelope enclosing all spectral sensitivity curves for non-amplifying silicon or germanium photodetectors.

Germanium photodetectors are nowadays only used for specialised applications. The lasers radiating on a wavelength in the range between 1 μm and $1 \cdot 6 \ \mu m$, for example YAG lasers on $1 \cdot 06 \ \mu m$, are spectrally well-matched to photodetectors germanium avalanche photodiode photodetectors. Figure 9.9 shows the relative spectral sensitivity $s(\lambda)_{rel}$ as a function of the radiation wavelength λ for the new obsolete germanium avalanche photodiode TIXL 68. The modulation frequency of the incident radiation is the variable. This photodiode had an antireflection coating for $\lambda = 1 \cdot 54 \ \mu m$.

Germanium photodetectors are being superseded more and more by silicon photodetectors.

The deliberate adjustment of the absolute spectral sensitivity $s(\lambda)$ of silicon junction photodetectors is of fundamental importance. The requirments for spectral

131

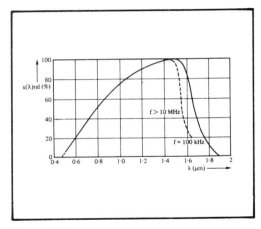

Figure 9.9
Relative spectral sensitivity s(λ)$_{rel}$ as a
function of the radiation wavelength λ
for the germanium avalanche photodiode
TIXL 68, with the modulation frequency
of the incident radiation as a variable.

performance can differ according to the
application. For selective radiometric
measurements, a linearised, flat spectral
sensitivity range is necessary, for photometric
applications the spectral photopic sensitivity
of the eye, V(λ), has to be simulated and
for general applications the theoretical
spectral sensitivity limits are desirable.

9.2.1
Schottky Barrier PIN Photodiodes

Within the spectral sensitivity range of non-
amplifying junction photodetectors, the
Schottky barrier PIN photodiodes have
the most linear sensitivity curve between
600 and 900 nm. These photodiodes
have a metal-semiconductor junction which
is formed by a very thin gold film (of the
order of 15 nm) being evaporated onto
N-doped silicon. If a voltage is applied to the
Schottky barrier PIN photodiode, with the
negative potential to the gold film and the
positive to the N-doped silicon, a space-
charge-free region is formed over the
whole junction. The charge carriers
produced by absorption can diffuse at a
relatively high speed through this intrinsic

layer. This is also accounts for the short
rise and fall times of these components.
Within the depletion region of about
1 μm, the Schottky barrier PIN photodiodes have a considerably higher absorption
coefficient than planar diffused silicon
junction photodetectors. Since shorter-
wavelength radiation penetrates less deeply
into silicon than longer-wavelength
radiation, Schottky-barrier PIN photodiodes are considerably more sensitive
than planar diffused silicon junction
photodetectors in the shorter-wavelength
absorption range. In the longer-wave
absorption range from about 800 nm, the
spectral sensitivity s(λ) of the Schottky
junction PIN photodiodes is somewhat
less than that of the planar diffused silicon
junction photodetectors. Here, the reflection
of the gold film has an effect ($\rho > 30\%$).
Figure 9.10 shows the spectral sensitivity of
Schottky barrier PIN photodiodes without
an antireflection coating.

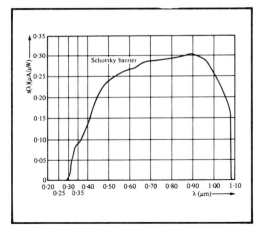

Figure 9.10
Absolute spectral sensitivity s(λ) as a
function of the radiation wavelength λ
for Schottky-barrier PIN photodiodes
without antireflection coating.

9.2.2
Planar diffused Si-photodiodes

Planar diffused silicon photodiodes have
very high quantum yields of about 70% to

132

90% in the range from 700 nm to 900 nm. The absolute spectral sensitivity s(λ) of these silicon junction photodetectors in this range is already very close to the theoretical sensitivity limit. At shorter radiation wavelengths the quantum yield Q(λ) and the blue-violet sensitivity falls off. So-called "dead zones" on the device surface, which are caused by surface recombination of diffusion atoms, prevent a higher quantum yield and higher sensitivities. To raise the blue-violet sensitivity, firstly the diffusion-induced surface defects can be reduced by improved technological processes, secondly the PN junction can be located very near the device surface and thirdly antireflection coatings can be evaporated on the surface *Figure 9.11* illustrates the spectral sensitivity s(λ) of a normal silicon photodiode and that of a silicon photodiode

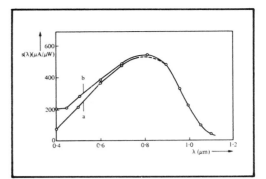

Figure 9.11
Absolute spectral sensitivity s(λ) as a function of the wavelength λ. Curve a: Normal silicon photodiode. Curve b: Silicon photodiode with increased blue-violet sensitivity

with increased blue-violet sensitivity.

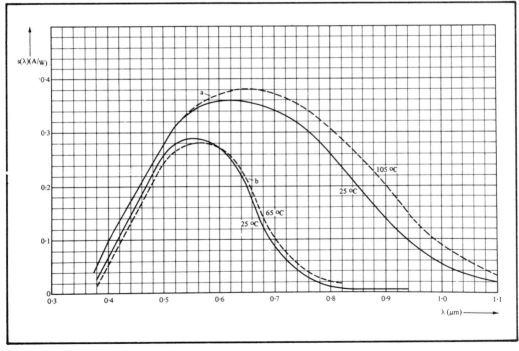

Figure 9.12
Absolute spectral sensitivity s(λ) as a function of radiation wavelength λ for silicon photodiodes produced by the CDI process. Curve a: for a deep junction. Curve b: for a shallow junction. The effect of temperature is also shown for both types.

9.2.3
Photodiodes by the CDI process

In the so-called CDI process (Collector Diffusion Isolation), the junction is even shallower (about $1 - 1.5$ μm), as compared with planar diffused photodiodes. The spectral sensitivity of these silicon junction photodiodes is thus displaced more into the shorter-wavelength region, so that these components are better matched to the sensitivity of the eye. As is shown by *Figure 9.12*, there are limits to this matching of sensitivity to the $V(\lambda)$ curve, since at the same time the absolute spectral sensitivity $s(\lambda)$ is reduced.

9.2.4
Effect of temperature on the spectral sensitivity of photodiodes

The spectral sensitivity $s(\lambda)$ can vary with increasing temperature. The sum of the energy of a photon and a phonon can give the energy difference needed for the transfer of an electron from the valency band into the conduction band. The spectral absorption and the spectral sensitivity range can therefore shift toward longer radiation wavelengths. *Figure 9.12* shows the change in the spectral sensitivity of silicon junction photodetectors, caused by the effect of temperature.

9.2.5
Possibilities for shifting the spectral sensitivity

Because of the dependence of their amplification factor M on the photo-current, junction photodetectors with internal amplification have a somewhat different absolute spectral sensitivity $s(\lambda)$ from non-amplifying junction photodetectors. In *Figure 9.13*, the typical relative spectral sensitivity $s(\lambda)_{rel}$ is shown as a function of the radiation wavelength λ for TIL 63-67 silicon phototransistors.

For certain applications, e.g., for the evaluation of the radiation from a

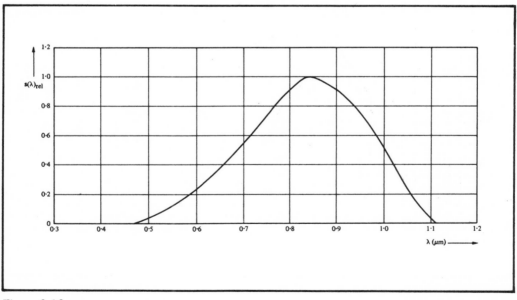

Figure 9.13
Typical relative spectral sensitivity as a function of the radiation wavelength λ for TIL 63-67 silicon phototransistors.

GaAs luminescence diode in the IR range, it is necessary, to shift the maximum of the sensitivity curve towards longer wavelengths. This is done firstly by

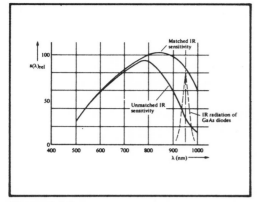

Figure 9.14
Relative spectral sensitivity s(λ)$_{rel}$ as a function of the wavelength for Si phototransistors. For the evaluation of the IR radiation from Si-doped GaAs luminescence diodes, Si photo-transistors with matched IR sensitivity must be used

suitable adjustment of the junction depth and also by application of an antireflection coating for $\lambda \approx 900$ nm to the component. *Figure 9.14* shows the effect of these measures

Apart from the above-mentioned technological possibilities, the spectral sensitivity of a photodetector can also be modified by externally attached optical filters. The makers of radiation measuring instruments either linearise the spectral sensitivity or they simulate the V(λ) curve, by placing a very expensive optical difference filter in front of a Schottky barrier PIN photodiode (see Section 6.5). Semiconductor manufacturers have also produced very low-priced silicon photodiodes which are reasonably well matched to the V(λ) curve, by fitting a green filter in the case of the silicon photodiode. Here too, the absolute spectral sensitivity s(λ) is reduced through the low transmission of IR by the filter. As an example of this, *Figure 9.15* shows the relative spectral sensitivity s(λ)$_{rel}$ as a function of the radiation wavelength λ

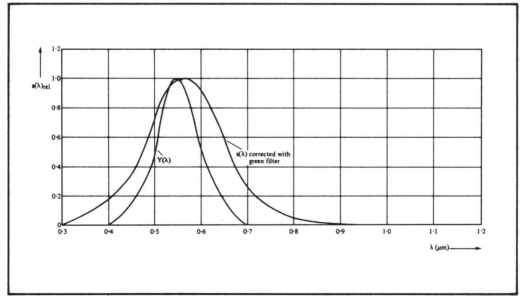

Figure 9.15
Relative spectral sensitivity s(λ)$_{rel}$ as a function of the radiation wavelength λ of the silicon photodiode Type TIL 77, produced with an attached green filter, in comparison with the photopic sensitivity curve of the eye, V(λ)

for the now obsolete silicon photodiode TIL 77, made with an attached green filter, in comparison with the photopic sensitivity of the eye $V(\lambda)$.

The possibilities described of affecting the absolute spectral sensitivity $s(\lambda)$ of a silicon junction photodetector within the theoretical sensitivity triangle, can be summarised as follows:

1
Variation of the absorption performance by the junction depth.

2
Reduction of surface recombination by technological methods.

3
Improvement of the absorption performance by antireflection coatings on the component surface.

4
Fitting of external optical filters.

5
Adjustment of the amplification factor M of junction photodetectors with internal amplification.

The quantum yield for monochromatic radiation $Q(\lambda)$ is a direct measure of the absolute spectral sensitivity $s(\lambda)$. The absolute spectral sensitivity $s(\lambda)$ can therefore be derived from the quantum yield for monochromatic radiation $Q(\lambda)$ in accordance with equation (9.4):

$$Q(\lambda) = \frac{I_{p,\lambda} \cdot h \cdot c_0}{\Phi_{e,\lambda} \cdot \lambda \cdot e} \qquad (9.4)$$

If the value of the constants from equation (9.5) and the wavelength λ of the monochromatic radiation used are inserted in this equation, there then remains, to define the absolute spectral sensitivity $s(\lambda)$ of a non-amplifying junction detector which is the ratio of the photocurrent produced $I_{p,\lambda}$ to the incident

monochromatic radiant power $\Phi_{e,\lambda}$. The relationship which applies for this

$$s(\lambda) = \frac{I_{p,\lambda}}{\Phi_{e,\lambda}} \qquad (9.6)$$

has already been stated as equation (9.6).

For the calculation of the absolute spectral sensitivity $s(\lambda)$ with the aid of the quantum yield for monochromatic radiation, the equations (9.4) and (9.5) have to be combined with equation (9.6). We obtain:

$$s(\lambda) = Q(\lambda) \cdot \frac{\lambda \cdot e}{h \cdot c_0}$$

$$= Q(\lambda) \cdot \lambda \cdot \frac{1}{1 \cdot 24} \cdot \frac{A}{W \cdot \mu m} \qquad (9.13)$$

9.3
Evaluation of radiation by non-amplifying junction photodetectors for mono-chromatic radiation sources

A non-amplifying junction photodetector evaluates an incident mono-chromatic radiation in accordance with its absolute spectral sensitivity. This is defined, for non-amplifying junction photodetectors, by the equation (9.6), which has already been defined.

The evaluation of radiation by the eye, described in Chapter 5, is equivalent in principle to that with non-amplifying junction photodetectors, if the relevant spectral sensitivity is taken into account. In order to understand the relationships which follow, the fundamentals from Chapter 5 will first be mentioned once more.

According to equation (5.103), the absolute spectral sensitivity $s(\lambda)$ corresponds to the equivalent absolute photopic sensitivity of the eye $K(\lambda)$:

$$s(\lambda) = K(\lambda) = K(\lambda)_{max} \cdot V(\lambda) \qquad (5.103)$$

The absolute sensitivity of the eye, $K(\lambda)$ is defined by equation (5.27) as the quotient of the output value $X_{v,\lambda}$ divided by the input value $X_{e,\lambda}$:

$$K(\lambda) = \frac{X_{v,\lambda}}{X_{e,\lambda}} \qquad (5.27)$$

Here, the output $X_{v,\lambda}$ is a corresponding photometric value for monochromatic light and the input, $X_{e,\lambda}$ is a spectral radiometric value equivalent to the input value.

Similarly to equation (5.27), the absolute spectral sensitivity $s(\lambda)$ of a non-amplifying junction photodetector corresponds to the ratio of the output value Y_λ to the input value $X_{e,\lambda}$:

$$s(\lambda) = \frac{Y_\lambda}{X_{e,\lambda}} \qquad (9.14)$$

This relationship is generally valid for non-amplifying junction photodetectors, since their output value Y_λ is proportional to the input value $X_{e,\lambda}$. In accordance with equation (9.6), the output value Y_λ corresponds to the photo-short-circuit current $I_{p,\lambda}$ of the junction photodetector, operated as a photocell, for the incident monochromatic radiation of wavelength λ. The sensitivity for junction photodetectors is also generally defined as the photocurrent sensitivity. To avoid misunderstandings, it should be mentioned, that an ideal non-amplifying junction photodetector shows the same sensitivity in both the photocell and photodiode modes.

The input value $X_{e,\lambda}$ in the denominator of equation (9.14) is either the monochromatic radiant power $\Phi_{e,\lambda}$ falling on the photodetector or the monochromatic radiant power $\Phi_{e,\lambda}/A$ related to a defined area A. In accordance with equation (2.40), the monochromatic irradiance $E_{e,\lambda}$ is thus obtained:

$$\frac{\Phi_{e,\lambda}}{A} = E_{e,\lambda} \qquad (2.40)$$

To calculate the absolute spectral sensitivity $s(\lambda)$, the values $I_{p,\lambda}$, $\Phi_{e,\lambda}$ or $E_{e,\lambda}$ are to be inserted in the general equation (9.14). Either the known equation (9.6) or the following relationship is obtained:

$$s(\lambda) = \frac{I_{p,\lambda}}{E_{e,\lambda}} \qquad (9.15)$$

The relative spectral sensitivity $s(\lambda)_{rel}$ is defined by the ratio of the absolute spectral sensitivity $s(\lambda)$ at the wavelength λ to the absolute spectral sensitivity $s(\lambda)_o$ at the reference wavelength λ_o:

$$s(\lambda)_{rel} = \frac{s(\lambda)}{s(\lambda)_o} \qquad (9.16)$$

If the maximum wavelength λ_{max} with maximum spectral sensitivity $s(\lambda)_{max}$ is inserted for λ_o in this expression, the equation (5.104), stated in Section 5.6, is obtained:

$$s(\lambda)_{rel} = \frac{s(\lambda)}{s(\lambda)_{max}} \qquad (5.104)$$

The absolute spectral sensitivity $s(\lambda)$ can be calculated by rearrangement of the equations (9.16) and (5.104):

$$s(\lambda) = s(\lambda)_{rel} \cdot s(\lambda)_o \qquad (9.17)$$

$$s(\lambda) = s(\lambda)_{rel} \cdot s(\lambda)_{max} \qquad (9.18)$$

The equation (9.18) corresponds to equation (5.103) for the analogous absolute spectral sensitivity of the eye, $K(\lambda)$.

9.4
Evaluation of radiation by non-amplifying junction photodetectors for chromatic radiation (mixed radiation)

A non-amplifying junction photodetector

evaluates incident radiation in accordance with its absolute sensitivity s. In the case of a non-amplifying junction photodetector, this corresponds, in analogy to equation (5.105), to the photometric radiation equivalent K of radiation source:

$$s = K = K(\lambda)_{max} \cdot V \qquad (5.105)$$

The photometric radiation equivalent K for chromatic radiation is determined by the equation (5.40):

$$K = \frac{K(\lambda)_{max} \cdot \int\limits_{\lambda 1}^{\lambda 2} X_{e,\lambda} \cdot V(\lambda) \cdot d\lambda}{\int\limits_{0}^{\infty} X_{e,\lambda} \, d\lambda} \qquad (5.40)$$

In this equation, the sensitivity values for the human eye can be replaced by the sensitivity values for non-amplifying junction photodetectors. Thus:

$$s_{rel} = V = \frac{\int\limits_{\lambda 1}^{\lambda 2} \int X_{e,\lambda} \cdot V(\lambda) \cdot d\lambda}{\int\limits_{0}^{\infty} X_{e,\lambda} \cdot d\lambda} \qquad (5.102)$$

$$s(\lambda) = K(\lambda) \qquad (5.103)$$

$$s(\lambda)_{rel} = V \qquad (5.104)$$

$$s = K \qquad (5.105)$$

$$s(\lambda)_{max} = K(\lambda)_{max} \qquad (9.19)$$

If the expressions from equations (5.104), (5.105) and (9.19) are inserted in equation (5.40), then the basic equation, equivalent to the latter, for the calculation of the sensitivity s of a non-amplifying junction photodetector is obtained:

$$s = \frac{s(\lambda)_{max} \int\limits_{\lambda 1}^{\lambda 2} X_{e,\lambda} \cdot s(\lambda)_{rel} \cdot d\lambda}{\int\limits_{0}^{\infty} X_{e,\lambda} \cdot d\lambda} \qquad (9.20)$$

If the expressions from equations (5.105), (5.103) and (5.37) are inserted in equation (5.40), then one obtains, for the sensitivity

$$s = \frac{\int\limits_{\lambda 1}^{\lambda 2} X_{e,\lambda} \cdot s(\lambda) \cdot d\lambda}{\int\limits_{0}^{\infty} X_{e,\lambda} \cdot d\lambda} \qquad (9.21)$$

Furthermore, by insertion of equations (5.102), (5.105) and (9.19) in equation (5.40), one obtains the relationship

$$s = s(\lambda)_{max} \cdot s_{rel} \qquad (9.22)$$

This expression is the analogous form of the equations (5.43) and (5.105) for the calculation of the photometric radiation equivalent K.

In the equations (5.40), (9.20) and (9.21), the value $X_{e,\lambda}$ occurs both in the numerator and the demoninator. It can therefore be replaced by its relative value, which characterises the radiation function S_λ of the radiation falling on the photodetector (see equation 2.66). The equation (9.20) then reads:

$$s = \frac{s(\lambda)_{max} \cdot \int\limits_{\lambda 1}^{\lambda 2} S_\lambda \cdot s(\lambda)_{rel} \cdot d\lambda}{\int\limits_{0}^{\infty} S_\lambda \cdot d\lambda} \qquad (9.23)$$

At the same time, equation (9.21) becomes:

$$s = \frac{\int\limits_{\lambda 1}^{\lambda 2} S_\lambda \cdot d(\lambda) \cdot d\lambda}{\int\limits_{0}^{\infty} S_\lambda \cdot d\lambda} \qquad (9.24)$$

In accordance with this expression, the sensitivity s of a non-amplifying junction photodetector is defined as the rate of

138

the output value, weighted both with the absolute spectral sensitivity distribution $s(\lambda) = f(\lambda)$ of the photodetector and with the radiation function $S\lambda$ of the incident radiation:

$$Y = \int_{\lambda 1}^{\lambda 2} S\lambda \cdot s(\lambda) \cdot d\lambda \qquad (9.25)$$

to the incident total radiation as the input value

$$X = \int_{o}^{\infty} S\lambda \cdot d\lambda \qquad (9.26)$$

The general equation for the calculation of the sensitivity s is thus given by (see Section 5.6):

$$s = \frac{Y}{X} \qquad (5.101)$$

If the spectral radiant flux $\Phi_{e,\lambda}$ is inserted in equation (9.21) for the value $X_{e,\lambda}$, then integration over the wavelength range which is of interest gives the photocurrent I_p in the numerator and the incident radiant power Φ_e in the denominator:

$$s = \frac{I_p}{\Phi_e} \qquad (9.27)$$

If the spectral irradiance $E_{e,\lambda}$ is inserted in equation (9.21) for the value $X_{e,\lambda}$, then integration over the wavelength range which is of interest gives the photocurrent I_p in the numerator and the irradiance E_e in the denominator:

$$s = \frac{I_p}{E_e} \qquad (9.28)$$

In contrast to equation (9.27), where the sensitivity s is related to the radiant power Φ_e, here it is related to the irradiance E_e.

As an example, Figure 9.16 shows the

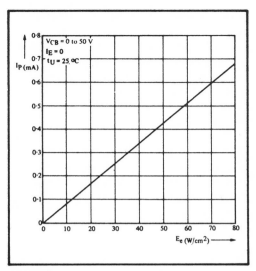

Figure 9.16
The function $I_p = f(E_e)$ for the collector-base junction of the phototransistor TIL 81.

typical linear characteristic of the function $I_p = f(E_e)$ for silicon photodiodes or photocells. The characteristic is applicable to the collector-base junction of the phototransistor TIL 81. In phtoodiode operation, the applied reverse voltage may have values up to 50 V. In photocell operation, the photocurrent I_p represents the short-circuit current.

9.5
Evaluation of radiation by amplifying junction photodetectors

An amplifying junction photo-detector evaluates incident radiation in accordance with its increased (amplified) absolute sensitivity for this incident radiation, s_M. The absolute sensitivity s_M of an amplifying photodetector is given by the product of the non-amplified absolute basic sensitivity s for the incident radiation and the amplification factor M:

$$s_M = s \cdot M \qquad (9.29)$$

For phototransistors, the amplification factor M corresponds to the DC current

139

gain B. Further, it can be assumed as an approximation, that the non-amplified absolute basic sensitivity s_{CB} corresponds to the absolute sensitivity s_{CB} of the collector-base junction acting as a photocell:

$$s_M = s_{CB} \cdot B \qquad (9.30)$$

The absolute sensitivity s_M of amplifying photodetectors is generally defined, corresponding to equation (5.101), as the ratio of the output value Y to the input value X_e.

$$s_M = \frac{Y}{X_e} \qquad (9.31)$$

For phototransistors, the output value Y is the collector current I_C. Other definitions, such as light current or photocurrent, for example, are incorrect and lead to confusion. As the input value X_e, a radiometric value should be selected, since phototransistors without spectral correction are sensitive both in the visible and the IR range. Corresponding to Section 9.6, the irradiance E_e is to be inserted as the radiometric input value X_e:

$$s_M = \frac{I_C}{E_e} \qquad (9.32)$$

For amplifying junction photodetectors, the amplification factor M depends on the input value X_e. In phototransistors, in particular, the DC current gain B is a function of the photocurrent I_{CB} of the collector-base photocell. Here, photo-current is defined as the control current occurring without the amplification effect, due only to charge-carrier generation, which would correspond, in the normal transistor, to the base current. The absolute sensitivity s_M for amplifying junction photodetectors is therefore also dependent on the input value X_e. For non-linear amplifying junction photodetectors, the absolute differential sensitivity s_d is stated. This

is given by the differential coefficient of the function $Y = f(X_e)$:

$$s_d = \frac{dY}{dX_e} \qquad (9.33)$$

But if the output value Y is proportional to the input value X_e, then the absolute differential sensitivity s_d can be made equal to the absolute sensitivity s_M. Thus in the linear case, the radiation evaluation of amplifying junction photodetectors can be carried out in accordance with Sections 9.3 and 9.4 for non-amplifying junction photodetectors, i.e.:

$$s_d = s_M = s \qquad (9.34)$$

Furthermore, with a non-linear relationship between the output value Y and the input value X_e, the absolute sensitivity s_M of amplifying photodetectors can also be stated by a finite interval between $X_{e2} - X_{e,1}$ and $Y_2 - Y_1$. The corresponding differential quotient then replaces the differential quotient in the equation (9.33). This statement of sensitivity is defined as the interval sensitivity

$$s_i = \frac{Y_2 - Y_1}{X_{e,2} - X_{e,1}} = \frac{\Delta Y}{\Delta X_e} \qquad (9.35)$$

For phototransistors, the function $Y = f(X_e)$ is shown in a graph as the characteristic curve $I_C = f(E_e)$ for ar applied collector-emitter voltage $V_{CE} = 5$ V and a case temperature of $t_G = 25°C$. *Figure 9.17* shows the typical function $I_C = f(E_e)$ for the well-known phototransistor TIL 81.

Generally, the function $I_C = f(E_e)$ is stated from an irradiance $E_e > 1$ mW/cm^2 for a given radiation function $S\lambda$.

In practice, irradiances $E_e < 1$ mW/cm^2 are also used. Here the user has to know either the typical function $B_{typ} = f(E_e)$ for the radiation function $S\lambda$ in question,

140

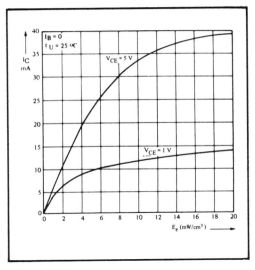

Figure 9.17
Collector current I_C as a function of the irradiance E_e for the photo-transistor TIL 81. The collector-emitter voltage V_{CE} is the variable. The irradiance is generated with the standard light A radiation.

or $B_{typ} = f(I_C)$ without taking account of a radiation function $S\lambda$.

The absolute sensitivity s_M of the photo-transistor can be calculated in accordance with the equations (9.29) to (9.32). For a point on the characteristic $B = f(I_C)$, the current gain B of a phototransistor with an external base connection is obtained by forming the ratio of the collector current I_C, measured with a given irradiance in phototransistor operation, to the photocurrent I_{CB}, measured with the same irradiance for the collector-base photocell working in the short-circuited mode.

$$B = \frac{I_C}{I_{CB}} \qquad (9.36)$$

As an example, the current gain B of the phototransistor TIL 81 will be determined. With the irradiance $E_e = 2 \cdot 5$ mW/cm² for standard light A, a collector current I_C = 9 mA and a photocurrent $I_{CB} = 30\ \mu A$ of the short-circuited collector-base photocell

were measured. From equation (9.36), the DC amplification is

$$B = \frac{9 \text{ mA}}{40\ \mu A} = 300$$

If the DC current gain B is determined for different collector currents I_C, the function $B = f(I_C)$ can be plotted in a graph as a characteristic curve. However, it is better to determine the relative DC current gain B_{rel} as a function of the collector current I_C, since this characteristic curve can be utilised for the whole corresponding family of phototransistors. We have:

$$B_{rel} = \frac{B}{B_{max}} \qquad (9.37)$$

Figure 9.18 shows the relative DC current gain B_{rel} as a function of the collector current I_C for the photo-transistor TIL 81. The maximum DC current gain $B_{rel,max}$ has been standardised to the value $B_{rel,max}$ = 100%. For opto-couplers, the relative DC current gain B_{rel} of the phototransistor is measured as a function of the collector current $I_{C(on)}$ when the GaAs light-emitting diode is switched on. A defined direct current I_F flows through the GaAs diode. Optocouplers are measured at an ambient temperature of $t_U = 25°C$. In Figure 9.19, the standardised DC current gain B_{rel} is shown as a function of the collector current $I_{C(on)}$ for the optocoupler TIL 112.

The measurement of the photocurrent sensitivity of phototransistors is usually carried out at an ambient temperature $t_U = 25°C$ for plastic transistors and a case temperature $t_G = 25°C$ for metal-encapsulated transistors. The Planck radiation used for this purpose, standard light A, causes heating of the wafer through the longer-wavelength IR raidation. This effect mainly occurs with phototransistors with high thermal resistances. Therefore the measurement

141

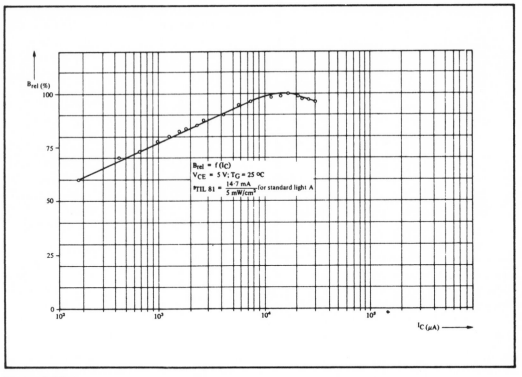

Figure 9.18
Relative DC current gain $B_{rel} = f(I_C)$ of the silicon phototransistor TIL 81

time is kept as short as possible, so that the measured result is as accurate as possible *Figure 9.20* shows the dependence of the collector current I_C on the ambient temperature t_U for the phtotransistor LS 400.

For phototransistors with external base connection, the DC current gain B can be measured, instead of with the photocurrent I_{CB} of the collector-base diode, alternatively with the base current I_B from an external current source. During this measurement the radiation-sensitive surface of the phototransistor is covered. With this method, the dependence of the collector current I_C on the applied collector-emitter voltage V_{CE}, that is, the function $I_C = f(V_{CE})$, can also be measured on the characteristic recorder. In this case, the base current I_B is the

variable. If the sensitivity s of the collector-base photodiode is known for the particular base current, (see *Figure 9.16*), the parameter I_B can be replaced by the irradiance E_e. As a specific example of this, *Figure 9.21* shows the typical function $I_C = f(V_{CE})$ of the silicon phototransistors TIL 602, TIL 610 and TIL 614 with the irradiance E_e as parameter.

In Section 9.2, it was explained that the absolute spectral sensitivity of an amplifying junction photodetector is also dependent on its particular amplification factor M. If, with monochromatic radiation, the output value Y_λ of an amplifying photodetector behaves proportionally to the spectral input value $X_{e,\lambda}$ and if the amplification factor M_λ or the DC current gain B_λ were independent of the output value Y_λ and of the input value $X_{e,\lambda}$

142

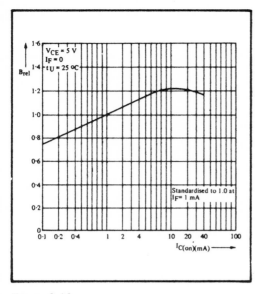

Figure 9.19
Standardised DC current gain B_{rel} as
a function of the collector current $I_{C(on)}$
with the luminescent diode conducting,
for the optocoupler TIL 112. The relative
DC current gain has been standardised to
the value $B_{rel} = 1$ for a current $I_F = 1$ mA
flowing through the GaAs diode.

(see equation 9.14), then the relative spectral
sensitivity $s(\lambda)_{M,rel}$ of amplifying junction
photodetectors corresponds to the basic
relative spectral sensitivity $s(\lambda)_{rel}$ (see
equation 5.104):

$$s(\lambda)_{rel} = \frac{s(\lambda)}{s(\lambda)_{max}} = s(\lambda)_{M,rel}$$

$$= \frac{s(\lambda)_M}{s(\lambda)_{M,max}} \qquad (9.38)$$

If equation (9.14) is inserted in equation
(9.38), we obtain:

$$s(\lambda)_{M,rel} = \frac{Y\lambda}{X_{e,\lambda}} : \frac{Y\lambda,max}{X_{e,\lambda,max}}$$

$$= \frac{Y\lambda \cdot X_{e,\lambda,max}}{X_{e,\lambda} \cdot Y\lambda,max} \qquad (9.39)$$

If, on the other hand, the amplification
factor $M\lambda$ or the DC current gain $B\lambda$
depends on the input value $X_{e,\lambda}$, then the
relative spectral sensitivity $s(\lambda)_{M,rel}$ can be

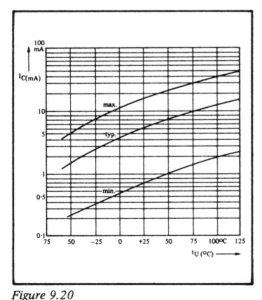

Figure 9.20
Function $I_C = f(t_U)$ for the phototransistor
LS 400. The applied collector-emitter
voltage is $V_{CE} = 5$ V.

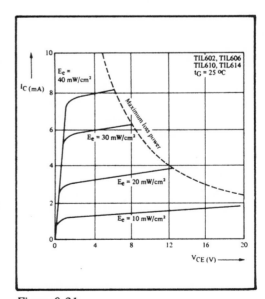

Figure 9.21
Typical function $I_C = f(V_{CE})$ of the silicon
phototransistors TIL 602, 606, 610, 614;
(The variable is the irradiance E_e)

143

measured by relation to equal output values of Y_λ and $Y_{\lambda,max}$.

$$Y_\lambda = Y_{\lambda,max} \qquad (9.40)$$

The relative spectral sensitivity $s(\lambda)_{M,rel}$ of amplifying junction photodetectors can be defined for equal output values Y_λ and $Y_{\lambda,max}$ as the ratio of the spectral radiometric input value $X_{e,\lambda,max}$ at the radiation wavelength λ_{max} with the maximum absolute spectral sensitivity $s(\lambda)_{M,max}$ to the spectral radiometric input value $X_{e,\lambda}$ at the radiation wavelength λ with the absolute spectral sensitivity $s(\lambda)_M$. By insertion of equation (9.40) in equation (9.39), one obtains:

$$s(\lambda)_{M,rel} = \frac{X_{e,\lambda,max}}{X_{e,\lambda}} \qquad (9.41)$$

9.6
Area-dependent sensitivity values for junction photodetectors

The sensitivity units s, $s(\lambda)$ and s_M are defined as the ratio of the output current I to the incident radiant power Φ_e or to the irradiance E_e. Here, a distinction is made between large-area and small area photodetectors.

Large-area junction photodetectors are mostly non-amplifying radiation detectors such as silicon solar cells and Schottky barrier PIN photodiodes. The manufacturer of such radiation detectors can give exact data on their photosensitive area A_E. If the receiving area is known, the incident radiant power can be calculated by the equation already defined (2.41).

$$\Phi_e = E_e \cdot A_E \qquad (2.41)$$

For large-area non-amplifying photo-detectors, the absolute photo-sensitivity is determined by the equation (9.27):

$$s = \frac{I_P}{\Phi_e} \qquad (9.27)$$

Its dimension is:

$$[s] = \frac{A}{W} = \frac{mW}{mW} = \frac{\mu A}{\mu W} \qquad (9.42)$$

Similarly, the absolute spectral photo-sensitivity $s(\lambda)$ is determined by the equation (9.6):

$$s(\lambda) = \frac{I_{P,\lambda}}{\Phi_{e,\lambda}} \qquad (9.6)$$

Its dimension is:

$$[s(\lambda)] = \frac{A}{W(\mu m^{-1})} = \frac{mA}{mW(\mu m^{-1})}$$

$$= \frac{\mu A}{\mu W(\mu m^{-1})} \qquad (9.43)$$

These sensitivities are defined accordingly as the specific photocurrent sensitivity s or the specific spectral photocurrent sensitivity $s(\lambda)$.

Small-area junction photodetectors can be amplifying radiation detectors, for example, such as silicon photodiodes, avalanche photodiodes and phototransistors. For these radiation detectors, the photosensitive area of their wafer, or their magnified photosensitive wafer area, due to built-in lenses or a plastic case which transmits radiation, is not exactly known.

The radiant power falling on the photo-detector can be calculated only with an excessive error by using equation (2.41). Therefore, the radiation falling perpendicularly on to a specified measurement plane is stated. The photo-detector device under test (e.g., a silicon photodiode) is located at the centre of the measurement plane in the normal direction.

Generally, the area of the measurement plane is specified at $A_{meas} = 1$ cm^2.

The incident radiant power is related to this area, so that instead of the radiant power Φ_e, the irradiance E_e is to be stated in the unit W/cm^2. For small-area non-amplifying junction photodetectors, the absolute photosensitivity s is determined by equation (9.28):

$$s = \frac{I_P}{E_e} \qquad (9.28)$$

Its dimension is:

$$[s] = \frac{A}{Wcm^{-2}} = \frac{mA}{mWcm^{-2}} = \frac{\mu A}{\mu Wcm^{-2}} \qquad (9.44)$$

Similarly, the absolute spectral photo-sensitivity $s(\lambda)$ is given by equation (9.15):

$$s(\lambda) = \frac{I_{P,\lambda}}{E_{e,\lambda}} \qquad (9.15)$$

Its dimension is:

$$[s(\lambda)] = \frac{A}{Wcm^{-2}(\mu m^{-1})} = \frac{mA}{mWcm^{-2}(\mu m^{-1})}$$

$$= \frac{\mu A}{\mu Wcm^{-2}(\mu m^{-1})} \qquad (9.45)$$

For small-area smplifying junction photo-detectors, the absolute sensitivity s_M is determined by the equation (9.32):

$$s_M = \frac{I_M}{E_e} \qquad (9.32)$$

Its dimension corresponds to that of equation (9.42). Similarly, the absolute differential sensitivity s_d is obtained from equation (9.33):

$$s_d = \frac{dY}{dX_e} = \frac{dI_M}{dE_e} \qquad (9.46)$$

Its dimensions corresponds to those of equation (9.44).

The interval sensitivity s_i is given by the equation (9.35):

$$s_i = \frac{Y_2 - Y_1}{X_{e,2} - X_{e,1}} = \frac{I_{M,2} - I_{M,1}}{E_{e,2} - E_{e,1}} \qquad (9.47)$$

The equation (9.44) also applies here for its dimensions:

The relative spectral sensitivity $s(\lambda)_{M,rel}$ is determined by the equation (9.41):

$$s(\lambda)_{M,rel} = \frac{X_{e,\lambda,max}}{X_{e,\lambda}} = \frac{E_{e,\lambda,max}}{E_{e,\lambda}} \qquad (9.48)$$

9.7
Actinic value of IR luminescence radiation from GaAs diodes for silicon junction photodiodes

The basic principles of actinic value were explained in Section 5.6. The sensitivity values needed for silicon photodetectors have been described in this chapter. The actinic value a(Z) of any given radiation spectrum Z for silicon junction photodetectors is the ratio of the sensitivity s(Z) when acted upon by this radiation to the sensitivity s(PA) when acted upon by the Planck standard light A radiation. The standard light A is always used as the reference radiation, since this form of radiation has become generally established for the sensitivity evaluation of silicon junction photodetectors.

For non-linear amplifying photodetectors, it will be assumed, for simplicity, that the actinic value will either be measured within a linear range or with equal output values, or will be calculated in accordance with equation (5.117) with Y(Z) = Y(N).

The actinic value of IR luminescence radiation from GaAs diodes for silicon junction photodetectors can be determined from equation (5.121):

$$a(Z) = \frac{s(Z)}{s(N)}$$

$$= \frac{\int S(Z)\lambda \cdot s(\lambda)_{rel} \cdot d\lambda \cdot \int S(N)\lambda \cdot d\lambda}{\int S(Z)\lambda \cdot d\lambda \cdot \int S(N)\lambda \cdot s(\lambda)_{rel} \cdot d\lambda}$$

(5.121)

The IR luminescence radiation from GaAs diodes is denoted by LIR. Its radiation function is therefore stated as follows:

$$S(Z)\lambda \simeq S(LIR)\lambda \qquad (9.49)$$

The reference radiation, the Planck radiation ',standard light A'', is denoted by PA, as in Section 5.6.2. Its radiation function is identified either by its distribution temperature $T_v = 2856$ K or by PA.

$$S(N)\lambda \simeq S\lambda,2856\ K = S(PA)\lambda \qquad (9.50)$$

The relevant sensitivities of a silicon junction photodetector are characterised in accordance with these radiation functions:

$$s(Z) \simeq s(LIR) \qquad (9.51)$$

$$s(Z)\lambda \simeq s(LIR)\lambda \qquad (9.52)$$

$$s(Z)\lambda,rel \simeq s(LIR)\lambda,rel \qquad (9.53)$$

$$s(N) \simeq s(PA) \qquad (9.54)$$

$$s(N)\lambda \simeq s(PA)\lambda \qquad (9.55)$$

$$s(N)\lambda,rel \simeq s(PA)\lambda,rel \qquad (9.56)$$

The actinic value of IR luminescence radiation from GaAs diodes for silicon junction photodetectors will be calculated in accordance with equation (5.121) with relative radiometric input values. Instead of $a(Z)$, it is defined as $a_e(LIR)$:

$$a(Z) \simeq a_e(LIR) \qquad (9.57)$$

Equations (9.49) to (9.57), inserted in equation (5.121), give a more convenient representation:

$$a_e(LIR) = \frac{s(LIR)_{rel}}{s(PA)_{rel}}$$

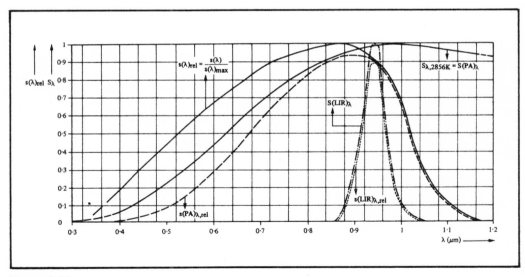

Figure 9.22
Radiation functions S(PA)λ of the standard light A, S(LIR)λ of IR luminescence radiation from the GaAs diode TIL 31 and the relative spectral sensitivity s(λ)rel, s(PA)λ,rel and s(LIR)λ,rel of a silicon photocell as functions of the wavelength λ

$$= \frac{\int S(LIR)\lambda \cdot s(\lambda)_{rel} \cdot d\lambda \cdot \int S(PA)\lambda \cdot d\lambda}{\int S(LIR)\lambda \cdot d\lambda \cdot \int S(PA)\lambda \cdot s(\lambda)_{rel} \cdot d\lambda} \qquad (9.58)$$

Integration of this expression is carried out graphically. For this, *Figure 9.22* shows the radiation functions $S(PA)\lambda$ of the standard light A and $S(LIR)\lambda$ of the IR luminescence radiation for the GaAs diode TIL 31, and also the relative spectral sensitivity $s(\lambda)_{rel}$ of a silicon photocell. The product of the ordinates $S(PA)\lambda$ and $s(\lambda)_{rel}$ gives relative spectral sensitivity $s(PA)\lambda,_{rel}$ for the standard light A. The product of the ordiantes $S(LIR)\lambda$ and $s(\lambda)_{rel}$ gives the relative spectral sensitivity $s(LIR)\lambda,_{rel}$ for the IR luminescence radiation of the GaAs diode TIL 31. The ratio of the measured area under the curve $s(PA)\lambda,_{rel}$ to the measured area under the surve $s(PA)\lambda$ gives the relative sensitivity $s(PA)_{rel}$. To give a clearer abscissa scale, the radiation function $S(PA)\lambda$ is only shown up to a wavelength $\lambda = 1 \cdot 2 \ \mu m$. The plani-metric measurement of the area under the curve $S(PA)\lambda$ was carried out up to a wavelength $\lambda = 4 \ \mu m$. This approximation is permissible, since up to $\lambda = 4 \ \mu m$, covers about 94% of the total radiation of standard light A and also since, in practical measure-ments, longer-wavelength IR radiation is naturally absorbed by the glass bulb of the radiation source.

The ratio of the measured area under the curve $s(LIR)\lambda,_{rel}$ to the measured area under the curve $S(LIR)\lambda$ gives the relative sensitivity $s(LIR)_{rel}$.

The actinic value $a_e(LIR)$ of luminescence radiation from the GaAs diode TIL 31 for a silicon photocell is, in accordance with equation (9.58), the ratio of the graphically determined relative sensitivity $s(LIR)_{rel}$ to the graphically determined relative sensitivity $s(PA)_{rel}$. Evaluation of the curves in *Figure 9.22* gives the following values:

$$s(PA)_{rel} = \frac{A_{s(PA)\lambda rel}}{A_{S(PA)\lambda}} = \frac{88 \ cm^2}{335 \ cm^2}$$

$$= 0 \cdot 263 \qquad (9.59)$$

$$s(LIR)_{rel} = \frac{A_{s(LIR)\lambda,rel}}{A_{S(LIR)\lambda}} = \frac{16 \cdot 5 \ cm^2}{18 \ cm^2}$$

$$= 0 \cdot 917 \qquad (9.60)$$

$$a_e(LIR) = \frac{s(LIR)_{rel}}{s(PA)_{rel}} = \frac{0 \cdot 917}{0 \cdot 263} = 3 \cdot 49 \qquad (9.61)$$

The actinic value of IR luminescence radiation from the GaAs diode TIXL 16 for the photocell with a relative spectral sensitivity as shown in *Figure 9.22* was also determined by measurement. The irradiance falling on the photocell was in one case 5 mW/cm² and in another case 20 mW/cm² for both of the two radiations PA and LIR. The photocurrent, I_p, measured with the photocell short-circuited, is shown in *Table 9.1*.

The actinic value $a_e(LIR)$ is:

a
for 5 mW/cm²:

Type of radiation	PA		LIR	
Irradiance E_e	5 mW/cm²	20 mW/cm²	5 mW/cm²	20 mW/cm²
Photocurrent I_p	3·5 mA	13·7 mA	9·27 mA	37·08 mA

Table 9.1
Measured photocurrents at various irradiances and with different types of radiation for the calculation of actinic value $a_e(LIR)$

147

$$a_e(LIR) = \frac{9 \cdot 27 \text{ mA}}{3 \cdot 5 \text{ mA}} = 2 \cdot 65 \qquad (9.62)$$

b

for 20 mW/cm^2:

$$a_e(LIR) = \frac{37 \cdot 08 \text{ mA}}{13 \cdot 7 \text{ mA}} = 2 \cdot 7 \qquad (9.63)$$

The different actinic values $a_e(LIR)$ arise through measurement errors, which are primarily due to the deviation of the radiation function $S\lambda$ of the filament lamp used from that of an ideal Planck radiator in the IR range, despite having the same colour temperature. The actinic value of IR luminescence radiation from GaAs diodes varies for Si phototransitors. The spectral sensitivity of a phototransistor also depends on the DC current gain B (see Section 9.5). *Figure 9.23* again shows, as for *Figure 9.22*, the radiation functions $S(PA)\lambda$ and $S(LIR)\lambda$ for the GaAs diode TIXL 26 and also the relative spectral sensitivities $s(\lambda)_{rel}$ and $s(PA)\lambda,_{rel}$. In contrast to *Figure 9.22*, almost the

whole radiation function $S(PA)\lambda$ is shown. The other functions are compressed into the first third of the graph. The function $s(LIR)\lambda,_{rel}$ has therefore not been shown. It corresponds approximately with the function $s(LIR)\lambda,_{rel}$ in *Figure 9.22*.

The actinic value of IR luminescence radiation from GaAs diodes being detected by phototransistors can be determined, for example, if the output variables have equal values. Measurement in this way is not always possible, so that in these cases the output variable is to be measured in each case with equal values of the input variable. The relative DC current gain B_{rel} of phototransistors is given for the collector currents $I_{C,(PA)}$ for the radiation PA and $I_{C,(LIR)}$ for the radiation LIR, measured with equal irradiance values in each case, by the characteristic curve $B_{rel} = f(I_C)$ (see *Figure 9.18*). The relative DC current gain for the incident radiation LIR is designated $B_{(LIR)rel}$ and for the incident radiation PA with $B_{(PA)\,rel}$. The quotient K_B of the relative DC current gain values for $B_{(PA)rel}$

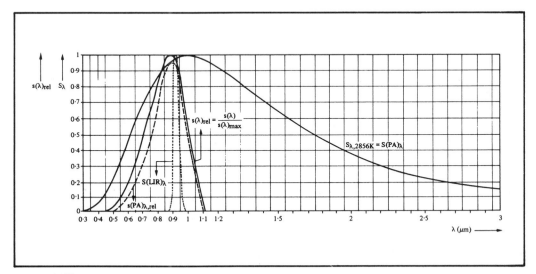

Figure 9.23
Radiation function S(PA)λ of standard light A, S(LIR)λ of IR luminescence radiation from the GaAs diode TIXL 26 and the relative spectral sensitivity s(λ $_{rel}$ and s(PA)λ,rel of a silicon phototransistor as functions of the wavelength λ

and $B_{(LIR)rel}$ is the correction factor for the collector current $I_{C(LIR)}$:

$$K_B = \frac{B_{(PA)rel}}{B_{(LIR)rel}} \qquad (9.64)$$

The product of the correction factor K_B and the collector current $I_{C(LIR)}$ gives the corrected collector current $I_{C(LIR)Kor}$, which would appear directly with the DC current gain $B_{(PA)}$, if this value were constant for all types of radiations.

$$I_{C(LIR)Kor} = K_B \cdot I_{C(LIR)} \qquad (9.65)$$

With this method, the non-linearity of the DC current gain B of a phototransistor can be eliminated from the actinic value calculation in many practical cases. But here, the conditions mentioned must be satisfied, i.e., the function $B_{rel} = f(I_C)$ must be known and furthermore the measured irradiances $E_{e(PA)}$ for the PA radiation and $E_{e(LIR)}$ for the LIR radiation must have exactly the same values without errors:

$$E_{e(PA)} = E_{e(LIR)} \qquad (9.66)$$

Thus the actinic value of IR luminescence radiation from GaAs diodes detected by silicon phototransistors is determined by the ratio of the corrected collector current $I_{C(LIR)C \text{ or } R}$ to the collector current $I_{C(PA)}$.

$$a_e(LIR) = \frac{s(LIR)_{Corr}}{s(PA)} = \frac{I_{C(LIR)Corr}}{I_{C(PA)}}$$
$$\qquad (9.67)$$

The actinic value of selective IR luminescence radiation with a maximum wavelength (λ_{max}) of 930 nm was measured for four LS600 silicon phototransistors. The IR luminescence radiation LIR with the maximum wavelength λ_{max} = 930 nm relates to the following GaAs diodes TIL 23, TIL 24, TIXL 12, TIXL 13, TIXL 14, TIXL 15, TIXL 16 and TIXL 26.

The sensitivities s(PA) and s(LIR) of the LS600 phototransistors were measured for both PA and LIR radiation with an irradiance E_e = 5 mW/cm^2.

In accordance with the statement on the device data sheet, the applied collector-emitter voltage must be V_{CE} = 5 V. In *Table 9.2*, the measured collector currents $I_{C(PA)}$ and $I_{C(LIR)}$ in each case, with the associated values of the relative DC current gain B_{rel}, are listed. The values of the relative DC current gain $B_{(PA)rel}$ and $B_{(LIR)rel}$ can be seen from *Figure 9.18*. The correction factor K_B was calculated from equation (9.64), the corrected corrected collector current $I_{C(LIR)Corr}$ from equation (9.65) and the actinic value $a_e(LIR)$ from equation (9.67). The mean actinic value $a_e(LIR)$ of IR luminescence radiation from the GaAs diodes listed for the four LS 600 phototransistors is:

$$a_e(LIR)_{typ} = \frac{2 \cdot 2 + 2 \cdot 1 + 2 \cdot 16 + 2 \cdot 47}{4}$$

$$= 2 \cdot 23 \qquad (9.68)$$

The typical correction factor K_B for the four phototransistors is calculated as follows:

$$K_{B,typ} = \frac{0 \cdot 869 + 0 \cdot 874 + 0 \cdot 8715 + 0 \cdot 854}{4}$$

$$= 0 \cdot 867 \qquad (9.69)$$

If no values are stated by the manufacturer for $a_e(LIR)_{typ}$ and $K_{B,typ}$, then the values obtained from equations (9.68) and (9.69) can be used for rough calculations when using IR luminescence radiation from Si-doped GaAs diodes with silicon transistors. When doing this, a rather large tolerance for the K_B value must be taken into account. With the typical value for a $a_e(LIR)_{typ}$ and the typical correction factor $K_{B,typ}$, the photocurrent sensitivity s(PA) for the standard light A, stated in the date sheet, can be converted

149

Phototransistor LS 600	1	2	3	4
$\dfrac{I_{C(PA)}}{mA} = s(PA) \cdot 5\ mW/cm^2$	0·26	0·24	0·21	0·152
$\dfrac{B_{(PA)rel}}{\%}$	63	62·5	61	58·5
$\dfrac{I_{C(LIR)}}{mA} = s(LIR) \cdot 5\ mW/cm^2$	0·66	0·58	0·52	0·44
$\dfrac{B_{(LIR)rel}}{\%}$	72·5	71·5	70	68·5
K_B	0·869	0·874	0·8715	0·854
$\dfrac{I_{C(LIR)Corr}}{mA}$	0·573	0·506	0·453	0·376
$A_e(LIR)$	2·2	2·1	2·16	2·47

Table 9.2
Calculation of the actinic value for silicon phototransistors

to the sensitivity s(LIR) for the IR radiation from Si-doped GaAs diodes.

The equations needed are obtained by rearrangement of the formulae (9.67) and (9.65):

$$I_{C(LIR)} = \frac{a_e(LIR) \cdot I_{C(PA)}}{K_B} \qquad (9.70)$$

$$s(LIR) = \frac{a_e(LIR) \cdot s(PA)}{K_B} \qquad (9.71)$$

Example calculation:
The collector current $I_{C(PA)}$ of the LS 613 phototransistor was measured for an incident radiation of standard light A with irradiance $E_{e(AP)} = 5\ mW/cm^2$, as

$I_{C(PA)} = 0·47\ mA$.

Its photocurrent sensitivity s(PA) is thus:

$$s(PA) = \frac{I_{C(PA)}}{E_{e(PA)}} = \frac{0·47\ mA}{5\ mW/cm^2} \qquad (9.72)$$

From equation (9.70) and the average values obtained from equations (9.68) and (9.69), the collector current $I_{C(LIR)}$ is calculated for an incident IR luminescence radiation from the GaAs diode SL 1183 (TIXL 16 with reflector) with the same irradiance.

$$I_{C(LIR)} = \frac{2·23 \cdot 0·47\ mA}{0·867} = 1·209\ mA$$
$$(9.73)$$

A collector current of

$I_{C(LIR)} = 1·4\ mA$ was measured.

Mainly because of the assumed correction factor $K_B = 0·867$, the calculation error still amounts to 13·6%. In practice, even

150

greater calculation errors must usually be taken into account. In general, simplified actinic value calculations are subject to relatively large tolerances. The calculation of actinic value by graphical integration becomes inaccurate through plotting errors on the individual functions, with their scatter and their planimetric measurement. Determination of the actinic value by direct measurement is only exact, if the radiation functions of the radiation sources used can be measured exactly and if no errors are made in the sensitivity measurements.

9.8
Dark current of junction photodetectors

The reverse current of a semiconductor diode consists essentially of two components: the reverse current caused by the presence of minority carriers and a leakage current with an approximately resistive (Ohmic) behaviour. In Si devices, the latter is very small and can usually be neglected. The reverse current due to minority carriers increases for small reverse voltages, in accordance with an exponential function of the reverse voltage.

$$I_R = I_{sat} \left(1 - e^{-\dfrac{V_R}{26 \text{ mV}}}\right) \qquad (9.74)$$

After an initial rise in the so-called starting range, the reverse current tends asymptotically towards a constant value (the reverse saturation current I_{sat}) so that above a reverse voltage of approx. 0.1 V the variable part of the exponential function can be neglected. Up to the breakdown of the junction, this range is sometimes called the "true" reverse range.

For semiconductor devices, the typical value of the reverse current at a temperature of 25°C is generally stated in the data sheet. Also, the reverse current also has an exponential relationship with the ambient temperature. Usually, $I_R = f(t)$ is shown on log/linear paper. The temperature is plotted linearly on the abscissa and the reverse current (or dark current) logarithmically on the ordinate. The relationship between I_R & t is as follows:

$$I_{R,t} = I_{R(25)} \cdot e^{K(t - 25^\circ C)} \qquad (9.75)$$

where:

$I_{R,t}$ = Reverse current at the specified temperature t,

$I_{R(25)}$ = Reverse current at 25°C (from data sheet)

K = Temperature coefficient of $0 \cdot 06$ to $0 \cdot 1$, average: $0 \cdot 082$

In this coordinate system, the function $I_{R,t}$ appears as a straight line, the slope being given by the temperature coefficient K and the parallel displacement of the lines by $I_{R(25)}$. In junction photodetectors, the reverse current is defined as the "dark current". This is subject to the proviso, that the ambient irradiance must be zero. In *Figure 9.24*, in principle, the dark current $I_D = f(t)$ can be read off for all nominal photodiodes and phototransistors. The temperature coefficient K serves as the variable. The value of the dark current $I_{D(25)}$ determines the scale of the ordinate. The $I_{D(25)}$ values (usually a second value is also stated at a higher temperature, e.g. $I_{D(80)}$) can be taken from the data sheet and transferred to the grpah. If these points are joined by a straight line, the function $I_D = f(t)$ is obtained for the corresponding device. The temperature coefficient K is only needed, if only one dark-current value is known. If K is not known, a mean value, e.g., $K = 0 \cdot 82$ can be used. This calculation of the dark current becomes unnecessary, if an appropriate graph is contained in the data sheet.

The examples given show $I_D = f(t)$ for the phototransistor TIL 78 in *Figure 9.25* and for the photodiode TIL 77 in *Figure 9.26*. It is also common, to indicate

151

Figure 9.24
The dark current I_D as a function of temperature t for junction photodetectors.

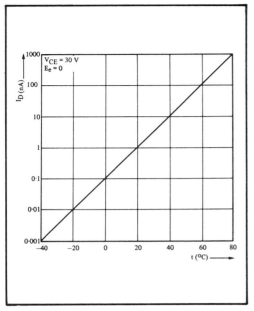

Figure 9.25
The dark current I_D as a function of temperature (t) for the phototransistor TIL 78

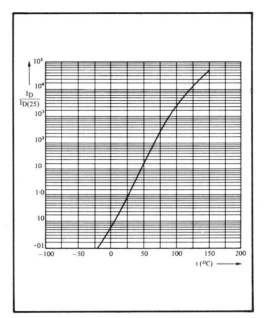

Figure 9.27
The dark current ratio $I_D/I_{D(25)}$ as a function of temperature (t) for the photodiode 1N2175

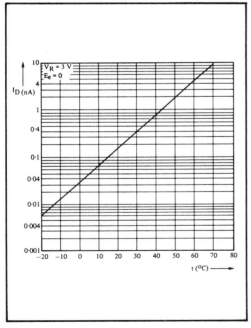

Figure 9.26
The dark current I_D as a function of temperature (t) for the photodiode TIL 77.

the normalised relationship $I_D/I_{D(25)}$ as shown in *Figure 9.27* for the photodiode 1N2175. In a typical emitter-sensor system, the dark current at the maximum case, or ambient, temperature of the photodetector should be less than the minimum useful photocurrent by at least a factor of 10.

9.8.1
Dark current of avalanche photodiodes

For avalanche photodiodes, the dark current $I_{D(25)}$ is normally stated separately in the data sheet as the leakage current in the device, the so-called bulk leakage current $I_{bu(25)}$, and as the leakage current on the device surface, the so-called surface leakage current $I_{su(25)}$, for a given applied reverse voltage V_R and an incident irradiance of $E_e = 0$. The dark current, $I_{D(25)}$, is the sum of the surface leakage current, $I_{su(25)}$, and the product of the photocurrent amplification

153

factor M multiplied by the bulk leakage current $I_{bu(25)}$:

$$I_{D(25)} = I_{su(25)} + (M \cdot I_{bu(25)}) \qquad (9.76)$$

The values for $I_{bu(25)}$ and $I_{su(25)}$ are determined by a relatively difficult measurement and a simple calculation:

The photosensitive surface of the wafer of an avalanche photodiode is arranged exactly at the focus of an optical system. The adjustment and the quality of the optical system must ensure, that the radiation at the focus does not fall on the whole wafer surface, but only on the central part of the active area of the wafer. The measurement is carried out with monochromatic radiation of wavelength $\lambda = 900$ nm. This radiation is modulated at 400 Hz. A radiant power of $\Phi_e \approx$ 10 nW is to fall on the active area of the wafer.

First, a reverse voltage V_R of about quarter of the typical breakdown voltage, $V_{BR,type}$, is applied to the avalanche photodiode to be measured, that is

$$V_R = \frac{1}{4} V_{BR,typ}$$

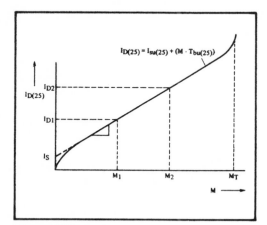

Figure 9.28
Dark current $I_{D(25)}$ as a function of the photocurrent amplification M. The measurement points I_{D1}, M_1 and I_{D2} lie in the linear part of the characteristic $I_{D(25)} = f(M)$

For the TIXL 56 silicon avalanche photodiode this reverse voltage is about 40 V and the photocurrent amplification has a value of M = 1. Then the signal current I_S is measured. As the next step, the reverse voltage V_R is increased, so that the signal current I_S increases to a higher, but freely selectable value, corresponding to the amplification factor M_1 (see *Figure 9.28*). The value of M_1 must, however, remain less than $M_T/2$ (with regard to M_T, see Section 9.1.1), otherwise the dark current and the noise superimposed on the signal current I_S would assume unusably high values. The incident radiant flux Φ_e is now interrupted and the value of the dark current I_{D1}, which now flows, is recorded.

Following this, the reverse voltage V_R is raised further, until the signal current I_S, and thus the amplification factor, have increased to twice the value; i.e.:

$$M_2 = 2 \cdot M_1 \qquad (9.77)$$

The incident radiant flux Φ_e is again interrupted, so that the value of the dark current which now occurs, I_{D2}, can be recorded. *Figure 9.28* shows the principle of this measurement. The surface leakage current $I_{su(25)}$ and the bulk leakage current $I_{bu(25)}$ are the calculated as follows:

$$I_{su(25)} = I_{D1} - (I_{D2} - I_{D1}) = 2I_{D1} - I_{D2}$$

$$(9.78)$$

$$I_{bu(25)} = \frac{I_{D2} - I_{D1}}{M_2 - M_1} = \frac{I_{D2} - I_{D1}}{M_1}$$

$$(9.79)$$

9.9
Sensitivity values of IR detectors

IR detectors, also called IR signal detectors, are understood to mean infra-red-sensitive receivers with comparable sensitivity and noise power for different spectral sensitivities. These detectors include, for example, very sensitive avalanche photo-diodes, PIN photodiodes, intrinsic photo-conductors and photomultipliers. For the noise voltage, which is of interest, it can be said in general, that as with most noise mechanisms it is roughly proportional to the square root of the bandwidth to be transmitted.

The signal-to-noise ratio, abbreviated as S/N ratio, is either the ratio of the rms signal current I_S to the rms noise current I_N or of the rms signal voltage V_S to the rms noise voltage V_N:

$$\frac{S}{N} = \frac{I_S}{I_N} = \frac{V_S}{V_N} \tag{9.80}$$

S is the rms value of the output signal of a detector or detector-amplification system. The detector is usually exposed to radiation with square-wave modulation. The pulse repetition frequency is either 800 Hz or 1000 Hz with a mark-space ratio of 1:1. The signal is measured selectively at f_{mod} and a defined narrow bandwidth, (e.g., $\Delta f = 5$ Hz) and is usually converted to a bandwidth of $\Delta f = 1$ Hz.

N is the rms value of the detector noise or the detector-amplifier noise. As a rule, 300 K background radiation falls on the detector. The noise is also measured selectively at f_{mod} and the defined bandwidth, (e.g., $\Delta f = 5$ Hz) and is usually converted to a bandwidth of $\Delta f = 1$ Hz.

The following data is indispensible for the measurement of S/N ratio:

The temperature of the black-body radiation or the wavelength of a mono-chromatic radiation, with which the most favourable S/N ratio occurs,

the electrical reference bandwidth,

the mid-band frequency,

the detector reference temperature or

the working temperature for cooled detectors,

the detector surface area A_D,

the incident irradiance E_e,

where applicable the angle of view and

the temperature of the aperture.

In order to compare different kinds of detectors, the term noise equivalent power (NEP) has been introduced. It is defined as the power of sinusoidally modulated radiation, which would produce the same rms output voltage in an imaginary ideal noise-free detector, as the real detector, without an input signal, delivers as noise voltage. The noise equivalent power, NEP, is the lower useful threshold of the detector, with a given detector surface area A_D, where the useful and noise signals are equal.

The S/N ratio is measured with the aid of a known, weak but still easily detectable incident radiant power, and the NEP is obtained approximately from the following formula:

$$NEP = E_e \cdot A_D \cdot \frac{1}{\sqrt{\Delta f}} \cdot (\frac{N}{S})$$

$$= \Phi_e \cdot \frac{1}{\sqrt{\Delta f}} \cdot (\frac{N}{S}) \tag{9.81}$$

In practice, the measurement unit for NEP is often stated in watts only, instead of in W/$\sqrt{Hz}$, when related to a bandwidth of $\Delta f = 1$ Hz.

If the irradiance is chosen for the calculation instead of the radiant power, the term *noise equivalent input power*, abbreviated NEI, is used.

For this:

$$NEI = \frac{NEP}{A_D} = E_e \cdot \frac{1}{\sqrt{\Delta f}} \cdot (\frac{N}{S}) \tag{9.82}$$

155

The reciprocal value of the NEP is the *detectivity* (detection capability), abbreviated D. The detectivity is a measure of the least detectable radiant power:

$$D = \frac{1}{NEP} = \frac{\sqrt{\Delta f}}{E_e \cdot A_D} \cdot (\frac{S}{N}) \qquad (9.83)$$

The measurement unit for detectivity is $\sqrt{Hz}/W$.

The NEP of a detector system is proportional to the square root of its surface area A_D. *The specific detectivity*, abbreviated D*, takes account of this factor

$$D^* = D \cdot \sqrt{A_D} = \frac{\sqrt{A_D}}{NEP} = \frac{1}{NEI \cdot \sqrt{A_D}} \qquad (9.84)$$

The reference bandwidth Δf of the specific detectivity is always 1 Hz. In practice, the measurement unit is often stated in cm/W instead of in cm . $\sqrt{Hz}/W$.

As well as the noise power and the detectivity, the *sensitivity s* of a detector is of interest. The sensitivity s is the ratio of the rms value of the signal voltage V_s to the rms value of a known radiant power falling on the detector:

$$s = \frac{V_s}{\Phi_e} = \frac{V_s}{E_e \cdot A_D} \qquad (9.85)$$

From this formula, the dimension of sensitivity is V/W. For the determination of sensitivity, the noise power, bandwidth and the modulation frequency when measuring with intermittent light (if f_{mod} is far enough below the limiting frequency of the detector and the measuring apparatus) are not needed. However, in sensitivity measurements, the distribution temperature of the radiation used, the detector temperature and the detector area A_D are to be defined. If the NEP or D* are known from the data

sheet and the noise voltage is measured, then, using the formulae (9.80), (9.81), (9.84), (9.85), the sensitivity can be calculated as follows:

$$s = \frac{V_N}{NEP \cdot \sqrt{\Delta f}} = \frac{D^* \cdot V_N}{\sqrt{A_D} \cdot \Delta f} \qquad (9.86)$$

The measurement conditions for the parameters NEP, NEI, D and D* are identified by figures in brackets, e.g.:

D* black-body — D*(500, 1000, 1)
D* spectral — D*(6, 1000, 1)
D* spectral max. — D*(λ_{max}, 1000, 1)

where 500 denotes T = 500 K of the black body as a radiation source.

1000 denotes f = 1000 Hz, pulse repetition frequency of the modulated radiation.

6 identifies the wavelength (λ = 6 μm) of the monochromatic radiation, with which D* D* was measured.

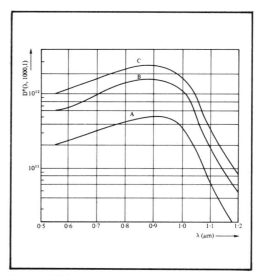

Figure 9.29
Specific detectivity D as a function of the wavelength for silicon photo-detectors with various detector surface areas: A = 0·02 cm^2, B = 0·2 cm^2, C = 1 cm^2.*

156

1 denotes $\Delta f = 1$ Hz as the converted bandwidth of the receiving amplifier, with which D* was measured at f_{mod}.

If the parameters NEP, NEI, D and D* are unknown for a Si photodetector, the NEP can be calculated approximately with special purpose calculators. Here, the current noise with the dark current I_D and the resistance noise with a load resistance R_L are taken into account for the corresponding Si photodetector.

Figure 9.29 shows the specific spectral detectivity D* of various silicon detectors as a function of wavelength λ. The variable is the detector surface area with A = 0·02 cm², B = 0·2 cm² and C = 1 cm². D* (2800, f,1) was converted to D*(λ_{max}) using the empirical value 5. In *Figure 9.30*, the specific detectivity D* (λ,450,1) in InAs detectors is shown as a function of the wavelength λ. The variable is the temperature T in degrees Kelvin.

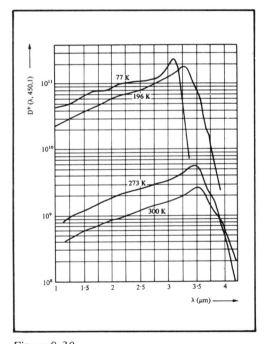

Figure 9.30
Specific detectivity D as a function of the wavelength λ for InAs photodetectors*

With sensitive detectors, the specific detectivity is mainly dependent on the background radiation (Blip); it can be improved by providing a cooled aperture in front of the detector. The restricted aperture angle is called the viewing field angle Θ. *Figure 9.31* shows the decrease

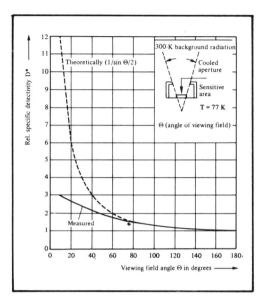

Figure 9.31
D = f(Θ) for InAs photodetectors*

in the relative specific detectivity with greater viewing field angles Θ for InAs detectors.

In *Figure 9.32*, the sensitivity s_{500} K is shown for Ge:Hg detectors as a function of the viewing field angle Θ at 5 K. The variable is the detector surface area A_D.

Radiation detectors have inertia (see rise and fall times) and their speed performance is determined by the *response time* or *time constant* (time response). The time constant is inversely proportional to the limiting frequency. It is determined when the modulated radiation falling on a detector surface causes a drop in the detector sensitivity of 3 dB as compared with the same radiation, but unmodulated.

157

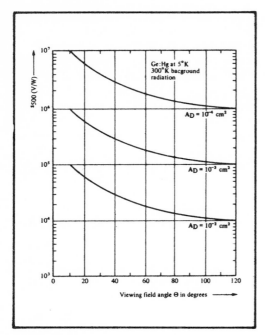

$$\tau = \frac{1}{\omega} = \frac{1}{2\pi f}$$

(9.87)

By means of the time constant of the detector, the maximum usable modulation frequency is determined, since above this both the sensitivity s and the quantum yield decrease. The theoretical measuring set-up for the determination of the time constant of silicon detectors, for example, is shown in *Figure 9.33*.

Figure 9.32
s = f(Θ) for mercury-doped germanium
(Ge:Hg) photodetectors

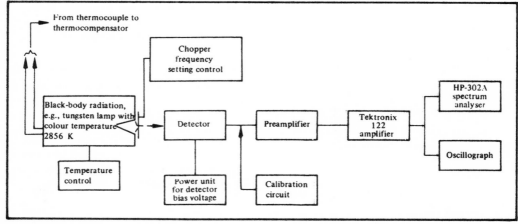

Figure 9.33
Circuit for measurement of the time constant

10
Parameters common to emitters and receivers

Parameters Common to Emitters and Receivers

10.1
Evaluation of the Radiation and Receiving Characteristics of Optoelectronic Components

Both the radiation intensity I_e of a radiation source and also the radiation sensitivity s of a detector are greatly dependent on the angle of emergence or of incidence of the radiation. This dependence is plotted in suitable graphs as the Radiant Intensity Distribution Curve $I_e = f(\varphi)$ or as the sensitivity distribution curve $s = f(\varphi)$. For this, polar coordinates are mostly used, but occasionally also Cartesian coordinate systems. A well-known example of this is the luminous intensity distribution curve of a filament lamp, illustrated in *Figure 10.1*, where the luminous intensity, which depends on the angle to the axis of summetry, is stated in cd.

On the same principle, *Figure 10.2* shows the emission conditions of a fluorescent tube. Here, the luminous intensity distribution curve is shown for a standardised luminous flux of 1000 lm for the tube.

From antenna and radar engineering in particular, it is generally well-known, that the efficiency of a transmitter and also of a receiver can be considerably improved by focussing the radiation into a beam. The distribution characteristics then form more or less narrow peaks.

In the field of optoelectronics, the functions $I_e = f(\varphi)$ and $s = f(\varphi)$ are of great importance for evaluating the performance of these components. These are characteristics of approximately spherical shape and also narrow patterns. The shape of the emission or reception characteristics is fixed by the mechanical construction of the components.

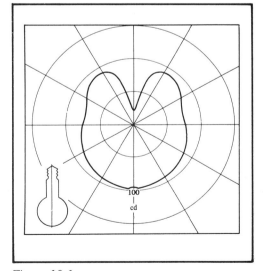

Figure 10.1
Luminous intensity distribution curve, with axial symmetry, of a filament lamp for 100 lm

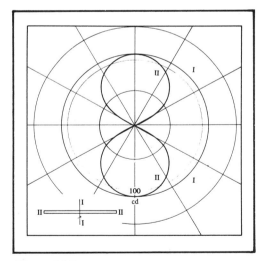

Figure 10.2
Luminous intensity distribution curve of a fluorescent lamp for 1000 lm. Plane I perpendicular to the axis of the lamp, plane II through the axis of the lamp

The case and its material play a decisive part here. The structural shape of the wafer also has an effect. A distinction is made here between flat and hemispherical dome wafers. Many wafers are embedded in an epoxy medium, through the shape and refractive index of which the radiation characteristic can be influenced in a planned manner. Lenses may be built to produce particularly narrow focussing of the maxima. It should be mentioned at this stage, that in the case of components with an asymetric characteristic, more or less pronounced secondary peaks can occur as well as the main peak, and must be taken into account in calculations and in the construction of optoelectronic systems.

For all components with a spherical radiation characteristic, Lambert's cosine law can be applied for the calculation. It is known from Chapter 3, that the intensity I_e of a Lambert radiator decreases, according to a cosine function, with increasing angle φ from the normal.

$$I_{e,\varphi} = I_{e,o} \cdot \cos\varphi \qquad (3.10)$$

A hemispherical space around the Lambert radiator is therefore not uniformly irradiated. A maximum radiant intensity is measured in the normal direction, and this decreases to zero, in proportion to $\cos\varphi$, with increasing angle φ.

Now, the Lambert radiator can have allocated to it a given solid angle Ω, which is so great, that a spherical cap with this solid angle, and assuming the radiant intensity I_e to be constant in space, receives the same total radiant power as the hemisphere defined by the cosine law. This means, the Lambert radiator is replaced, in the imagination, by a radiator, which emits with its full intensity I_{max} within a given solid angle Ω and emits no radiation outside this solid angle.

The solid angle can be determined mathematically as follows: The original

hemispherical surface is broken down, as shown in *Figure 10.3*, into annular zones. If the zones are sufficiently narrow, then the area of such a zone is:

$$dA = 2\pi \cdot r \cdot \sin\varphi \cdot d\varphi \cdot r \qquad (10.1)$$

By integration of this expression from 0 to $90°$, and taking account of the elementary radiation laws (2.18) and (3.1), the relationship

$$\Phi_e = I_e \cdot \Omega = I_e \cdot \frac{A}{r^2} \cdot \Omega_0 \qquad (10.2)$$

is obtained and, with equation (3.10), the following integral for the radiant power:

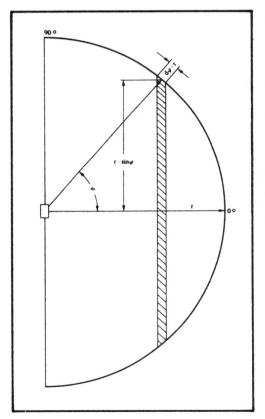

Figure 10.3
Derivation of the solid angle in the case of the Lambert radiator

162

$$\Phi_e = \frac{\Omega_0}{r^2} \cdot \int\limits_0^{90^\circ} 2\pi \cdot r \cdot \sin\varphi \cdot I_{e,o} \cdot \cos\varphi \, d\varphi \cdot r$$

$$= 2\pi \cdot I_{e,o} \cdot \Omega_0 \cdot \int\limits_0^{90^\circ} \sin\varphi \cdot \cos\varphi \cdot d\varphi$$

$$= 2\pi \cdot I_{e,o} \cdot \Omega_0 \cdot [\frac{\sin^2\varphi}{2}\Big|^{90^\circ}] = \pi \cdot I_{e,o} \cdot \Omega_0$$

$$(10.3)$$

Division of the power Φ_e, which is assumed to be constant, by the maximum radiant intensity $I_{e,o}$ gives a solid angle

$$\Omega = \frac{\Phi_e}{I_{e,o}} = \pi \cdot \Omega_0 = \pi \quad \text{sr} \qquad (10.4)$$

for the Lambert radiator.

This result has already been pointed out in (3.11). The expression Ω_0 or 1 sr has no unit in itself, but is treated mathematically like a constant in the formulae.

By means of *Figure 10.4*, it will now be shown geometrically that a hollow sphere corresponding to the Lambert radiation characteristic is irradiated with equal intensity at all points. For every surface element of this hollow sphere, conditions, such as are shown in *Figure 3.9*, apply, where both the angle φ and also the distance a are variables:

$$\Phi_e, dA_{E1} = f(\varphi, \text{a}) \qquad (10.5)$$

Since the normal to the emitter and the normal to the receiver always intersect at the centre of the sphere, they form the two equal sides of an isosceles triangle, the base of which is the ray under consideration. For reasons of symmetry, the angles φ_S and φ_E must therefore also be equal:

$$\varphi_S = \varphi_E = \varphi \qquad (10.6)$$

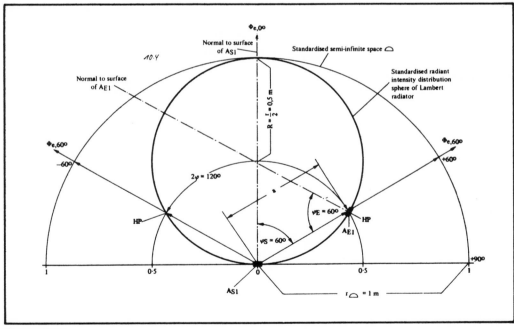

Figure 10.4
Standardised radiant intensity distribution sphere of a very small plane area A_{S1} of the Lambert radiator, radiating on one side in the standardised semi-infinite space with unit radius r = 1 m

With the Lambert radiator, the distance 'a' is clearly dependent on the angle φ, namely:

$$a = r \cdot \cos\varphi \qquad (10.7)$$

If the results of equations (10.6) and (10.7) are inserted in (3.22), we obtain:

$$d^2\Phi_e = L_e \frac{dA_E \cdot \cos\varphi \cdot dA_S \cdot \cos\varphi}{(r \cdot \cos\varphi)^2} \cdot \Omega_o \qquad (10.8)$$

Since the only variable in this equation, $\cos\varphi$, can be cancelled out, pure constants are obtained and the differential notation can therefore be dispensed with:

$$\Phi_e = L_e \cdot \frac{A_E \cdot A_S}{r^2} \cdot \Omega_o \qquad (10.9)$$

In this equation, Φ_e is no longer dependent on the angle and is therefore a constant value for the hollow sphere under consideration.

Now let us again consider the hemisphere with radius r. According to the cosine law, at an angle $\varphi = 60°$, the radiant power obtained is:

$$\Phi_{e,60°} = \Phi_{e,o} \cdot \cos 60° = 0 \cdot 5 \cdot \Phi_{e,o} \qquad (10.10)$$

Since, for semi-infinite space, only half the radiant power is emitted here, in comparison with the power in the normal direction, the corresponding points on the distribution curve (a complete sphere with radius r/2) are designated as the *half-power points*, abbreviated HP.

Next, the solid angle Ω is examined. It is obtained by consideration of the spherical cap within the half-power points with $2\varphi = 120°$. According to equation (3.4):

$$\Omega = 2\pi(1 - \cos 60°) \cdot \Omega_o = \pi \cdot \text{sr} \qquad (10.11)$$

If this result is compared with (10.4), then it is seen, that the boundary of the solid angle, in the case of the Lambert radiator, goes exactly through the half-power points of its distribution curve.

It has already been pointed out, that many optoelectronic components have asymmetric directional characteristics. An exact calculation of the solid angle Ω_K which applies here is only possible, if the asymmetric characteristic can be described unambiguously by an equation of the form

$$I_{e,\varphi} = f(\varphi) \cdot I_{e,o} \qquad (10.12)$$

In practice, an attempt will therefore be made, to find a function approximately the same as the characteristic, which can be calculated with a minimal expenditure of mathematical effort. Functions of the form

$$I_{e,\varphi} = \cos(n \cdot \varphi) \cdot I_{e,o} \qquad (10.13)$$

are suitable for this purpose, n being a number depending on the degree of narrowness of the distribution.

As an example, *Figure 10.5* shows, in addition to the Lambert radiator with n = 1, two maxima with n = 3 and n = 10. Calculation of the solid angle is carried out on the same principle as for the Lambert radiator, and the following integral is obtained:

$$\Omega_K = 2\pi \cdot \Omega_o \cdot \int_o^{\frac{90°}{n}} \sin\varphi \cdot \cos(n\varphi) \cdot d\varphi$$

$$= 2\pi \cdot \Omega_o \cdot [-\frac{\cos(1+n)\varphi}{2(1+n)} - \frac{\cos(1-n)\varphi}{2(1-n)}]_o^{\frac{90°}{n}}$$

$$(10.14)$$

Calculation of this integral for a few selected values of n gives the results stated in *Table 10.1*.

n	Ω_K/sr	$\varphi\Omega/^{-\circ}$	$\varphi_{HP}/^{\circ}$
1·5	1·50	40·4	40
2	0·867	30·5	30
2·5	0·562	24·4	24
3	0·393	20·4	20
4	0·222	15·3	15
5	0·143	12·2	12
6	0·0993	10·2	10
8	0·0559	7·65	7·5
10	0·0358	6·12	6

Table 10.1
Results obtained from equation (10.14)
for a few selected values of n

Also, from the solid angle Ω_K according to equation (3.4), the angles $\varphi\Omega$, which limit the cone for Ω_K, have been determined. At the same time, the angle for the half-power points, at which the radiated power has fallen to half, has also been stated. As can be seen, with these peaks, $\varphi\Omega$ and φ_{HP} are no longer exactly equal, as was the case with the Lambert radiator. The small differences are not due to inaccuracies in calculation, but actually exist. Of course, they are very small (of the order of 1 to 2%), so that in practice,

in the case of asymmetric radiation maxima which are similar to the functions discussed here, the solid angle can be calculated directly with the aid of the half-power points.

Solid angle calculations are mainly of interest for components which emit radiation. In the data sheets, statements on the radiated power Φ_e are generally found. In the case of asymmetric characteristics, the total power, including any secondary peaks which may be present, is often stated, but often also only the power of the main area alone.

Nowadays, statements of the radiant intensity $I_{e,o}$ in the normal direction are most often quoted and it is to be expected, that this parameter will be stated in future in all data sheets.

The radiation characteristics, which define the radiant intensity as a function of the angle φ in suitable graphs, generally show relative values, related to the value in the principal direction of emission. In evaluating the graphs, it must not be forgotten, that as a result of various tolerances, which are unavoidable in

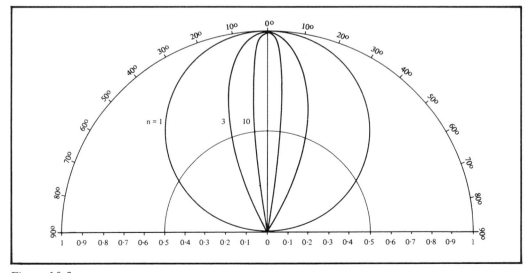

Figure 10.5
Functions of the form a = cos $n\varphi$ in polar coordinates

165

mass production, the final results can sometimes differ from the values obtained theoretically. In particular, even small tolerances of the radiation distribution curve can cause a considerable reduction of the irradiance E_e falling on a photodetector. Further details on tolerances will follow in Section 10.2.

The problems mentioned will be illustrated by two practical examples: *Figure 10.6* shows the radiation emission characteristic of the GaAs diode TIXL 27. Here, we are concerned with an approximately spherical pattern, with which the calculation of the solid angle Ω_K can be carried out with sufficient accuracy from the half-power points. As can be seen from the graph, the rays through the half-power points have an angle $\varphi = 67°$ to the major axis. With equation (3.4), one therefore obtains:

$$\Omega_K \approx 2\pi(1-\cos 67°) \cdot \Omega_0 = 3 \cdot 826 \text{ sr}$$
$$(10.15)$$

In the data sheets of the TIXL 27, a minimum radiant power $\Phi_{e,min} = 15$ mW is stated. This gives a minimum radiant intensity $I_{e,o,min}$ in the direction of the major axis of:

$$I_{e,o,min} = \frac{\Phi_{e,min}}{\Omega_K} = \frac{15 \text{ mW}}{3 \cdot 826 \text{ sr}}$$

$$= 3 \cdot 92 \text{ mW/sr} \qquad (10.16)$$

As a further example, the results of a practical measurement on the GaAs diode TIL 31 will be discussed.

The emitter-receiver distance was selected at r = 20 cm and a solar cell, with an active area $A_E = 3 \cdot 6$ cm^2 and a sensitivity s = 0·51 A/W, was used as the detector. The radiation of the TIL 31 is first observed on a projection screen with an IR viewer. Here, the projection of the wafer is very clearly visible as a square with a length of side of 1 = 4·3 cm. At a somewhat

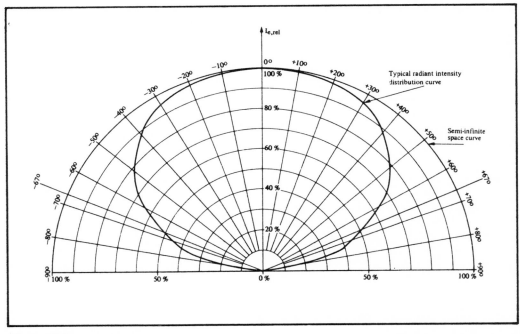

Figure 10.6
Calculation of the solid angle Ω_K in the case of the GaAs diode TIXL 27. This diode has a modified T05 can. The projecting cylindrical cover is removed. The flat wafer is embedded in an epoxy dome

greater distance around this square, a number of rings, which arise from the secondary peaks, can also be seen quite faintly. At first, these were not taken into account.

The solid angle of the main maximum can be calculated with sufficient accuracy on the basis of the water projection found, from (3.1):

$$\Omega_K = \frac{A_K}{r^2} \cdot \Omega_0 = \frac{4 \cdot 3^2 \, cm^2}{20^2 \, cm^2} \; sr = 0 \cdot 0462 \; sr \tag{10.17}$$

From the converted formula (3.4), the angle of the rays through the half-power points of the idealised main peak to the normal is obtained:

$$\varphi = arc \; cos \; (1 - \frac{\Omega_K}{2\pi}) = arc \; cos \; (1 - \frac{0 \cdot 0462}{2\pi})$$

$$= 7^\circ \tag{10.18}$$

The measuring instrument connected to the fully irradiated solar cell showed a short-circuit current $I_P = 160 \; \mu A$. From this the radiant power falling on the solar cell was calculated with equation (9.27):

$$\Phi_{e,E} = \frac{I_P}{s} = \frac{160 \; \mu A}{0 \cdot 51 \; W/A} = 313 \cdot 7 \; \mu W \tag{10.19}$$

Further, (2.40) gives:

$$E_e = \frac{\Phi_{e,E}}{A_E} = \frac{313 \cdot 7 \; \mu W}{3 \cdot 6 \; cm^2} = 87 \cdot 14 \frac{\mu W}{cm^2} \tag{10.20}$$

Since the angle of incidence of the rays at the edge onto the solar cell is still very small, the $cos \; \varphi_E$ in equation (3.25) can be made equal to 1 without any great error, and the radiant intensity of the main area of the distribution can be calculated as follows:

$$I_e = \frac{E_e \cdot r^2}{\Omega_o} = 87 \cdot 14 \cdot 20^2 \frac{\mu W}{sr} = 34 \cdot 86 \frac{mW}{sr} \tag{10.21}$$

The radiant power of the main peak is obtained from (2.18) as

$$\Phi_e = I_e \cdot \Omega_K = 34 \cdot 86 \frac{mW}{sr} \cdot 0 \cdot 0462 \; sr$$

$$= 1 \cdot 61 \; mW \tag{10.22}$$

It has been mentioned above, that secondary maxima are present as well as the main peak. In order to include their radiant power as well, the solar cell was now arranged directly in front of the TIL 31. In this case, evaluation of this greatly simplified measurement gave a total radiant power $\Phi_{e,tot}$ 7 mW. From this it can be seen, that only about 23% of the total power is concentrated in the main peak, while the remainder is scattered in a fan-like pattern by the secondary maxima.

The calculation of radiation characteristics which differ considerably from the previous idealised curves is rather more difficult. For this, it is convenient to apply graphical methods. In the Rousseau diagram, the radiation pattern is divided into sufficiently small angular areas and integrated graphically. A second method describes a breakdown into constant solid-angle zones, which can also be integrated graphically.

10.2
Optical Tolerances of optoelectronic components

In the previous section 10.1, it has already been pointed out, that optoelectronic components can be subject to relatively large tolerances. Distinctions are made here between:

1
Tolerances on the electrical parameters.

167

2

Tolerances of the total radiated power Φ_e of semiconductor radiation sources.

3

Tolerances of the basic photosensitivity of semiconductor junction photodetectors. Here, all external or built-in optical parts such as lenses, mirrors or reflectors are to be assumed to be dismantled from the photodetector.

4

Installation tolerances of optoelectronic components on the part of the user.

5a

Tolerances of the emission peak of semiconductor radiation sources.

5b

Tolerances of the receiving peak of junction photodetectors.

The electrical tolerances and the tolerances on radiant power and photosensitivity of optoelectronic components should be overcome by an optimised circuit design. The radiant power tolerances can be eliminated in many cases by adjusting the flux current I_F through the radiating diode. In simple applications, the photosensitivity tolerances can be compensated for by gain adjustments on the following amplifier. The user should examine all measures of circuit design and also keep the mechanical installation tolerances of the optoelectronic components as small as possible, in order to be able to use lower priced, mass-produced standard types of optoelectronic component instead of the more expensive selected types.

The tolerances of the radiation peak of` semiconductor radiation sources and the tolerances of the receiving peak of junction photodetectors must be dealt with in great detail, since these tolerances can have a very drastic effect on the output signal from the detector.

10.2.1

Effect of wafer centering, lens quality, distance from lens to wafer, refractive index of epoxy resin and shape of dome and case

The highest-quality optoelectronic components, which are mostly used in commercial electronic systems, have metal cases with flat windows. Metal cases are preferred, since they have better thermal resistance $R_{th\,JC}$ and are easy to mount on heat sinks. Flat windows are installed intentionally, so that the user can design the focussing of the radiant power optimally for his needs, with simple lenses or high-quality reflector, glass fibre or lens combinations.

Epoxy resin domes are not used, since otherwise these components cannot be used for temperatures above 100°C. The GaAs power diodes are constructed with a dome wafer for a better radiation yield.

Components with flat windows often have emission or reception maxima with wide aperture angles. The deviation of the principal direction of emission or reception from the normal is mainly determined only by the wafer centering. Here, the tolerances are not great, since the spatial distribution alone gives a drop in power of a few percent in the normal direction. In addition, the aperture angle of such maxima is usually so large, that despite the tolerances, the aperture angles of attached lenses or lens systems are covered almost without a decrease in power. As an example, *Figure 10.7* shows a distribution curve of an optoelectronic component with a tolerance deviation. The aperture angle between the half-power points is about 135°. The drop in power in the normal direction as compared with the main emission or reception direction is aobut 4%.

"Low-cost" GaAs power diodes have wafers with flat geometry and are installed in a metal case. For a better

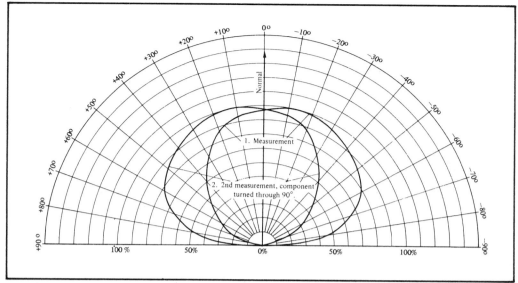

Figure 10.7
Radiant intensity or sensitivity distribution of an optoelectronic component affected by tolerance. The aperture angle is about 135°, the drop in power in the normal direction about 4%

radiation yield (see Section 6.6 and Chapter 11), an epoxy-resin dome or lens is fitted on the wafer. Here, somewhat greater radiant intensity distribution tolerances are to be expected. They are mainly determined by the wafer centering, the refractive index of the epoxy resin and the shape of the epoxy dome. *Figure 10.8* shows a radiant-intensity distribution curve, subject to tolerance, for a "Low cost" GaAs power diode. The aperture angle betwen the half-power

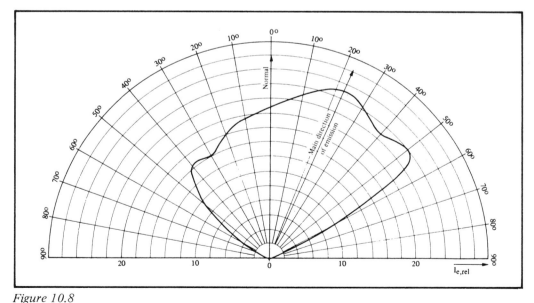

Figure 10.8
Radiant intensity distribution of a "Low-cost" GaAs Power diode, subject to tolerance

169

points is about $110°$, while the drop in radiant intensity in the normal direction, as compared with the main direction of radiation, is around 15%. Despite the larger radiant intensity distribution tolerances of such GaAs diodes, the aperture angles of attached lenses are covered without significant losses of radiant power.

Low-power GaAs diodes also have, in general, wafers with a flat geometry. They are installed in metal cases with flat windows, if close radiant intensity distribution tolerances and high operating temperatures are called for. Epoxy resin droplets are applied to flat wafer surfaces, if somewhat higher radiation yields and radiant powers are desired. The radiant intensity distribution tolerances then increase. For focussing the emission peak of GaAs diodes or the reception peak of junction photodetectors, i.e., if high radiant intensities or high sensitivities are required, glass lenses are fixed on the metal case. The optical quality of these very small lenses is low as compared with the usual single lenses. The distance from the wafer to the lens is subject to the assembly tolerances. Therefore the distribution tolerances increase once more. The wafers of "Low cost" low-power GaAs diodes or junction photo-detectors are completely embedded in epoxy-resin cases. Here the distribution tolerances are increased yet again.

10.2.2
Effects of internal case reflections and wafer geometry

The emission or reception peaks of opto-electronic components are also affected by internal case reflections and the wafer geometry. Flat wafers have either a cathode bond (GaAs diodes) or a base and an emitter bond (junction photodetectors) on their surface. A bond can sometimes be located in the centre of the wafer. The

output power distribution can then show dips in the normal direction. This effect can also occur through the shape of epoxy resin cases. *Figure 10.9* shows a distribution curve with a dip in the principal direction of emission. Discontinuities in radiant power in the radiant intensity distribution profile can be detected visually with an IR viewing device, by observing the radiation from a GaAs diode falling on a projection screen. Corresponding to the radiation distribution shown in *Figure 10.9*, a more weakly reflecting circle will be projected in the normal direction, with a more strongly reflecting annular area surrounding it.

Optoelectronic components in metal transistor cases with fitted lenses can show additional undesired secondary maxima due to internal reflections from the case, as well as a closely focussed peak in the main direction. In *Figure 10.10*, the radiant intensity distribution of a low-power GaAs diode with unwanted secondary peaks is shown. If this radiation in the normal direction falls perpendicularly on a projection screen, then a strongly reflecting circular area will be projected in the main direction of radiation and two surrounding less strongly reflecting annular areas at intervals from it.

Furthermore, the radiant intensity distribution of a GaAs diode can be assessed visually with the IR viewer by a second method. For this, transparent paper is stretched out flat in the normal direction to the GaAs diode. The radiant intensity distribution on the transparent plane can be seen very impressively and clearly, since it corresponds to the radiant intensity distribution in the polar coordinate system.

With epoxy resin and metal cases for low-power GaAs diodes, the radiation also emerges through the base of the package through internal reflections in the case. For epoxy resin cases, this is clearly under-standable, since the base of the case is

170

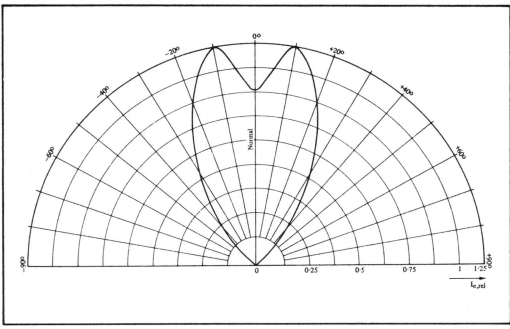

Figure 10.9
Radiant intensity or sensitivity distribution of an optoelectronic component affected by tolerance. The distribution profile shows a dip in radiant intensity or sensitivity in the normal direction

transparent to IR radiation. With metal cases, the radiation emerges through the plastic insulation of the connecting wires, which, although opaque to light, sometimes transmits radiation even in the very near IR range. This effect is mentioned, since

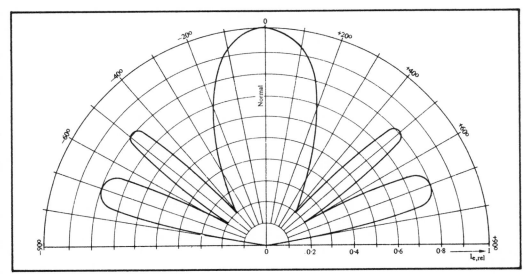

Figure 10.10
Radiant intensity distribution of a GaAs diode with secondary maxima

171

in extreme cases it could cause cross-talk interference with adjacent radiation links.

GaAs diodes which are overloaded from the point of view of loss power can also change their radiant intensity distribution. Dead areas can then form in the wafer. With an IR viewer, it is easy to ascertain, whether only part of the wafer is still emitting radiation. *Figure 10.11* shows the wafer of a GaAs diode overloaded in this way. Its emission decreases more and more from the centre to the left-hand side. It is equally possible that only a small corner area will emit radiation.

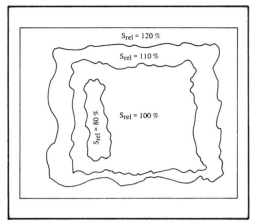

Figure 10.12
Sensitivity distribution over the active wafer area of a junction photodetector

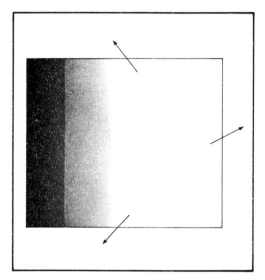

Figure 10.11
Emitting wafer of an overloaded GaAs diode with dead zones

For the enlarged observation of the wafer of an IR-emitting GaAs diode, an IR viewer with an additional lens or with an external lens will be needed. For GaAs diodes with built-in, strongly refracting lenses, it is simpler to observe the wafer. Here, the radiation from these GaAs diodes is already focussed so that the wafer can be shown as an image on a flat projection screen and observed with the IR viewer. Wafer defects, bonds and bonding wires are then clearly visible. If the radiance of a GaAs diode is considered over all

active parts of the wafer area, it cannot be regarded as constant. In the same way, the sensitivity of a junction photodetector is not constant over all the active parts of the wafer area. *Figure 10.12* shows, in simplified form, the sensitivity distribution over the active wafer area of a photodiode. Here the sensitivity increases towards the outer zones. Various people have investigated the sensitivity distribution of Si photodetectors over the photosensitive area. From what has been said so far, the reasons for the distribution tolerances of low-power GaAs diodes with closely focussed emission peaks or of junction photodetectors with closely focussed reception peaks can be summarised. The factors which are responsible are the wafer centering, the quality of built-in lenses, the distance from lens to wafer, the refractive index of the epoxy resin, the shape of the epoxy dome or case, bonds on the wafer and bonding wires, internal case reflections and last but not least the radiance distribution over the active wafer area of GaAs diodes or the sensitivity distribution over the active wafer area of junction photodetectors. The stated causes for the tolerances clearly show, that to achieve a uniform mode of representation, the radiant intensity of GaAs diodes and the

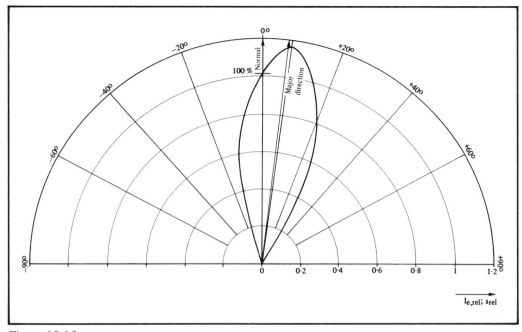

Figure 10.13
Radiant intensity distribution curve or sensitivity distribution curve of a "cross-eyed"
optoelectronic component. The principal direction deviates from the normal by 9.5°

sensitivity of junction photodetectors must always be measured in the normal direction instead of the main direction of emission or reception. *Figure 10.13* reproduces the distribution curve of a "cross-eyed" optoelectronic component. The principal direction deviates from the normal by 9.5°. Here, the sensitivity of a detector or the radiant intensity of a GaAs diode is 15% higher in the main direction than in the normal direction.

10.2.3
Tolerance levels

Generally, data on the minimum, typical or maximum sensitivity for junction photodetectors only relates to the normal direction of the radiant power data of GaAs diodes only relates to the corrected emission peak in the normal direction. Thus the only case which can occur is that a junction photodetector shows a higher sensitivity in the principal reception direction or a GaAs diode shows a higher radiant power or radiant intensity peak in the principal emission direction. The sensitivities for junction photodetectors or the radiant power values for GaAs diodes stated in data sheets mainly relate to the distribution curves illustrated. Closely focussed distribution maxima are sometimes represented with idealised distribution curves.

Users often set very strict tolerance requirements, which are not attainable in mass production, especially on optoelectronic components with built-in lenses or with epoxy-resin cases. It should be strictly noted that due to the inherant tolerances of these components comapred to devices without epoxy encapsulations and lenses they will have less optical precision. The built-in lenses only represent a poor substitute for external and optimally selected individual lenses.

173

In electro-optical precision instruments or systems, optoelectronic components with flat windows are mainly used in conjunction with attached external lenses or objectives. For reasons of space, in optoelectronic reading heads, or for reasons of cost, e.g., for consumer applications, external individual lenses cannot be used in many cases. Here, at least, taking account of the tolerances, one can have resources to selected optoelectronic components with lenses.

The selection types needed, e.g., for optoelectronic components with lenses, can be determined in a rough tolerance calculation in conjunction with radiant intensity and sensitivity measurements for the components in question.

a

For roughly-selected GaAs diodes with lenses:
Radiant power tolerances

$$\frac{\Phi_{e,max}}{\Phi_{e,min}} \approx 3 \qquad (10.24)$$

Estimated emission peak tolerances in the measurement direction

$$\frac{\Omega_{K,max}}{\Omega_{K,min}} \approx 2 \qquad (10.25)$$

The radiant intensity tolerance is then:

$$\frac{\Phi_{e,max}}{\Phi_{e,min}} \cdot \frac{\Omega_{K,max}}{\Omega_{K,min}} \approx 3 . 2 = 6 \qquad (10.26)$$

b

For preselected Si photodetectors with lenses:

The difference in the photocurrent sensitivity amounts to

$$\frac{s_{max}}{s_{min}} \approx 3 \qquad (10.27)$$

c

Installation tolerances:
The deviation of the opposing optical axes in each case for the receiver and emitter, which are caused by assembly and by temperature-dependent stresses in the installation support material, depend on the precision of the mechanical structure. The overall installation tolerances are estimated by the inclusion of a factor 2.

d

Transmission medium:
In many cases, the tolerances of the transmission medium are not taken into account. However, in an atmosphere containing water, dirt and oil, e.g., for machine controls, these tolerances are by no means negligible.

f

Overall tolerance factor:
The product of the factors from a to d gives the overall tolerance factor of

$$a . b . c . d \approx 6 . 3 . 2 . 1 \approx 36 \qquad (10.28)$$

An overall tolerance factor of 36 corresponds, for example, to a photo-current in the detector of

$I_{p,min} = 0 \cdot 1$ mA and $I_{p,max} = 3 \cdot 6$ mA

or as a second example of

$I_{p,min} = 5$ mA and $I_{p,max} = 180$ mA

(see also *Figure 10.15*).

Depending on the application and the selection of components, the overall tolerance factor can be even greater or also less. For very short-range radiation links, the overall tolerance factor can be reduced by the amount of the installation tolerances, since the emitter irradiates the whole detector.

GaAs diodes usually show the greatest tolerances, since, as already discussed, the semiconductor manufacturer only guarantees the minimum radiant power. The radiant intensity tolerances of GaAs diodes can be reduced by about a factor of 2 by adjusting the diode forward current I_F. Radiant

174

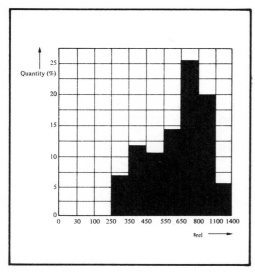

Figure 10.14a
Percentage numbers of phototransistors
as a function of the relative photocurrent
sensitivity for a photo-transistor type of
batch A

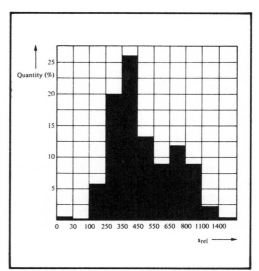

Figure 10.14b
Percentage numbers of phototransistors
as a function of the relative photocurrent
sensitivity for a phototransistor type of
batch B

intensity tolerances can also be taken into
account by the photodetector amplifier,
by varying the gain. With high tolerance
requirements, it is convenient, if the
user, during his goods-inward inspection, at
least selects the GaAs diodes supplied for
the relative minimum, typical and maximum
. radiant intensity.

GaAs diodes with low radiant intensities are
grouped for Si photodetectors with high
radiant intensity for Si photodetectors
with low sensitivity.

For mass production, it should be taken into
account, that the distribution of the
radiant intensity values for a GaAs diode
type or the distribution of the sensitivity
values for a Si photodetector type can
vary from batch to batch within the
data sheet tolerances. For one phototransistor type, the selection of which only
related to an uppe photocurrent sensitivity
limit, the relative photocurrent sensitivity
s_{rel} is illustrated in *Figure 10.14a* for
Batch A and in *Figure 10.14b* for Batch B.

A further reduction of the overall
tolerance factor is achieved, by choosing
closely selected photodetectors with
sensitivity differences of

$$\frac{s_{max}}{s_{min}} = 1 \cdot 5 - 2$$

Here, a large number of selected Si-photo-
detectors, developed by Texas Instruments,
are available to the user. To achieve the
smallest overall tolerance factors, the
GaAs diodes and Si photodetectors
recorded or selected in the customer's
goods inward inspection are combined as
matching pairs in each case.

Competent manufacturers of electronic
systems have already used this method
for years. In general, in the many simple
visible light or applications, which only
call for a Yes-No decision, relatively large
tolerance factors are acceptable with
appropriate circuit design. Here, a signal
is only initiated when a fixed irradiance
is exceeded or not exceeded.

175

10.3
Coupling characteristics, Transmission Ratios and Contrast Current Ratios of Very Small Photocell Junctions.

10.3.1
Coupling Characteristics and Transmission Ratios

Very small opto isolators are, as the name implies, light or IR sensitive devices with very short distances between source and detector. They also include the optoelectronic reading heads and optocouplers. The maximum source-detector distance corresponds to the photometric limiting distance. For this, an opto isolator can be designed with a theoretical accuracy of 1%. For shorter distances, practical calculation of a very small opto isolator performance is hardly possible. For this reason, the semiconductor manufacturer states the coupling characteristic between GaAs diodes and Si photodetectors. Here, the emitter and the detector face one another in the same optical axis. The coupling characteristic is represented in a graph as a function of the photodetector output signal in relation to the source-detector distance. *Figure 10.15* shows, as an example, the theoretical coupling characteristic for very small optoisolators or optoelectronic reading heads. The possible "On" collector current $I_{C(On)}$ is illustrated as a function of the source-detector distance for typical pair combinations with the GaAs diodes TIL 24 and the phototransistors LS 600.

The expression "On" relates to the switched-on or conducting state of the GaAs diode. The $I_{C(On)}$ tolerances correspond approximately to the tolerance calculation state in Section 10.2. For a distance of r = 0·25 inch, a typical "On" collector current of $I_{C(On),typ} = 2$ mA is obtained. This output current is high enough to drive either subsequent amplifiers or interface circuits. The following Figures 10.16 to 10.19 show typical coupling

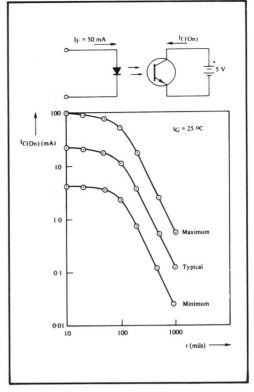

Figure 10.15
Basic coupling characteristic for very small opto isolators and optoelectronic reading heads. The possible "On" collector current $I_{C(On)}$ is illustrated as a function of the source-detector distance r for a typical pair combination with a GaAs diode TIL 24 and a Si phototransistor LS 600. 1000 mils = 1 inch = 24·4 mm

characteristics for very small opto isolators and optoelectronic reading heads from selected source-detector component families, some of which have the same packages.

The components used for *Figure 10.19*, the GaAs diode TIL 31 and the phototransistor TIL 81, have a metal encapsulated in the transistor TO 46 metal can to which a glass lens is fitted. Such optoelectronic components are often operated without a heat-sink. This has the result, that the highest GaAs

176

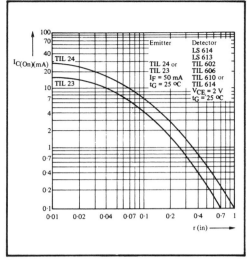

Figure 10.16
Typical coupling characteristic for source-detector combinations with GaAs diodes of types TIL 23, TIL 24 and Si photo-transistors of types LS 613, LS 614, TIL 602, TIL 606, TIL 610 and TIL 614. The "On" collector current $I_{C(On)}$ is shown as a function of the distance r

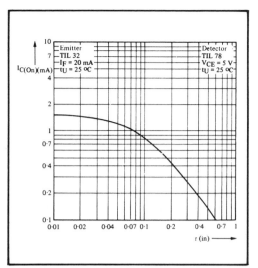

Figure 10.18
Typical coupling characteristic for a source-detector combination with the GaAs diode TIL 32 and the SI photo-transistor TIL 78. The "On" collector current $I_{C(On)}$ is shown as a function of the distance r

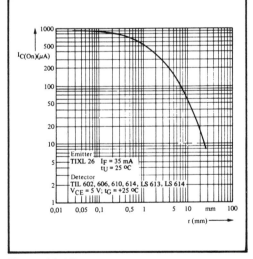

Figure 10.17
Typical coupling characteristic for source-detector combinations with the GaAs diode TIXL 26 and Si phototransistors of Types LS 613, LS 614, TIL 602, TIL 606, TIL 610 and TIL 614. The "On" collector current $I_{C(On)}$ is shown as a function of the distance r

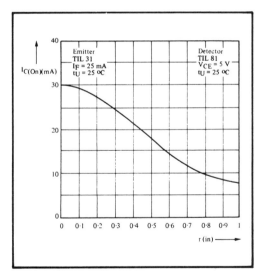

Figure 10.19
Typical coupling characteristic for a source-detector combination with the GaAs diode TIL 31 and the Si photo-transistor TIL 81. The "On" collector current $I_{C(On)}$ is shown as a function of the distance r

177

diode forward current $I_{F,max}$ for an ambient temperature $t_U = 25°C$ may only have a value, because of the loss power, of about a quarter of the maximum permissible value stated in the data sheet for a case temperature $t_G = 25°C$. However in comparison, the diode forward current $I_{F,max}$ at an ambient temperature $t_U = 25°C$ for the GaAs diode TIL 32 (which is equivalent to the TIL 31 except for its plastic case) may only be about a fifth of the maximum permissible value stated for the TIL 31 at $t_G = 25°C$. The relative radiant power $\Phi_{e,rel}$ as a function of the ambient temperature t_U for the GaAs diode TIL 31 can therefore also be read off approximately from the $\Phi_{e,rel} = f(t_U)$ graph of the GaAs diode TIL 32.

The absolute radiant power Φ_e of the TIL 31 can be calculated in simplified form with this relative value and the absolute TIL 32 data-sheet value for any desired ambient temperatures t_U. With a case temperature of $t_G = 80°C$, according to the TIL 31 data sheet, the diode forward current must have a value of $I_{F,t_G,80} = 0$ mA.

Now, the current drift factor per $°C$ ambient temperature, $\Delta I_F/t_U$, can be calculated for the GaAs diode TIL 31. This current drift factor is the ratio of the maximum permissible diode forward current $I_{F,max,t_U,25}$ for the ambient temperature $t_{U,25} = 25°C$ to the ambient temperature difference $\Delta t_{U,max-25}$ between the maximum permissible ambient temperature $t_{U,max}$ and $t_{U,25}$. In the absence of a statement in the data sheet, the maximum permissible case temperature $t_{G,max}$ is to be used as the maximum permissible ambient temperature $t_{U,max}$, since $t_{U,max}$ may not be greater than $t_{G,max}$. We have:

$$\frac{\Delta I_F}{t_U} = \frac{I_{F,max,\,t_{U,25}}}{\Delta t_{U,max-25}} \qquad (10.29)$$

The values for the equations (10.29 and 10.30) can be taken from the data sheet for the GaAs diode TIL 31. We obtain:

$$\Delta t_{U,max-25} = t_{U,max} - t_{U,25}$$

$$= 80°C - 25°C = 55°C \qquad (10.30)$$

$$I_{F,max,t_U,25} \approx \frac{I_{f,max,t_G,25}}{4}$$

$$= \frac{200\ \text{mA}}{3} = 50\ \text{mA} \qquad (10.31)$$

$$\frac{\Delta I_F}{t_U} = \frac{I_{F,max,t_U,25}}{\Delta t_{U,max-25}} = \frac{50\ \text{mA}}{55°C}$$

$$= 0.91\frac{\text{mA}}{°C} \qquad (10.32)$$

For the GaAs diode TIL 31, the maximum forward current $I_{F,max,t_U,50}$ can now be calculated as an example for an ambient temperature of $t_{U,50} = 50°C$.

$$\Delta t_{U,50-25} = t_{U,50} - t_{U,25}$$

$$= 50°C - 25°C = 25°C \qquad (10.33)$$

$$I_{F,max,t_U,50} = I_{F,max,t_U,25}$$

$$- (\Delta t_{U,50 - 25} \cdot \frac{\Delta \cdot I_F}{t_U}) \qquad (10.34)$$

$$I_{F,max,t_U,50} = 50\ \text{mA}$$

$$- \frac{25°C \cdot 0.91\ \text{mA}}{°C} = 27.25\ \text{mA} \qquad (10.35)$$

In order to be able to use the graph, stated for an ambient temperature of $t_U = 25°C$ for the GaAs diode TIL 31 and the phototransistor TIL 81, with the "On" collector current as a function of the source-detector distance (see *Figure 10.19*), a correction factor must be calculated, corresponding to the different temperatures and the diode forward currents.

178

From the data-sheet graph $\Phi_{e,rel} = f(I_F)$ for the type TIL 31, a relative increase in the radiant power of about

$$\Delta\Phi_{e,rel} = +10\%$$

is obtained, corresponding to this almost linear function, for the calculated forward current $I_{F,max,t_U,50} = 27.75$ mA, in comparison with the current $I_{F,t_U,25}$ = 25 mA used in *Figure 10.19* at an ambient temperature $t_U = 25°C$.

The ambient temperature difference $\Delta t_{U,50-25}$ between the required ambient temperature $t_{U,50} = 50°C$ and the temperature $t_{U,25} = 25°C$ stated in equation (10.30) is $50°C - 25°C = 25°C$.

The effect of the ambient temperature difference $t_{U,50-25}$ on the change in the "On" collector current of the phototransistor TIL 81 can be read off from a function $I_{C(On)} = f(t_U)$. Sufficient accuracy is obtained by taking the average of the values read off from *Figures 10.23* and *10.24*. The change in the "On" Collector current is approximately

$$\Delta I_{C(On)} \text{ for } t_U = (50-25) = -5\% \quad (10.36)$$

The correction factor for the function $I_{C(On)} = f(r)$ which is shown with an ambient temperature of $t_U = 25°C$ in *Figure 10.19* is:

$$K_f = \Delta\Phi_{e,rel} + |\Delta I_{C(On)}| \text{ for } t_U$$

$$= (50-25) \ °C \quad (10.37)$$

$$K_f = +10\% + (-5\%) = +5\% \quad (10.38)$$

The consequent slightly higher "On" collector current only changes the DC gain B of the phototransistor TIL 81 to an insignificant extent, so that ΔB can be neglected.

For the source-detector combination with the GaAs diode TIL 32 and the photo-

transistor TIL 78, the calculation of the "On" collector current is simpler, since the figures in the data sheet relate to the ambient temperatures and the coupling characteristic to the data sheet test conditions. With the exception of the TIL 32/TIL 78 combination, almost every representation of the coupling characteristic requires a separate evaluation for the individual needs. In addition, every calculation should be confirmed by measurements. The coupling characteristics stated apply in principle for all those Si phototransistors, which are in the same photocurrent sensitivity range as that of the types stated. Only phototransistors which differ greatly in actinic value form an exception. For combinations with a specified source-detector distance r or for ready-assembled combinations (e.g., optocouplers), the "On" photocurrent $I_{P(On)}$ or the "On" collector current $I_{C(On)}$ of the photodetector is of interest as a function of the diode current I_F of the GaAs diode. In a graph, these two values are each shown as a function of the GaAs diode forward current I_F, in most cases for equal source-detector case temperatures of $t_G = 25°C$ or an ambient temperature of $t_U = 25°C$. The ratio of the "On" photocurrent $I_{P(On)}$ or the "On" collector current $I_{C(On)}$ to the GaAs diode forward current I_F is designated as the transmission ratio or transmission factor:

$$u = \frac{I_{P(On)}}{I_F} \quad (10.39)$$

$$u = \frac{I_{C(On)}}{I_F} \quad (10.40)$$

The typical transmission ratio of a given source-detector combination, or an opto-coupler can be determined for the desired GaAs diode forward current I_F from the graph $I_{P(On)} = f(I_F)$ or $I_{C(On)} = f(I_F)$, together with the equations (10.39) and (10.40). As an example, *Figure 10.20* shows the "On" collector current $I_{C(On)}$ as a function of the GaAs diode current I_F

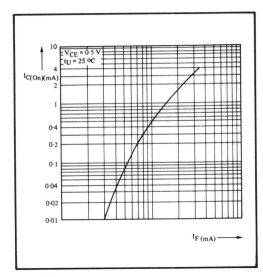

Figure 10.20
"On" collector current $I_{C(On)}$ as a function of the GaAs diode current I_F for the TIL 138 optocoupler. This typical function corresponds to a separately constructed source-detector combination with the same cooling conditions, with the GaAs diode TIL 32 and the Si phototransistor TIL 78, for a distance of r = 3·2 mm

for the TIL 138 optocoupler. For a GaAs diode current of I_F = 10 mA, the transmission ratio is

$$u = \frac{I_{C(On)}}{I_F} = \frac{0·45 \text{ mA}}{10 \text{ mA}} \equiv 4·5\% \qquad (10.41)$$

The transmission ratio for optocouplers is also dependent on the transmission medium such as glass, epoxy resin or gas. Furthermore, the operating mode, which can sometimes be selected, plays a part. The transmission ratio in photodiode operation is about 1%, in phototransistor operation about 100% and in photo-Darlington operation up to 500%. Figure 10.21 shows the "On" collector current $I_{C(On)}$ as a function of the GaAs diode forward current I_F for the optocouplers TIL 102 and TIL 130. For a GaAs diode forward current of I_F = 10 mA, the transmission ratio of the optocoupler is:

$$\text{TIL } 102 : u = \frac{I_{C(On)}}{I_F} \quad \frac{6·5 \text{ mA}}{10 \text{ mA}} \equiv 65\% \qquad (10.42)$$

$$\text{TIL } 103 : u = \frac{13 \text{ mA}}{10 \text{ mA}} \equiv 130\% \qquad (10.43)$$

$$\text{TIL } 112 : u = \frac{2 \text{ mA}}{10 \text{ mA}} \equiv 20\% \qquad (10.44)$$

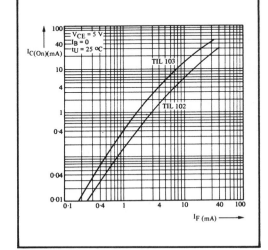

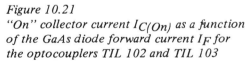

Figure 10.21
"On" collector current $I_{C(On)}$ as a function of the GaAs diode forward current I_F for the optocouplers TIL 102 and TIL 103

The transmission ratio and thus the "On" receiver output current of a photodetector is temperature-dependent. At low temperatures the radiant power of the GaAs diode increase, while at high temperatures the "On" collector current of the phototransistor rises. Depending on the selected combination of components, the temperature drifts of the source and detector compensate for each other within a certain temperature range. In this, only small changes in the "On" collector current of the phototransistor occur. Very small optoisolators have almost equal source and detector case temperatures or equal source and detector ambient temperatures. This makes it possible for

180

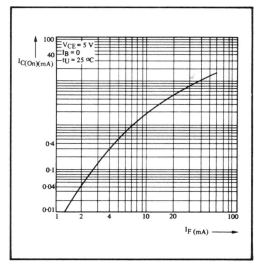

Figure 10.22
"On" collector current $I_{C(On)}$ as a function of the GaAs diode forward current I_F for the optocoupler TIL 112

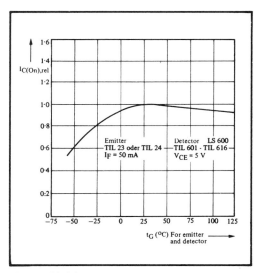

Figure 10.23
Dependence of the "On" collector current $I_{C(On)}$ of a Si phototransistor from the LS 600 series or TIL 600, produced by the radiation of the GaAs diodes TIL 23 or TIL or TIL 24, on the source and detector case temperatures t_G, which are equal. The relative "On" collector current has been normalised to the value $I_{C(On),rel}$ = 1 for a case temperature of t_G = 25°C

the typical detector output current to be shown for the whole combination in one graph as a function of the component case temperature t_G or the component ambient temperature t_U.

Figure 10.23 shows the normalised "On" collector $I_{C(On),rel}$ as a function of the case temperature t_G for source-detector combinations of the GaAs diodes TIL 23, TIL 24 and Si phototransistors of the type families LS 600 and TIL 600. The normalised "On" collector current $I_{C(ON),rel}$ is shown as a function of the ambient temperature t_U for the optocouplers TIL 102 and TIL 103 in *Figure 10.24* and for the optocoupler TIL 112 in *Figure 10.25*.

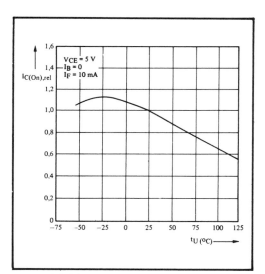

Figure 10.24
Normalised "On" collector current $I_{C(ON),rel}$ as a function of the ambient temperature t_U for the TIL 102 and TIL 103 optocouplers

10.3.2
Contrast ratio

A very important criterion, related to the detector output signal, is the contrast ratio. This is defined as the ratio of the "On" collector current $I_{C(On),N}$ (as a result of useful radiation falling on the

181

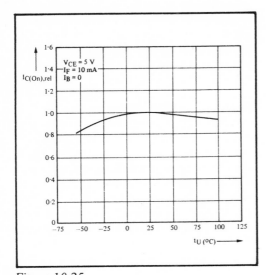

Figure 10.25
Standardised "On" collector current
$I_{C(On),rel}$ *as a function of the ambient*
temperature t_U *for the optocoupler*
TIL 112

phototransistor through a marking hole in
a card or tape material) to the "On"
collector current $I_{C(On),S}$ (as a result of
stray or background radiation falling on
the phototransistor through a card or tape
material).

For stable optical scanning conditions, the
contrast ratio should be as large as possible.
A contrast current ratio $I_{C(On),N}/I_{C(On),S}$
= 50/1 can still be simply evaluated by an
electrical circuit. A lower contrast current
ratio gives response levels in very narrow
ranges, so that the circuit design becomes
more difficult. The contrast current ratio
is determined by several factors, which
interact on one another:

a
Source and detector parameters,

b
Distance between source and detector,

c
Location of the punched card or tape
between the source and detector,

d
Aperture and its arrangement between
the source and detector,

e
Opacity of the material to be scanned
(degree of opacity to radiation).

f
Coupling medium (if any)

g
Cross-talk between individual channels.

For every application, the contrast
current ratio should be calculated
roughly with the factors listed and
confirmed by practical measurements.
The GaAs diodes selected should still
show an adequate radiant intensity for
their energisation, for example with
TTL circuits, with a small forward
current of about I_F = 15 mA. The
phototransistor selected should still
produce a reasonable "On" collector
current $I_{C(On)}$ for driving the
subsequent amplifier, TTL or interface
circuits.

To achieve optimum contrast current
ratios, hole masks of opaque material,
card or tape material of high opacity
and screening tubes for each channel
to avoid optical cross-talk are very decisive
factors. If a high contrast current ratio
is specified for card or tape material with
low opacity, then as the most effective
protection against stray radiation and optical
cross-talk, each optoelectronic component
will be installed in a spearate tube with a
mask fitted to it. The tube and mask will
advantageously be anodised or sprayed matt
black. *Figure 10.26* shows the contrast
current ratio $I_{C(On),N}/I_{C(On),S}$ of
different card and tape materials as a
function of the GaAs diode current I_F. The
source-detector distance r is the variable.
The data recorded relates to a mechanical
arrangement, which had been equipped with
separate tubes and hole masks, with an
aperture diameter of 0·9 mm, for each

182

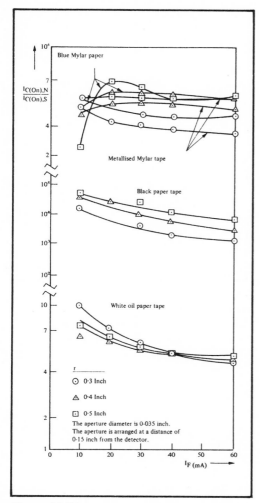

Figure 10.26
Contrast current ratio $I_{C(On),N}/I_{C(On),S}$
of various card and tape materials as
functions of the GaAs diode current I_F,
with the source-detector distance as
variable

channel, in order to eliminate cross-talk.
Because of its low opacity, the white oil
paper shows the lowest contrast current
ratios. With black paper, contrast current
ratios of greater than 1000:1 are
already achieved. The Mylar materials
give the most favourable values.

The "On" collector current of reflection
photocell units is stated for a reflecting

surface of known reflectivity and its
distance from the photocell unit.
Among others, the following are used as
reflection surfaces:

a

Neutral white paper with 90% diffuse
reflection (e.g., Eastman Kodak),

b

Mylar magnetic tape (Registered Trade
Mark of Du Point),

Aluminium foil 0·001 inch thick, which is
used as strips for the start and end of
magnetic tapes.

For the TIL 139 reflection reading head,
the reflection surfaces listed are arranged
at a distance of $r = 0·150$ inch in each case.
The "On" collector current $I_{C(On)}$ is
measured for a current I_F of 40 mA flowing
through the GaAs diode TIL 32, a collector-
emitter voltage of $V_{CE} = 5$ V applied to
the phototransistor TIL 78 and at an
ambient temperature of $t_U = 25°C$. During
this, no radiation from the surroundings may
fall on the reflecting surface or directly on
the phototransistor. The "On"
collector current of the reflection reading
head TIL 139 for the various reflecting
surfaces is:

	$I_{C(On),min}$	$I_{C(On),typ}$
White Kodak paper	10 μA	15 μA
Mylar magnetic tape	5 μA	7 μA
Aluminium foil	100 μA	130 μA

Very small optoisolators or optoelectronic reading heads should be rigid
and mechanically stable, since the mechanical
construction has a very decisive effect on
the coupling characteristic and thus on the
output signal of an electro-optical system.
Phototransistors are usually preferred as
photodetectors because of their internal

183

gain. As a consequence of their slim glass rod housing, the phototransistors of the LS 400 family are suitable for installation in tubes. The phototransistors of the LS 600 type and TIL 600, and also the GaAs diodes TIL 23 and TIL 24 are obtainable in small pill housings. These pill housings permit easy mounting and convenient soldering on double-coated circuit boards, which, as usual, should be free of grease and sprayed with a soldering lacquer. These component families contain small glass lenses, so that no additional fibre-optics or external lenses are needed for focussing the GaAs diode radiation on the phototransistor.

10.4
Half-power and Half-value Points

The half-power point, HP, marks the two points on a distribution curve, where the power has fallen by 3 dB in comparison with the maximum power P_{max}.

$$P_{HP} = \frac{P_{max}}{2} \qquad (10.45)$$

For photodetectors, the half-value point HP denotes the angle of incidence Φ with respect to the normal, at which the photocurrent sensitivity s_{HP} has fallen by 50% in comparison with the photocurrent sensitivity s_o in the normal direction.

$$s_{HP} = \frac{s_o}{2} \qquad (10.46)$$

The designation "half-power point" is inappropriate for photodetectors, since a drop in power by 0·5 times corresponds to a drop in sensitivity of only 0·707 times.

For radiation sources, the half-power point HP determines the emission angle, Φ, with respect to the normal, at which the radiant intensity $I_{e,HP}$ has fallen by 50% in comparison with the radiant intensity $I_{e,o}$ in the normal direction.

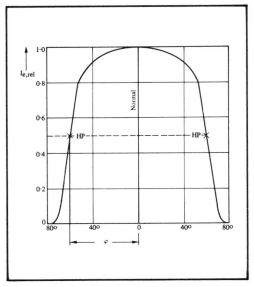

Figure 10.27
Radiant intensity distribution curve with the half-power points HP at $\varphi = 58·5°$, in the Cartesian coordinate system for the GaAs diode TIXL 06.

$$I_{e,HP} = \frac{I_{e,o}}{2} \qquad (10.47)$$

Figure 10.27 shows the radiant intensity distribution curve with the half-power points HP in the Cartesian coordinate system for the GaAs diode TIXL 06 and *Figure 10.28* shows the curve in the polar coordinate system for the GaAs diode TIXL 16. The angle φ is measured between the half-power point, HP, and the normal. The angle between the two half-power points is also called the aperture angle Θ.

$$\Theta = 2\varphi \qquad (10.48)$$

The aperture angles for the two GaAs diodes are:

TIXL 06: $\Theta = 115°$
TIXL 16: $\Theta = 150°$

For emission spectra, the half-power point, HP, marks the wavelengths λ_{HP}, at which the spectral radiant power $\Phi_{e,\lambda,HP}$ only shows the half value as compared with the

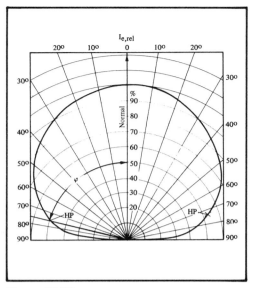

Figure 10.28
Radiant intensity distribution curve with the half-power points HP at φ = 75°, in the polar coordinate system for the GaAs diode TIXL 16

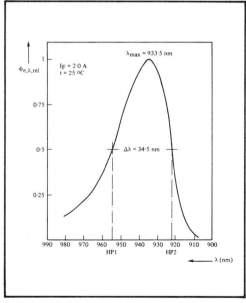

Figure 10.29
Relative spectral radiant power $\Phi_{e,\lambda,rel}$ as a function of the wavelength λ for the Si-doped GaAs diode TIXL 16. The half-power points lie at λ_{HP1} = 956 nm and λ_{HP2} = 911 nm

maximum spectral radiant power $\Phi_{e,\lambda,max}$ emitted at the maximum wavelength λ_{max}.

$$\Phi_{e,\lambda,HP} = \frac{\Phi_{e,\lambda,max}}{2} \qquad (10.49)$$

Figure 10.29 shows the relative spectral radiation power $\Phi_{e,\lambda,rel}$ as a function of the wavelength λ for the Si-doped GaAs diode TIXL 16. The maximum wavelenght is λ_{max} = 933·5 nm.

The half-power points occur at

$$\lambda_{HP\,1} = 956 \text{ nm}$$

and $\quad \lambda_{HP\,2}$ = 911 nm.

The bandwidth $\Delta\lambda$ is:

$$\Delta\lambda = \lambda_{HP1} - \lambda_{HP2} \qquad (10.50)$$

$$\Delta\lambda = 956 \text{ nm} - 911 \text{ nm} = 45 \text{ nm} \qquad (10.51)$$

For the lifetime of light-emitting or luminescent diodes, the half-power point HP is used to define the operating time, at which the total radiant power of an IR-emitting diode or the total luminous power of light-emitting diodes has fallen to 50%, as compared with the initial value. One definition relates the initial value to the radiant or luminous power of unused

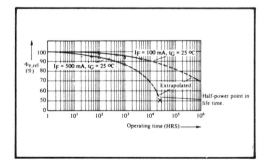

Figure 10.30
Basic life-time curve of a GaAs diode. The relative radiant power $\Phi_{e,rel}$ is shown as a function of the operating time. For the parameter I_F = 500 mA, the life is characterised by the half-power point at 30 000 hours

components. Other definitions which are often used omit the relatively high initial ageing (5% to 20%) during the first 20 – 100 operating hours. In these, the initial value is related to an operating time after twenty to one hundred hours.

Figure 10.30 shows the basic life curve of a luminescent diode. The relative radiant power $\Phi_{e,rel}$ is shown as a function of the operating time. The value of the forward current I_F affects the lifetime considerably. The half-power point for the device life-time, with I_F = 500 mA and t_G = 25°C, lies around 30 000 hours. After this time in operation, the luminescent diode is by no means destroyed, but still works satisfactorily but with correspondingly lower radiant power.

Figure 10.31 shows (with the forward

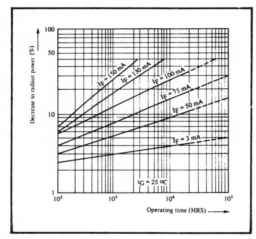

Figure 10.31
Decrease in the radiant power of a "Low Power" GaAs diode as a function of the operating time, with I_F as a variable

current I_F as a variable) the percentage decreases in the radiant power as a function of the operating time. This very clearly shows the slight decrease in the radiant power for small forward currents I_F and the high decrease in the radiant power for large values of I_F.

10.5
Dynamic Data

The dynamic data is understood to refer to the switching time performance of electronic components. *Figure 10.32* shows the theoretical test circuit for measurement of the switching time of source-detector combinations. A GaAs diode is operated in the pulsed mode. The pulsed radiation falls on a photo-transistor. The input pulse and the output pulse can be observed on an oscilloscope to determine the actual switching time.

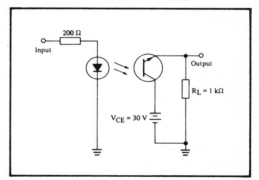

Figure 10.32
Test circuit for measurement of the swtiching time performance for the TIL 138 miniature optoisolator. The amplitude of the input pulse is adjusted for an "On" collector current of $I_{C(On)}$ = 500 µA. The pulse generator is to have a specification of: Z_{out} = 50Ω, t_r < 100 ns, t_f ≤ 100 ns, mark-space ratio ≈ 1:1

In *Figure 10.33*, the most important times are illustrated:

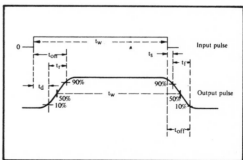

Figure 10.33
Definition of switching times

186

The *switch-on time* t_{on} is the time, in which the amplitude rises from 0% to 90% of the final value.

The *switch-off time* t_{off} is the time in which the amplitude falls from 100% to 10% of the final value.

The *delay time* t_d is the time, in which the amplitude rises from 0% to 10% of the final value.

The *storage time* t_s is the time, in which the amplitude falls from 100% to 90% of the final value.

The *rise time* t_r is the time, in which the amplitude rises from 10% to 90% of the final value.

The *fall time* t_f is the time, in which the amplitude falls from 90% to 10% of the final value.

The *pulse width* t_w is given by the time interval which lies between 50% of the amplitude of the rising edge and 50% of the amplitude of the falling edge.

Separate measurement of the switching times of GaAs diodes or of photo-detectors necessitate that, in the first case the measuring photodetector and in the second case the GaAs diode as the measurement emitter, must have considerably shorter switching times than the test device. Fast avalanche photodiodes are suitable as photodetectors for GaAs:Zn, GaAs:Si, GaAsP and GaP diodes. The GaAs-Zn diodes are suitable for use as measurement emitters for commercial Si photodiodes and Si phototransistors.

The switching time characteristics of non-amplifying junction photodetectors is determined, down to rise times of 10 ns, by the RC performance. In practice, the rise time will be determined by

$$t_r = 2{\cdot}2 \, . \, R \, . \, C \qquad (10.52)$$

In this expression, as an approximation the value of the total component capacitance from the relevant data sheet can be inserted for C.

Figure 10.34 shows as an example, the

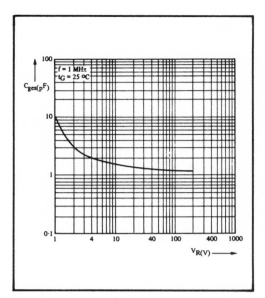

Figure 10.34
Total capacitance C as a function of the applied blocking voltage V_R for the Si avalanche photodiode TIXL 56.

total capacitance C in pF as a function of the applied blocking voltage V_R for the Si avalanche photodiode TIXL 56.

As a further simplification, the connected load resistance R_L can be inserted for the resistance value R. From equation (10.52), a photodiode capacitance of 100 pF and a load resistance of 1 kΩ give rise time of

$$t_r = 2{\cdot}2 \, . \, 100 \, . \, 10^{-12} \, \text{As/V} \, . \, 10^3 \, \text{V/A}$$

$$= 220 \text{ ns} \qquad (10.53)$$

Commercially available Si photodiodes have rise times from about $t_r = 100$ ns to 10 μs. Si-PIN photodiodes, on the other hand, have rise times of about $t_r = 10$ ns. Si avalanche photodiodes have even better values, their rise times lie around $t_r = 0{\cdot}35$ ns.

The switching times of phototransistors are mainly determined by the Miller capacitance between collector and base. This Miller capacitance is dependent on the DC gain

187

B. Phototransistors have rise times of about 5 to 50 μs.

The switching times of GaAs diodes are substantially determined by the lifetime of the charge carriers in the crystal. Red-emitting GaP diodes with the wavelength λ_{max} = 690 nm, which have the PN junction formed by growing a Zn and O-doped P-layer, in the liquid epitaxy process, onto a Te-doped N-type epitaxial GaP layer (double-epitaxy diode), have average rise and fall times of $(t_r + t_f)/2$ = 200 ns to 500 ns. Components made by the vapour epitaxy process show shorter times.

The green-emitting GaP diodes with the wavelength λ_{max} = 565 nm have average rise and fall times $(t_r + t_f)/2$ = 50 ns. The same value applies for the yellow-emitting GaP diodes with the wavelength λ_{max} = 590 nm.

On the other hand, the red-emitting GaAsP diodes with the wavelength λ_{max} = 650 nm

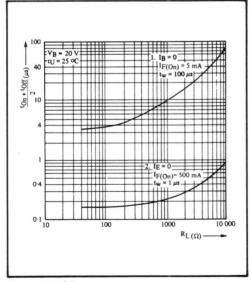

Figure 10.35
Average switch-on and switch-off times $(t_{on} + t_{off})/2$ as a function of the load resistance R_L for the optocouplers TIL 102 and TIL 103. Curve 1 applies for phototransistor operation, curve 2 for photodiode operation

have a somewhat shorter rise and fall time of $(t_r + t_f)/2$ = 1 ns to 50 ns.

IR-emitting GaAs diodes with the wavelength λ_{max} = 890 nm, which have the PN junction formed by the liquid epitaxy process by counter-doping the melt with zinc during the growing of the epitaxial layer, have average rise and fall times of $(t_r + t_f)/2$ = 2 ns to 10 ns.

IR-emitting GaAs diodes with the wavelength λ_{max} = 940 nm, which have their PN junction formed by the liquid epitaxy process from the Si-doped melt, which behaves amphotically in the GaAs, have average rise and fall times of $(t_r + t_f)/2$ = 500 ns.

In general, it is also usual to state the average switch-on and switch-off times in data sheets. As an example of this, *Figure 10.35* gives the average switch-on and switch-off times as a function of the load resistance R_L for the optocouplers TIL 102 and TIL 103.

Furthermore, the dynamic characteristic of a component is marked by the *frequency limit* f_g. This denotes the frequency, at which the power has fallen to a half, as compared with its maximum value. With photodetectors, however, the sensitivity is stated, not with a power but with a current or voltage per incident unit of irradiance. Thus, on a constant load resistance R_L, according to equation (10.54), a fall in power by 0·5 times corresponds to a voltage reduction of 0·707 times and a current reduction of 0·707 times.

$$0\cdot5\ P = 0\cdot5\ U \cdot J = \sqrt{0\cdot5}\ U \cdot \sqrt{0\cdot5}\ I$$

$$= 0\cdot707\ U \cdot 0\cdot707\ I \qquad (10.54)$$

The limiting frequency of a photo-detector is thus that frequency, at which the sensitivity is only 0·707 times the maximum sensitivity.

As the following transformation shows, the

limiting frequency f_g and the rise time t_r can be related to one another. Starting from

$$t_r = 2 \cdot 2 \, . \, R \, . \, C$$

The product RC is the time constant

$$\tau = R \, . \, C \tag{10.55}$$

The time constant τ is the reciprocal of the angular frequency ω.

$$\tau = \frac{1}{\omega} \tag{10.56}$$

The angular frequency ω corresponds to the angular velocity of a rotating vector, in radians

$$\omega = \frac{2\pi}{T} = 2\pi f \tag{10.57}$$

The cycle duration T_g at the limiting frequency f_g is obtained from the equation

$$T_g = \frac{1}{f_g} \tag{10.58}$$

If the equations from (10.54) to (10.57) are inserted in the equation (10.52), then with the cycles duration (period) T_g, the rise time t_r can be calculated:

$$t_r = 2 \cdot 2 \, . \, R \, . \, C = 2 \cdot 2 \, . \, \tau \frac{2 \cdot 2}{\omega} = \frac{1 \cdot 1}{\pi} \, . \, T_g \tag{10.59}$$

$$t_r = 0 \cdot 35 \, T_g \tag{10.60}$$

$$t_r = \frac{0 \cdot 35}{f_g} \tag{10.61}$$

$$f_g = \frac{0 \cdot 35}{t_r} \tag{10.62}$$

A phototransistor with a limiting frequency of $f_g = 100$ kHz thus has a rise time of

$$t_r = \frac{0 \cdot 35}{10^5 \ Hz} = 3 \cdot 5 \ \mu s \tag{10.63}$$

For a photodiode with a rise time of $t_r = 0 \cdot 1 \ \mu s$, a limiting frequency of

$$f_g = \frac{0 \cdot 35}{0 \cdot 1 \ \mu s} = 3 \cdot 5 \ MHz \tag{10.64}$$

is obtained.

For Si-doped GaAs diodes, the limiting frequency f_g lies around 1 MHz, while for Zn-doped GaAs diodes it is about 100 MHz. With Si phototransistors, a frequency limit between 50 kHz and 200 kHz can be expected.

For photodetectors and optocouplers the whole frequency response is usually stated in the data sheet. The sensitivity s or the "On" collector current $I_{C(On)}$ is shown in a graph as a function of the modulation frequency f_m of the radiation falling on the photodetector. *Figure 10.36* shows the frequency response

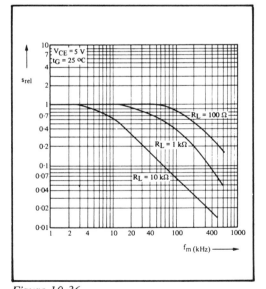

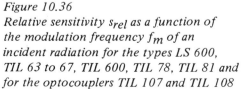

Figure 10.36
Relative sensitivity s_{rel} as a function of the modulation frequency f_m of an incident radiation for the types LS 600, TIL 63 to 67, TIL 600, TIL 78, TIL 81 and for the optocouplers TIL 107 and TIL 108

of the phototransistor families LS 600,
TIL 63 to 67, TIL 600, TIL 78, TIL 81
and the optocouplers TIL 107 and TIL 108.

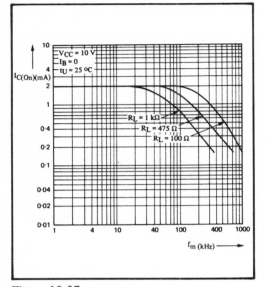

Figure 10.37
"On" collector current as a function of
the modulation frequency f_m of the
GaAs diode current for the optocouplers
TIL 111 and TIL 112

In Figure 10.37, the "On" collector
current $I_{C(ON)}$ is shown as a function of
the modulation frequency f_m of the GaAs
diode current for the optocouplers TIL 111
and TIL 112.

Figure 10.38 shows the measurement
circuit for recording the frequency
response of GaAs power diodes. A load-
independent bias current I_F is applied
to the GaAs diode through a series resistance.
A low-impedance generator is capacitatively
coupled to the GaAs diode. The modulation
current I_M through the GaAs diode is to
be kept constant during the measurement
of the frequency response. I_M must not
exceed twice the value of bias current I_F.

The modulation current can be monitored
on the oscilloscope. In order to avoid
possible electrical cross-talk between
the emitter and detector, a large
emitter-detector distance can be selected.
To focus the radiation from the GaAs diode,
a lens can be arranged in the beam path
between the emitter and detector. The
Si avalanche diode TIXL 56 which is used
as the measuring photodetector is operated

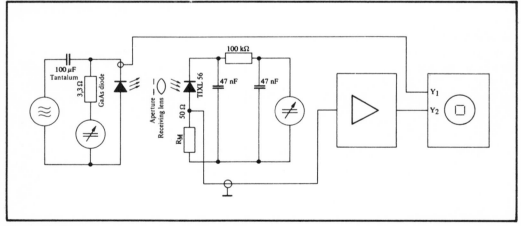

Figure 10.38
Measurement circuit for recording the frequency response of GaAs power diodes. The
following values are to be set: a) GaAs diode: Modulation current I_m = 100 mA (peak to
peak), independent of frequency; bias current I_F = 55 mA constant. b) Photodetector
TIXL 56: Dark current with applied voltage $V_R \approx 170$ V set to $I_D = 0.5$ μA; reverse current
with the aperture open $I_R = 10$ μA

190

with optimum gain. For this, with the applied reverse voltage $V_R \approx 70$ V, a dark current of approx. $0.5\,\mu A$ is set. Although a further increase in the applied blocking voltage V_R gives a higher amplification factor M, at the same time it gives a higher current noise.

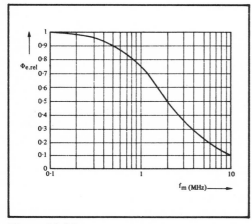

Figure 10.39
Frequency response of the GaAs diode TIXL 13. Relative radiant power $\Phi_{e,rel}$ as a function of the modulation frequency f_m

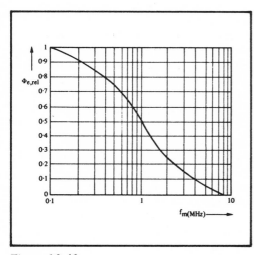

Figure 10.40
Frequency response of the GaAs diode TIXL 16. Relative radiant power $\Phi_{e,rel}$ as a function of the modulation frequency f_m

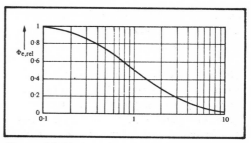

Figure 10.41
Frequency response of the GaAs diode TIXL 27. Relative radiant power $\Phi_{e, rel}$ as a function of the modulation frequency f_m

With the unmodulated radiation at first falling on the avalanche diode through an aperture set in the beam path, the reverse current I_R is to be set to $I_R \approx 10\ \mu A$. The input voltage signal appearing on the measuring resistor $R_M = 50$ is amplified and observed on the oscilloscope. The Input voltage is a measure of the modulated radiant power of the GaAs diode. Figures 10.39 to 10.41 show the frequency response of Si-doped GaAs power diodes.

10.6
Reliability of optoelectronic semiconductor components

The term "reliability of a component" covers component lifetime, changes which occur in the parameters during life and the probability of the occurrence of a total failure.

The life of a semiconductor component depends, among other factors, on the electrical loss power, the selection of the working point, the current and power density in the wafer (important for pulse operation), the junction temperature and the overall thermal loading. The life of optoelectronic components relates firstly to the operating condition and secondly to the storage condition. The reliability data should include all important parameters. (It is unnecessary to go further into the

191

reliability prediction of Si phototransistors, since it corresponds to that for ordinary Si transistors).

The life test data on luminescent diodes relate to the half-power life point (see Section 10.4). Thermal fatigue, resulting in wafer breakage, occurs less often in GaAs diodes than for other semiconductor devices. Efforts are made to work with relatively low junction temperatures t_J, so that a high radiation yield is achieved. The half-power point of the lifetime is mainly determined by the junction temperature t_J and the diode forward current I_F. Different current densities can occur in the crystal, so that different local thermal loadings occur. With increasing forward current, the active region can displace itself more towards the contact point, which can also lead to failure of the luminescent diodes (see *Figure 10.11*). In addition, for pulse operation of luminescent diodes, the permissible current and power density in the bonding wire is important. High temperatures can cause changes in the epoxy resin used for lens domes or housings.

By improved technological processes, suitable component designs, careful component assembly and through the selection of suitable system supports and plastics, the life of luminescent diodes has been considerably increased.

For protection against environmental effects, GaP wafers and Si-doped GaAs wafers, made by liquid epitaxy, are mounted with the P-junction side of the wafer system on the system support. The system support is either a metal base of the transistor case, a mounting bolt, in power luminescent diodes, or the end of a massive square connecting wire in plastic-encapsulated components. Since the electrical loss power mainly causes a heat build-up at the anode, the mounting of heat-sinks is only meaningful at the system support. For example, the mounting of the threaded bolt on a chassis is very convenient for

the power luminescent diodes TIXL 12 to TIXL 15, while the GaAs diode TIL 31, mounted in a TO-46 case, needs a heat-sink which pushes onto the case.

It was possible to improve the life of Zn-doped GaAs diodes and GaAsP diodes considerably by use of the well-proven planar technique.

With the data-sheet operating values, the minimum life of luminescent diodes is about 50,000 hours. For many GaAs diodes, the typical lifetime is over 100,000 hours. After this time, the radiant or liminous power has fallen to 50% of the initial value. Otherwise these components still work satisfactorily (see Section 10.4). With the present state of technology, the probability of the occurrence of a total failure of a luminescent diode is comparable with that of silicon components. The reliability of a component is considerably influenced by the manufacturing process.

The quality control department is responsible throughout for the component yield. It provides for the necessary direct interventions during the manufacturing process, in order to guarantee the planned and required reliability and quality of the components produced. Quality control includes the inspection of raw materials on receipt, the necessary inspections and tests during the critical production stages and the final testing and release of the finished product. The manufacturing steps and the frequent inspections and tests, which are carried out by the manufacturer of optoelectronic components, are illustrated in the flow diagrams in *Figure 10.42* for the production of Si phototransistors of the pill-case family LS 600 and in *Figure 10.43* for the production of GaAs diodes of the pill case types TIL 23 and TIL 24.

In the following sections, life tests on 11 000 phototransistors of the type family LS 600, with ten million component operating hours and on GaAs

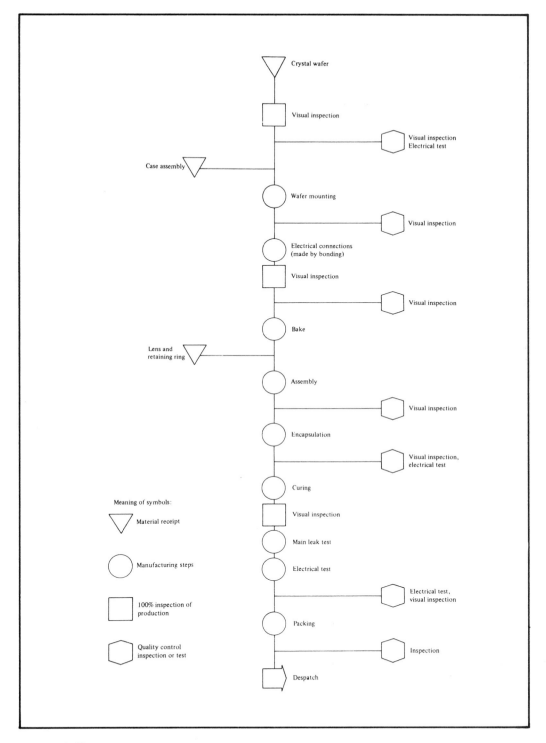

Crystal wafer

Visual inspection

Visual inspection
Electrical test

Case assembly

Wafer mounting

Visual inspection

Electrical connections
(made by bonding)

Visual inspection

Visual inspection

Bake

Lens and
retaining ring

Assembly

Visual inspection

Encapsulation

Visual inspection,
electrical test

Curing

Visual inspection

Main leak test

Electrical test

Electrical test,
visual inspection

Packing

Inspection

Despatch

Meaning of symbols:

Material receipt

Manufacturing steps

100% inspection of
production

Quality control
inspection or test

Figure 10.42
Typical production flow diagram for Si phototransistors of the pill-case type LS 600 family.

193

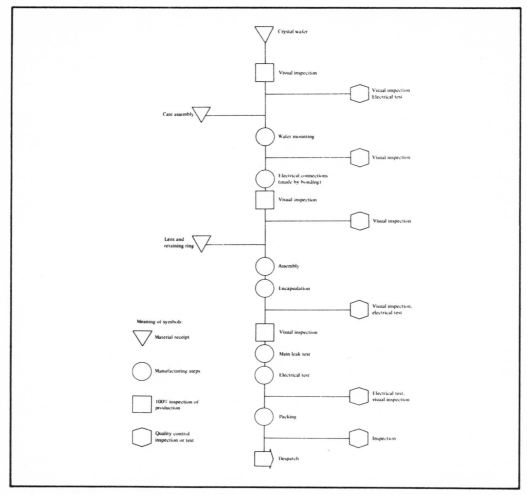

Crystal wafer

Visual inspection

Visual inspection
Electrical test

Case assembly

Wafer mounting

Visual inspection

Electrical connections
(made by bonding)

Visual inspection

Visual inspection

Lens and
retaining ring

Assembly

Encapsulation

Visual inspection,
electrical test

Meaning of symbols:

Visual inspection

Material receipt

Main leak test

Electrical test

Manufacturing steps

Electrical test,
visual inspection

100% inspection of
production

Packing

Inspection

Quality control
inspection or test

Despatch

Figure 10.43
Production flow diagram for GaAs diodes of the pill-case types TIL 23 and TIL 24

diodes of the type families TIL 23 and TIL 24 with over two million component operating hours are described. It is expressly pointed out, that only previously used components were used for these. The tests listed are typical of TI sensor and emitter products. The test results have been summarised by means of graphs and tables. In addition, various mechanical and temperature stress tests are listed.

Test 1: Operational life test for phototransistors at an ambient temperature of $t_U = 25°C$ for 1000 hours.

The irradiance E_e falling on every phototransistor, with the applied collector-emitter voltage of $V_{CE} = 10$ V, gave a total loss power for each phototransistor of $P_V = 50$ mW. The dark current I_D and the collector current I_C were measured in each case after 0, 250, 500 and 1000 hours of operation. As the criteria of failure a maximum dark current of $I_{Dmax} = 0·2$ μA and a $\pm 20\%$ deviation of the collector current from the original limit were laid down. In all, 3210 phototransistors were tested to these criteria, while 6 failures were recorded. The samples tested were taken from batches, which contained

194

over 1,050,000 phototransistors.

Figures 10.44 and 10.45 and Table 10.2 show the evaluation of this operating life test at $t_U = 25°C$.

Test 2: Operating life test for phototransistors with an ambient temperature of $t_U = 55°C$ for 1000 hours.

The irradiance E_e falling on each phototransistor LS 600, with the applied collector-emitter voltage of $V_{CE} = 10$ V, gave a loss power for each phototransistor of $P_V = 50$ mW. The dark current I_D and the collector current I_C were measured after an operating time of 0, 168 and 1000 hours. As the criteria of failure a maximum dark current of $I_{Dmax} = 0.2 \mu A$ and a $\pm 40\%$ collector-current deviation from the original limit were laid down. In all, 3356 phototransistors were tested to these criteria, while 11 failures were recorded. The samples tested were taken from batches,

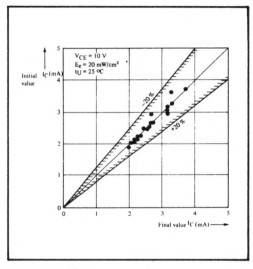

Figure 10.44
Change in the initial collector current value after 500 hours operation for 3210 phototransistors of the LS 600 family. During this operating life test, the ambient temperature was $t_U = 25°C$, the applied collector-emitter voltage was $V_{CE} = 10$ V and the loss power for each phototransistor was $P_V = 50$ mW

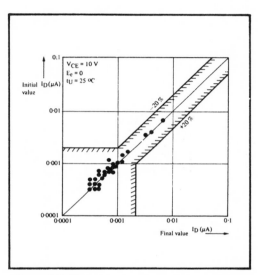

Figure 10.45
Change in the initial dark current value after 500 hours operation for 3210 phototransistors of the LS 600 family. During this operating life test, the ambient temperature was $t_U = 25°C$, the applied collector-emitter voltage was $V_{CE} = 10$ V and the loss power for each phototransistor was $P_V = 50$ mW

				Failures due to parameter deviations		
Units tested	Unit hours	Total failures	Total	Failure rate in % per 1000 h		Mean time to failure
				60% probability	90% probability	
3210	2 847 000	0	6	0.20	0.33	700 000 h

Table 10.2
Evaluation of the operational life test after an operating time of 1000 hours for the phototransistor family LS 600. The ambient temperature was $t_U = 25°C$.

195

which contained over 2,064,000 photo-transistors.

The results and the evaluation of this operational life test, carried out at $t_U = 55°C$ are shown in *Figurures 10.46* and *10.47* and in *Table 10.3*.

Test 3: High-temperature storage test for the phototransistor family LS 600.

The phototransistors were stored in the oven at $t_U = 150°C$ for 500 and for 1000 hours. The dark current I_D and the collector current I_C were measured after

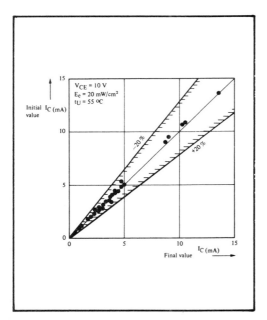

Figure 10.46
Change in the initial collector current value after 1000 hours operation for 3356 phototransistors of the LS 600 family. During this operational life test, the ambient temperature was $t_U = 55°C$, the applied collector-emitter voltage V_{CE} = 10 V and the loss power per phototransistor $P_v = 50$ mW

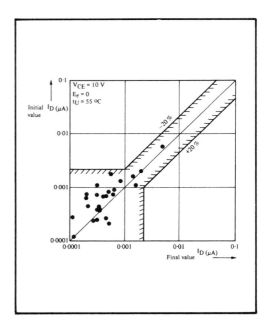

Figure 10.47
Change in the initial dark current value after 1000 hours operation for 3356 phototransistors of the LS 600 family. During this operational life test, the ambient temperature was $t_U = 55°C$, the applied collector-emitter voltage V_{CE} = 10 V and the loss power per phototransistor $P_v = 50$ mW.

Units tested	Unit hours	Total failures	Total	Failures due to parameter deviations		Mean time to failure
				Failure rate in % per 1000 h		
				60% probability	90% probability	
3356	3356000	0	11	0·36	0·49	300 000 h

Table 10.3
Evaluation of the operating life test after 1000 hours operation for the phototransistor family LS 600. The ambient temperature was $t_U = 55°C$

a storage time of 0, 250, 500 and 1000 hours. As the criteria of failure a maximum dark current of $I_{Dmax} = 0.2\ \mu A$ and a 20% collector current deviation from the original limit were laid down. In all, 1829 phototransistors were tested to these criteria, while six failures were recorded. The samples tested were taken from batches, which contained over 745,000 phototransistors.

The *Figures 10.48* and *10.49* and *Table 10.4* show the evaluation of the high-temperature storage test carried out at $t_U = 150°C$.

Test 4: Long-term reliability test on phototransistors LS 600.

The long-term reliability of the phototransistor family LS 600 is illustrated by

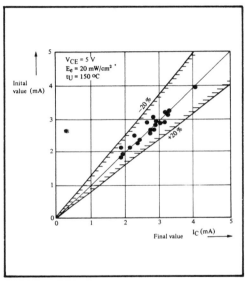

Figure 10.48
Change in the initial collector current value after a storage time of 5000 hours for 1829 phototransistors of the LS 600 family. During storage, the ambient temperature was $t_U = 150°C$

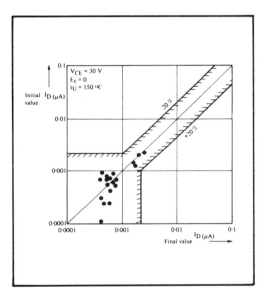

Figure 10.49
Change in the initial dark current value after a storage time of 500 hours for 1829 phototransistors of the LS 600 family. During storage, the ambient temperature was $t_U = 150°C$

| | | | | Failures due to parameter deviations | | |
| | | | | Failure rate in % per 1000 h | | |
Units tested	Unit hours	Total failures	Total	60% probability	90% probability	Mean time to failure
1 829	963 500	0	0	0·78	1·1	160 000 h

Table 10.4
Evaluation of the high-temperature storage test after a storage duration of 1000 hours. During storage, the ambient temperature was $t_U = 150°C$

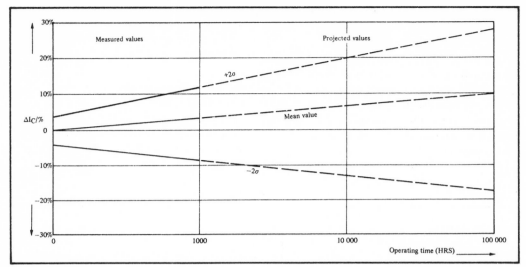

Figure 10.50
Change in collector current ΔI_C as a function of the operating time in a long-term reliability test at $t_U = 25°C$ for the phototransistor family LS 600

the change in collector current ΔI_C as a function of the operating time, in *Figure 10.50* for an ambient temperature of $t_U = 25°C$ and in *Figure 10.51* for an ambient temperature of $t_U = 55°C$. Here, all test results during a three-year reporting period were evaluated. The projected discrepancy limits are based on an exponential failure distribution.

Test 5: High-temperature reverse blocking voltage test on LS 600 phototransistors.

This high-temperature reverse blocking voltage test was carried out for 1000 hours with an applied collector-emitter reverse voltage of $V_{CE} = 45$ V in a dark oven with an ambient temperature of $150°C$. The dark current I_D, the breakdown

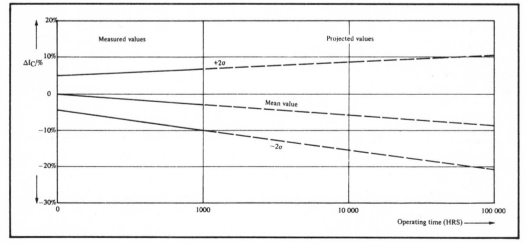

Figure 10.51
Chance in collector current ΔI_C as a function of the operating time in a long-term reliability test at $t_U = 55°C$ for the phototransistor family LS 600

198

				Failures due to parameter deviations		
Units tested	Unit hours	Total failures	Total	Failure rate in % per 1000 h		Mean time to failure
				60% probability	90% probability	
3 320	3 320 000	0	11	0·38	0·5	300 000 h

Table 10.5
Evaluation of the high-temperature reverse blocking voltage test on phototransistors of the
LS 600 family after a storage time of 1000 hours. During the storage time, the ambient
temperature was $t_U = 150°C$ and the applied collector-emitter blocking voltage $V_{CE} = 45 V$

voltage V_{CEO} and the collector current I_C during irradiation were measured after a storage time of 0, 168 and 1000 hours. As the criteria of failure, a maximum dark current of $I_{Dmax} = 0·2 \mu A$ and a 20% deviation of the collector current from the original limit were laid down. In all, 3320 phototransistors were tested to these criteria, while eleven failures were recorded.

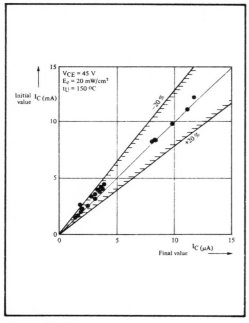

Figure 10.52
Change in the initial collector current value
after a storage time of 1000 hours for
3320 phototransistors of the LS 600 family.
During the storage time, the ambient
temperature was $t_U = 150°C$ and applied
collector-emitter blocking voltage $V_{CE} = 45 V$

The samples tested were taken from batches, which contained over 2,027,000 phototransistors.

The results and the evaluation of this high-temperature blocking voltage test, carried out at $t_U = 150°C$, are shown in *Figures 10.52* and *10.53* and in *Table 10.5*.

Test 6: Operational Life Test on GaAs diodes.

The GaAs diode families TIL 23 and TIL 24 were also subjected to various operational life tests. The results of one operational life test are summarised in *Table 10.6*. In this, the GaAs diodes were tested at various ambient temperatures with different forward current I_F. The evaluation relates to the average percentage deviation in radiant power for the test conditions stated in the data sheet ($I_F = 50 mA$, $t_U = 25°C$).

Test 7: Operational life test on GaAs diodes.

For the GaAs diode families TIL 23 and TIL 24, further 1000-operating hour life tests were carried out at ambient temperatures of $t_U = 25°C$ and $t_U = 55°C$. The diode forward currents were: $I_{F1} = 10 mA$; $I_{F2} = 30 mA$ and $I_{F3} = 50 mA$. The radiant power was measured with a solar cell after 0, 168, 500 and 1000 hours. The diode forward voltage was also measured at the same time intervals. No significant deviations of the diode forward voltages were detected. In all, 77 components were tested with $I_{F1} = 10 mA$ and 98

Quantity	Test conditions	Test time (hours)	$\Delta \Phi_e$	Total failures
104	$I_F = $ 10 mA; $t_U = 25^\circ$C	1000	+1·680%	0
104	$I_F = $ 30 mA; $t_U = 25^\circ$C	1000	−0·420%	0
104	$I_F = $ 50 mA; $t_U = 25^\circ$C	1000	−4·089%	0
104	$I_F = $ 10 mA; $t_U = 55^\circ$C	168	+1·488%	0
104	$I_F = $ 30 mA; $t_U = 55^\circ$C	168	−0·4761%	0
104	$I_F = $ 50 mA; $t_U = 55^\circ$C	168	+1·470%	0
100	$I_F = $ 10 mA; $t_U = 25^\circ$C	500	+1·856%	0
100	$I_F = $ 30 mA; $t_U = 25^\circ$C	500	−1·177%	0
100	$I_F = $ 50 mA; $t_U = 25^\circ$C	500	−1·167%	0
100	$I_F = $ 100 mA; $t_U = 25^\circ$C	1000	−1·188%	0

Table 10.6
Summary of a life test for the GaAs diode family TIL 23 and TIL 24. $\Delta \Phi_e$ *relates to the average percentage deviation in radiant power for the test conditions stated in the data sheet* ($I_F = 50$ mA, $t_U = 25^\circ$C)

components with $I_{F2} = 30$ mA. In these tests, no significant deviation of the radiant power of greater than −10% was found.
A total of 96 components were tested with the current $I_{F3} = 50$ mA, among which four GaAs diodes exceeded a −20% deviation (maximum −27%).

Furthermore, operational life tests, extended to 4000 hours, were carried out with 300 GaAs diodes. These confirm the extrapolated curves in *Figures 10.54* to *10.59.*

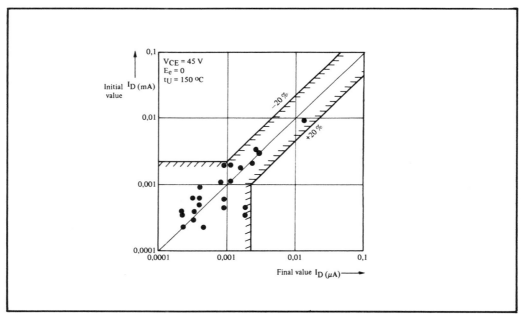

Figure 10.53
Change in the initial dark current value after a storage time of 1000 hours for 3320 phototransistors of the LS 600 family. During the storage time the ambient temperature was $t_U = 150^\circ$C and the applied collector-emitter voltage $V_{CE} = 45$ V

200

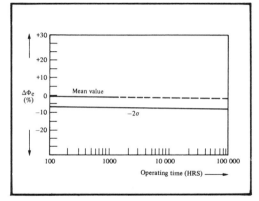

Figure 10.54
Percentage change in radiant power $\Delta\Phi_e$ as a function of the operating time for the GaAs diode families TIL 23 and TIL 24. During the operational life test, the ambient temperature was $t_U = 25°C$ and the diode forward current $I_{F1} = 10\ mA$

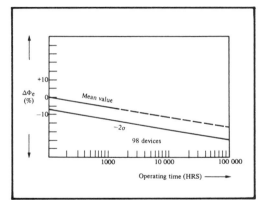

Figure 10.56
Percentage change in radiant power $\Delta\Phi_e$ as a function of the operating time for the GaAs diode families TIL 23 and TIL 24. During this operational life test, the ambient temperature was $t_U = 25°C$ and the diode forward current $I_{F3} = 50\ mA$

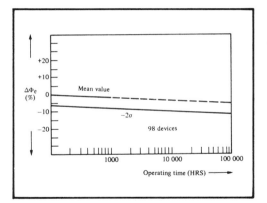

Figure 10.55
Percentage change in radiant power $\Delta\Phi_e$ as a function of the operating time for the GaAs diode families TIL 23 and TIL 24. During this operational life test, the ambient temperature was $t_U = 25°C$ and the diode forward current $I_{F2} = 30\ mA$

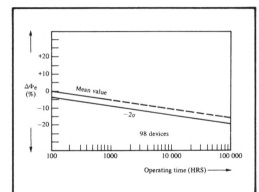

Figure 10.57
Percentage change in radiant power $\Delta\Phi_e$ as a function of the operating time for the GaAs diode families TIL 23 and TIL 24. During this operational life test, the ambient temperature was $t_U = 55°C$ and the diode forward current $I_{F1} = 10\ mA$

201

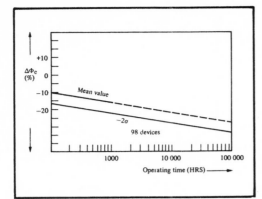

Figure 10.58
*Percentage change in radiant power $\Delta\Phi_e$
as a function of the operating time for the
GaAs diode families TIL 23 and TIL 24.
During this operational life test, the ambient
temperature was $t_U = 55°C$ and the diode
forward current $I_{F2} = 30\ mA$*

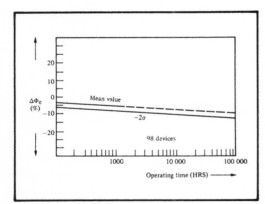

Figure 10.59
*Percentage change in radiant power $\Delta\Phi_e$
as a function of the operating time for the
GaAs diode families TIL 23 and TIL 24.
During this operational life test, the ambient
temperature was $t_U = 55°C$ and the diode
forward current $I_{F3} = 50\ mA$*

Test 8: Storage temperature test on
GaAs diodes.

In addition to the previous tests, a storage
temperature test was carried out for the
GaAs diode families TIL 23 and TIL 24.

172 components were stored at an ambient
temperature of $t_U = 85°C$ for 1000 hours.
In this test, only two GaAs diodes showed a
change in radiant power greater than -5%
(maximum -13%). The forward voltage
did not change in any component.

Test 9: Operational life test under
extreme temperature conditions.

In some applications, it is not the life, but
the reliability under extreme stress conditions
which is the decisive factor for a component.
For example, for semiconductor devices
in motor vehicles, only operating lives of
the order of 3000 hours are called for.
However, the environmental conditions
necessitate reliability over an extreme
temperature range of about $t_U = -40°C$ to
$t_U = +120°C$. *Table 10.7* shows results on
temperature shock tests from $t_U = -40$ to
$+130°C$ on nine optical detectors with the
GaAs diodes TIL 23 and the phototransistors
TIL 81. In this test, the components were
operated outside the permissible data-sheet
conditions. The phototransistors and the
GaAs diodes of the individual light-links
were inserted in the holes provided in circuit
boards. They were not specially optically
aligned. The arrangement was of very stable
construction. The source-detector distance
(lens to lens) was 6 mm. Only previously
unused components were used. The diode
forward current was $I_{F,\text{TIL 23}} = 10\ mA$,
and the applied collector-emitter voltage
$V_{CE,\text{TIL 81}} = 8\ V$.

Test 10: Environmental conditions test.

The environmental condition tests carried
out for the phototransistors LS 600 and
the GaAs diodes TIL 23 are shown in
Table 10.8. The product test specimens
showed no total failures and only one
parametric failure. Here it must be
mentioned, that these test conditions
imposed by customers do not represent
the most severe component stresses which
can be tolerated. In most cases, the
component can withstand considerably
more severe loading conditions.

202

Measurement series	Temp. (°C)	Time	Light-link No. and collector current $I_{C(On)}$								
			1	2	3	4	5	6	7	8	9
1	25	13.00	900 µA	545 µA	615 µA	1 mA	550 µA	1·02 mA	630 µA	1·25 mA	410 µA
	−40	14.00	475 µA	180 µA	490 µA	800 µA	500 µA	690 µA	500 µA	1·07 mA	260 µA
	+130	14.20	965 µA	1·2 mA	620 µA	940 µA	485 µA	990 µA	600 µA	970 µA	550 µA
2	−40	14.40	460 µA	180 µA	490 µA	810 µA	500 µA	725 µA	495 µA	1·1 mA	275 µA
	+130	15.00	960 µA	1·18mA	610 µA	930 µA	480 µA	970 µA	590 µA	960 µA	540 µA
3	−40	15.20	460 µA	175 µA	485 µA	785 µA	490 µA	700 µA	475 µA	1·06 mA	265 µA
	+130	15.40	955 µA	1·18 mA	610 µA	930 µA	480 µA	970 µA	585 µA	960 µA	540 µA
4	−40	16.00	450 µA	175 µA	490 µA	800 µA	490 µA	710 µA	480 µA	1·07 mA	272 µA
	+130	16.20	940 µA	1·16 mA	610 µA	930 µA	480 µA	965 µA	580 µA	950 µA	540 µA
5	−40	16.40	Not evaluate						µ		
	+130	17.00	940 µA	1·16 mA	610 µA	930 µA	480 µA	965 µA	580 µA	950 µA	540 µA
	25	8.00	820 µA	495 µA	610 µA	970 µA	530 µA	1·01 mA	600 µA	1·2 mA	410 µA

Table 10.7
Temperature shock tests of nine light-links with GaAs diodes TIL 23 and phototransistors TIL 81. Parameters: $I_{F,TIL\,23} = 10\ mA$; $V_{CE,TIL\,81} = 8\ V$; $r = 6\ mm$

MIL-STD-750 Test methods	Environmental condition test	Quantity		Parametric changes and total failures	
		LS 600	TIL 23	LS 600	TIL 23
1051	Temperature cycles: 5 cycles, 30 min, $+40^\circ$C to $+100^\circ$C	2100		0	
	5 cycles, 30 min, -65°C to $+125^\circ$C	126		0	
	5 cycles, 30 min, -40°C to $+100^\circ$C		110		0
	5 cycles, 30 min, -65°C to $+150^\circ$C		50		0
1056	Temperature shock: 5 cycles, Condition A	126	50	0	0
1021	Moisture resistance	126	50	0	0
2016	Mechanical shock test: 1000 g, 5 each axis, 0·5 ms	126		0	
	1500 g, Z_1 axis, 0·5 ms		146		0
2056	Vibration test, variable frequency: 10 g	126		0	
	20 g		146		0
2006	Constant acceleration: 1000 g, 1 min	126		0	
	2000 g, 1 min Z_1		146		0
2046	Vibration fatigue: 10 g	126		0	
1001	Air pressure: 15 mm Hg, 45V	126		0	
2026	Solderability: 240°C, 3 min	126		0	
1071	Airtight enclosure: Test condition E	2100	390	0	1

Table 10.8
Results of environmental condition tests for the Si phototransistors LS 600 and the GaAs diodes TIL 23

11
Parameters of Luminescence diodes

Parameters of Luminescence Diodes

11.1
Quantum efficiency

The quantum efficiency Q_D of a luminescence diode is the quotient of the number of photons n_{ph} emerging per unit time divided by the number of electrons n_E flowing through the diode in unit time:

$$Q_D = \frac{n_{Ph}}{t} : \frac{n_E}{t} = \frac{n_{Ph}}{n_E} \qquad (11.1)$$

This shows, that the quantum efficiency of a radiation emitter, as compared with that of a radiation detector, which was defined in equation (9.1), is defined in exactly the opposite way.

The quantum efficiency Q_D of a luminescent diode, emitting almost monochromatically, is calculated in simplified form with the maximum wavelength λ_{max}.

Number of photons:

$$n_{Ph} = \frac{\Phi_{e,\lambda} \cdot t}{h \cdot \nu} \qquad (9.2)$$

$$n_{Ph} = \frac{\Phi_{e,\lambda} \cdot \lambda_{max} \cdot t}{h \cdot c_o} \qquad (11.2)$$

Number of electrons:

$$n_E = \frac{I_F \cdot t}{e} \qquad (11.3)$$

The equation (11.2) and (11.3) are inserted in equation (11.1) and the constants are simplified as in equation (9.5):

$$Q(\lambda)_D = \frac{\Phi_{e,\lambda} \cdot \lambda_{max} \cdot t \cdot e}{h \cdot c_o \cdot I_F \cdot t}$$

$$= \frac{\Phi_{e,\lambda} \cdot \lambda_{max}}{I_F} \cdot \frac{10^6 \text{ A}}{1 \cdot 24 \text{ mW}} \qquad (11.4)$$

Here, the determination of $\Phi_{e,\lambda}$ is difficult, since the total radiation power $\Phi_{e,\lambda}$ has to be included.

If the total radiated power $\Phi_{e,\lambda}$ is measured in accordance with the arrangement described in Section 12.7, with a Si solar cell, then the value of $\Phi_{e,\lambda}$ must be calculated from the short-circuit current I_p and the quantum efficiency $Q(\lambda)_{SC}$ of the solar cell by means of the equation (9.4)

$$Q(\lambda)_{SC} = \frac{I_p \cdot h \cdot c_o}{\Phi_{e,\lambda} \cdot \lambda \cdot e}$$

$$= \frac{I_p}{\Phi_{e,\lambda} \cdot \lambda} \cdot \frac{1 \cdot 24 \text{ mW}}{10^6 \text{ A}} \qquad (9.4)$$

The quantum efficiency $Q(\lambda)_D$ of a luminescence diode can be calculated as the ratio of the short-circuit current I_p of the solar cell to the product of the forward current I_F of the luminescence diode and the quantum efficiency $Q(\lambda)_{SC}$ of the solar cell. For this, equation (9.4) is inserted in equation (11.4). The following is obtained:

$$Q(\lambda)_D$$

$$= \frac{I_p \cdot \lambda_{max}}{Q(\lambda)_{SC} \cdot \lambda \cdot I_F} \cdot \frac{1 \cdot 24 \text{ mW} \cdot 10^6 \text{ A}}{10^6 \cdot \text{A} \cdot 1 \cdot 24 \text{ mW}}$$

$$= \frac{I_p}{I_F \cdot Q(\lambda)_{SC}} \qquad (11.5)$$

Example:
The quantum efficiency $Q(\lambda)_D$ of the GaAs power diode TIXL 27 is to be determined with the following parameters:

$I_{F,TIXL27}$ = 300 mA; $I_{P,SC}$ = 10·5 mA

$Q(\lambda)_{SC}$ = 70%

We obtain:

$$Q(\lambda)_{TIXL27} = \frac{I_P}{I_F \cdot Q(\lambda)_{SC}}$$

$$= \frac{10\cdot5 \text{ mA}}{300 \text{ mA} \cdot 0\cdot7} = 0\cdot05 = 5\% \qquad (11.6)$$

The quantum efficiency depends on the construction of the luminescence diode. In this connection, especially, any total reflection inside the wafer system considerably reduces the quantum efficiency.

The semiconductors which are suitable for mass production of luminescence diodes have relatively high refractive indices. Gallium arsenide, for example, has a refractice index of n = 3.6. The radiation emitted from a point near the junction will be reflected, if the angle between the normal to the surface and the incident ray is greater than the angle for total reflection.

By suitable wafer geometries, the angle of incidence of a ray at the crystal-to-air boundary can be reduced and thus the total reflection can be decreased. (See Section 6.6). Flat wafer geometries have a small emission angle for each surface element, since, in accordance with equation (6.32), the critical angle for GaAs is only α_G = $16\cdot1^\circ$.

$$\sin\alpha_G = \frac{n_{Luft}}{n_{GaAs}} = \frac{1}{3\cdot6} = 0\cdot278 \rightarrow \alpha_G = 16\cdot1^\circ$$

$$(6.32)$$

Above the critical angle, total reflection occurs. *Figures 11.1 to 11.6* show the emission conditions for flat wafer geometries. For diffused GaAs wafers, the geometry shown in *Figure 11.2* has proved its worth. For example, the GaAs diode TIXL 06 is constructed from such

wafers. In the TIXL 06, an epoxy dome is dispensed with, since as a high-quality diode it has to have a very good emission characteristic and a high temperature tolerance.

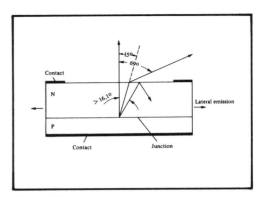

Figure 11.1
Flat epitaxial or diffused wafer. This produces an unfavourable quantum efficiency, since, as a result of total reflection, the greater proportion of the radiation cannot leave the system. The radiation of the surface elements far away from the centre point undergoes almost complete total reflection

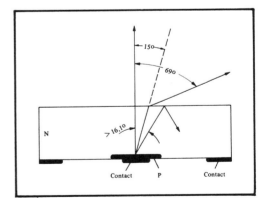

Figure 11.2
Flat diffused wafer. Its quantum efficiency is somewhat better, as compared with that in Figure 11.1. Radiation is only produced in the central part. The losses of the surface elements located at the outer edges are eliminated

208

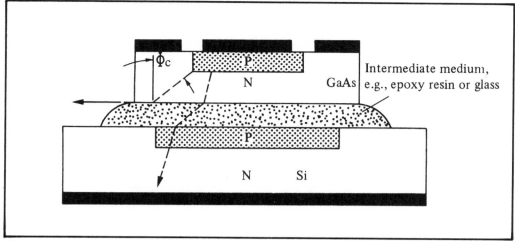

Figure 11.3
Principle of construction of an optocoupler with diffused GaAs diode and diffused photodiode. The quantum efficiency of the GaAs diode is raised further by the intermediate medium with n ≈ 1.5

With Si-doped GaAs wafers produced by liquid epitaxy, greater quantum efficiencies are achieved, in comparison with diffused wafers. *Figure 11.4* shows a typical Si-doped GaAs wafer made by the liquid epitaxy process. Here, the bonds can also be located in the outer regions. For example, GaAs diode TIXL 27 has four bonds, one in each corner.

For the diode TIXL 27, the critical angle α_G has been improved, by setting an epoxy dome directly on the wafer. The epoxy resin has a refractive index of approx. $n = 1.5$. The new critical angle α_{G1} of a radiating wafer surface is then:

$$\sin\alpha_{G1} = \frac{n_{Epoxyd}}{n_{GaAs}} = \frac{1.5}{3.6} = 0.417 \rightarrow \alpha_{G1}$$

$$= \underline{\underline{24.6°}} \qquad (11.7)$$

Thus, for the approximately Lambertian radiation, the following improvement ratio K_V is obtained:

$$K_V = \frac{1-\cos\alpha_{G1}}{1-\cos\alpha_G} = \frac{1-\cos24.6°}{1-\cos 16.1°}$$

$$= \frac{1-0.91}{1-0.96} = 2.25 \qquad (11.8)$$

For high-quality GaAs power diodes, the quantum efficiency is improved by a more expensive procedure. The possibility of emergence of the radiation emitted from a point near the junction is greater, if a ray strikes a spherical surface. Therefore the wafers are formed into hemispheres. In addition, the effective radiating area is restricted, by mesa etching, to a small central area. If the ratio of the radiating area to the total base area of the wafer is optimally designed, the "point radiation" undergoes no total reflection, at the hemispherical GaAs-air boundary. A considerably greater radiant power emerges from the crystal, since the surface angles of emergence of almost all rays are less than the critical angle α_G. *Figures 11.7* and *11.8* show the emission conditions of domed wafers.

Furthermore, the quantum efficiency of GaAs diodes can be improved, if an antireflection coating is evaporated onto the wafer surface.

209

The quantum efficiency of red-emitting GaAsP diodes is affected, among other factors, by their own absorption. The

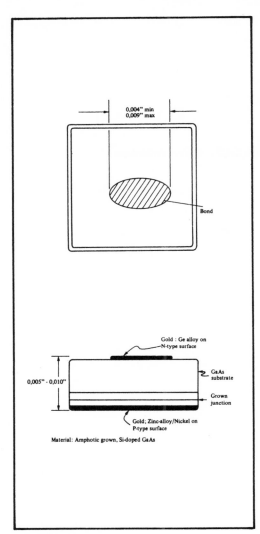

Figure 11.4
Flat, Si-doped GaAs wafer, produced by liquid epitaxy. For protection against environmental effects, the P-side of the junction is mounted on the system substrate. The N-connection is located in the centre of the upper surface of the wafer. This kind of wafer is used in many "low-cost" products. Often, an epoxy dome is also fitted. The quantum efficiency of these wafers, produced by the liquid epitaxy process is greater than that of diffused wafers

luminous power produced by recombination is sufficient to re-excite charge carriers. For good quantum and luminous efficiencies, therefore, the junction is formed very closely under the surface. As an example, *Figure 11.9* shows a wafer of a GaAsP luminescent diode, while the wafer of a monolithic display element shown in *Figure 11.10.*

In the following, the typical quantum efficiency $Q(\lambda)_D$ will be calculated for various kinds of luminescence diodes by the formula

$$Q(\lambda)_D = \frac{\Phi_{e,\lambda,typ} \cdot \lambda_{max}}{I_F} \cdot \frac{10^6 \, A}{1 \cdot 24 \, mW}$$

$$(11.4)$$

In these calculations, the data-sheet test conditions were taken as a basis.

a
Luminescence diode TIXL 27.
Si-doped GaAs wafer, flat wafer geometry, epoxy-resin dome, $\lambda_{max} = 0 \cdot 94 \, \mu m$.

$$Q(\lambda)_D = \frac{20 \, mW \cdot 0 \cdot 94 \, \mu m}{300 \, mA} \cdot \frac{10^6 \, A}{1 \cdot 24 \, mW}$$

$$= 5\% \qquad (11.9)$$

b
Luminescence diode TIXL 12
Si-doped GaAs dome wafer, with mesa etching, $\lambda_{max} = 0 \cdot 93 \, \mu m$.

$$Q(\lambda)_D = \frac{50 \, mW \cdot 0 \cdot 93 \, \mu m}{300 \, mA} \cdot \frac{10^6 \, A}{1 \cdot 24 \, mW}$$

$$= 12 \cdot 5\% \qquad (11.10)$$

c
Luminescence diode TIL 31
Si-doped GaAs wafer, flat wafer geometry, epoxy-resin dome, $\lambda_{max} = 0 \cdot 94 \, \mu m$

$$Q(\lambda)_D = \frac{6 \, mW \cdot 0 \cdot 94 \, \mu m}{100 \, mA} \cdot \frac{10^6 \, A}{1 \cdot 24 \, mW}$$

$$= 4 \cdot 5\% \qquad (11.11)$$

210

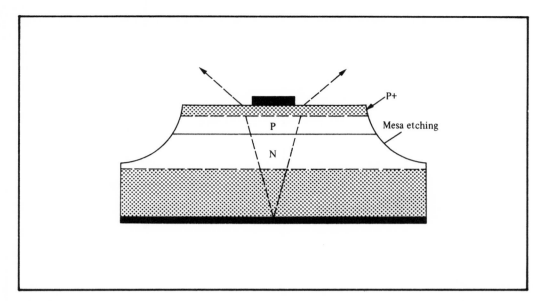

Figure 11.5

Flat wafer produced by liquid epitaxy. Improvement of the quantum efficiency by the convex mesa sides. The indirect radiation is reflected and some of it can emerge through the upper surface. The lifetime of the device is limited by the presence of the unprotected junction

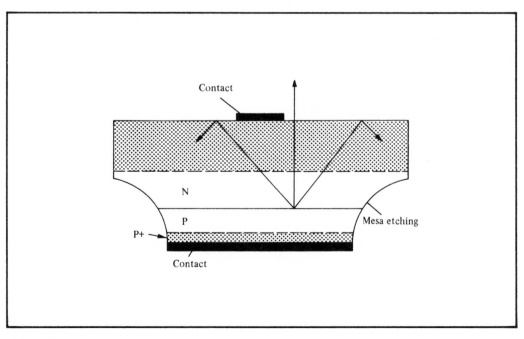

Figure 11.6

Flat wafer, produced by liquid epitaxy. For protection against environmental effects, the P-side of the junction is mounted on the system substrate

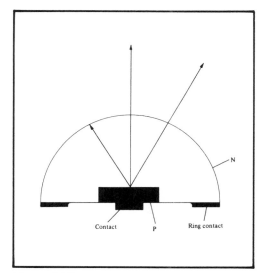

Figure 11.7
Diffused dome wafer, Its quantum efficiency corresponds approximately to that of flat wafers produced by liquid epitaxy

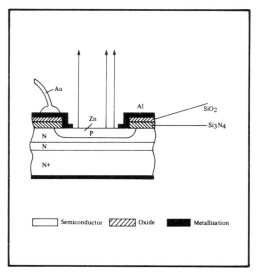

Figure 11.9
Diffused wafer for GaAsP light-emitting diodes or Zn-doped GaAs diodes

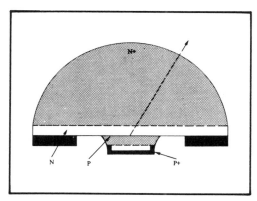

Figure 11.8
Si-doped GaAs dome wafer, produced by liquid epitaxy, with convex mesa junction sides. Through the liquid epitaxy process, the hemispherical wafer configuration and the optimum dimensioning of the effective radiating area in relation to the total base area, the quantum efficiency reaches the highest values so far attainable

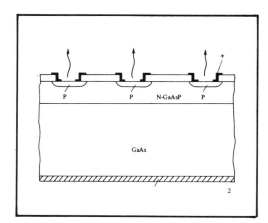

Figure 11.10
Diffused wafer for monolithic display elements

d

Luminescence diode TIXL 06 Si-diffused GaAs wafer, flat wafer geometry, $\lambda_{max} = 0.91\ \mu m$

$$Q(\lambda)_D = \frac{1.2\ mW \cdot 0.91\ \mu m}{500\ mA} \cdot \frac{10^6\ A}{1.24\ mW}$$

$$= 0.18\% \tag{11.12}$$

212

e

Luminescence diode TIL 24
Si-doped GaAs wafer, flat wafer geometry,
$\lambda_{max} = 0.93\ \mu m$.

$$Q(\lambda)_D = \frac{2\ mW \cdot 0.93\ \mu m}{50\ mA} \cdot \frac{10^6\ A}{1.24\ mW}$$

$$= 3\% \tag{11.13}$$

f

Luminescence diode TIXL 20
GaAsP dome wafer, mesa etching, $\lambda_{max} = 0.85\ \mu m$

$$Q(\lambda)_D = \frac{15\ mW \cdot 0.85\ \mu m}{200\ mA} \cdot \frac{10^6\ A}{1.24\ mW}$$

$$= 5.1\% \tag{11.14}$$

g

Red light-emitting diode TIL 210
GaAsp wafer, flat wafer geometry, epoxy
resin case, $\lambda_{max} = 0.65\ \mu m$.

$$Q(\lambda)_D = \frac{25\ \mu W \cdot 0.65\ \mu m}{20\ mA} \cdot \frac{10^6\ A}{1.24\ mW}$$

$$= 0.065\% \tag{11.15}$$

h

Yellow light-emitting diode
GaAsP wafer, flat wafer geometry,
epoxy resin case, $\lambda_{max} = 0.59\ \mu m$.

$$Q(\lambda)_D = \frac{10\ \mu W \cdot 0.59\ \mu m}{50\ mA} \cdot \frac{10^6\ A}{1.24\ mW}$$

$$\approx 0.01\% \tag{11.16}$$

i

Amber-coloured light-emitting diode
GaAsP wafer, flat wafer geometry, epoxy
resin dome, $\lambda_{max} = 0.610\ \mu m$.

$$Q(\lambda)_D = \frac{10\ \mu W \cdot 0.61\ \mu m}{50\ mA} \cdot \frac{10^6\ A}{1.24\ mW}$$

$$\approx 0.01\% \tag{11.17}$$

k

Red light-emitting diode
GaP wafer, flat wafer geometry, epoxy
resin case, $\lambda_{max} = 0.69\ \mu m$.

$$Q(\lambda)_D = \frac{225\ \mu W \cdot 0.69\ \mu m}{40\ mA} \cdot \frac{10^6\ A}{1.24\ mW}$$

$$\approx 0.3\% \tag{11.18}$$

l

Green light-emitting diode
GaP wafer, flat wafer geometry, epoxy
resin case, $\lambda_{max} = 0.56\ \mu m$.

$$Q(\lambda_D) = \frac{14\ \mu W \cdot 0.56\ \mu m}{40\ mA} \cdot \frac{10^6\ A}{1.24\ mW}$$

$$\approx 0.015\% \tag{11.19}$$

m

Yellow light-emitting diode
GaP, flat wafer geometry, epoxy resin case,
$\lambda_{max} = 0.589\ \mu m$.

$$Q(\lambda)_D = \frac{5\ \mu W \cdot 0.589\ \mu m}{50\ mA} \cdot \frac{10^6\ A}{1.24\ mW}$$

$$\approx 0.005\% \tag{11.20}$$

The calculated quantum efficiencies apply
with metal cans for the case temperature
t_G and with epoxy resin packages for the
ambient temperature t_U. At higher
temperatures, the quantum efficiency falls,
but at lower temperatures it rises.

11.2
Thermal calculations

11.2.1
Basic principles

The mathematical treatment of thermal
calculations with semiconductors is carried
out with equations similar to those for Ohm's
law in elementary electrical theory.

The higher the elctrical energy converted into heat per unit time, the so-called loss power P_{tot}, the more will the temperature of the component rise and give up its thermal energy to the surroundings by a heat flow.

According to the laws of thermodynamics the amount of a heat transfer through a transmission medium depends both on the temperature gradient Δt and also on the characteristics of the medium, its thermal resistance R_{th}.

Analogy with Ohm's law gives the following comparisons:

$$\Delta t \equiv V \; ; P_{tot} \equiv I \; ; R_{th} \equiv R \qquad (11.21)$$

Thus the "Ohm's law of thermodynamics" can be written as follows:

$$R_{th} = \frac{\Delta t}{P_{tot}} \qquad (11.22)$$

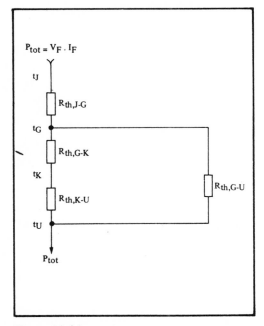

Figure 11.11
Thermal equivalent circuit for junction semiconductor

For the junction semiconductor, the total thermal resistance $R_{th,J\text{-}G}$ is made up of a series and parallel connection of individual thermal resistances. (See *Figure 11.11*).

In the following calculations, the following symbols will be used:

t_J = Junction temperature
t_G = Case temperature
t_K = Heat-sink temperature
t_U = Ambient temperature
$R_{th,J\text{-}G}$ = Thermal resistance, junction to case
$R_{th,G\text{-}K}$ = Thermal resistance, case to heat-sink
$R_{th,K\text{-}U}$ = Thermal resistance, heat-sink to ambient
$R_{th,G\text{-}U}$ = Thermal resistance, case to ambient
$R_{th,J\text{-}U}$ = Thermal resistance, junction to ambient

With plastic-encapsulated luminescence diodes, the thermal resistance $R_{th,J\text{-}U}$ is of interest.

For luminescence diodes in metal cans without heat sinks, the thermal resistance $R_{th,J\text{-}G}$, $R_{th,G\text{-}U}$ and $R_{th,J\text{-}U}$ are significant.

$$R_{th,J\text{-}U} = R_{th,J\text{-}G} + R_{th,G\text{-}U} \qquad (11.23)$$

With luminescence diodes in metal cans and with heat-sinks, the thermal resistance $R_{th,J\text{-}G}$, $R_{th,G\text{-}K}$, $R_{th,K\text{-}U}$ and $R_{th,J\text{-}U}$ are of interest. In this case, the thermal convection and radiation resistance $R_{th,G\text{-}U}$ can usually be neglected.

$$R_{th,J\text{-}U} = R_{th,J\text{-}G} + R_{th,G\text{-}K} + R_{th,K\text{-}U} \qquad (11.24)$$

The individual thermal resistances will be calculated in accordance with equation (11.22) as follows:

$$R_{th,J\text{-}G} = \frac{t_J - t_G}{P_{tot}} \qquad (11.25)$$

214

$$R_{th,G\text{-}U} = \frac{t_G - t_U}{P_{tot}} \qquad (11.26)$$

$$R_{th,J\text{-}U} = \frac{t_J - t_U}{P_{tot}} \qquad (11.27)$$

$$R_{th,G\text{-}K} = \frac{t_G - t_K}{P_{tot}} \qquad (11.28)$$

$$R_{th,K\text{-}U} = \frac{t_K - t_U}{P_{tot}} \qquad (11.29)$$

11.2.2
Determination of the thermal resistance R_{th} of luminescence diodes

The thermal resistance of a luminescence diode can be determined by measurement. The method used is based on a correlation, starting from the fact, that with constant diode forward current, I_F, the diode forward voltage V_F varies inversely with the junction temperature t_J.

In an oven, with the luminescence diodes operated with a constant forward current of $I_F = 5$ mA, the diode forward voltage V_F is recorded as a function of the junction temperature t_J. For each luminescence diode, a characteristic curve $V_F = f(t_J)$ is plotted. Following this, each luminescence diode is mounted on a large heat-sink and the electrical connections are made to the thermal resistance measuring test set. The temperature of the heat-sink is kept constant.

The measuring test set contains a circuit to interrupt the diode forward current for pulse durations of $t_p = 100$ μs with a mark-space ratio of approx. 100:1. During the pulse, the diode forward voltage V_F is measured. The junction temperature t_J is determined by means of the prepared characteristic curve $V_F = f(t_J)$.

The junction temperature t_J is determined for each diode for several values of the

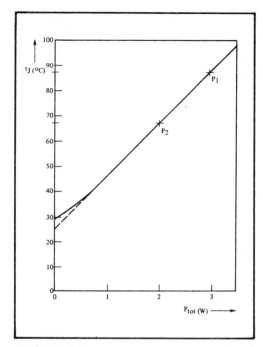

Figure 11.12
Junction temperature t_J as a function of the loss power P_{tot} for a GaAs power diode. The thermal resistance between the junction and the case is approx. $R_{th,J\text{-}G} = 20°C/W$

forward current I_F. Then t_J is plotted on a graph as a function of the loss power P_{tot}. *Figure 11.12* shows a function $t_J = f(P_{tot})$ recorded by this method for a GaAs power diode. In the linear part of this curve, the thermal resistance $R_{th,J\text{-}G}$ can be calculated in simplified form by forming the differential quotient. In this example it is:

$$R_{tj,J\text{-}G} \approx \frac{t_{J1} - t_{J2}}{P_{tot,1} - P_{tot,2}} = \frac{88°C - 68°C}{3\,W - 2\,W}$$

$$= 20\,\frac{°C}{W} \qquad (11.30)$$

The thermal resistance $R_{th,J\text{-}G}$ determined with this measurement method must not be used for the guaranteed operation of a luminescence diode in accordance with the data sheet. The maximum permissible loss

215

power P_{max} for low temperatures up to
$t = 25°C$ and the highest permissible
temperature t_{max} at $P_{max} = 0$ are stated in
the data sheet by the semiconductor
manufacturer, after quantum efficiency and
radiant efficiency measurements and after
extensive life and reliability tests. *Figure
11.13* shows the usual representation of the

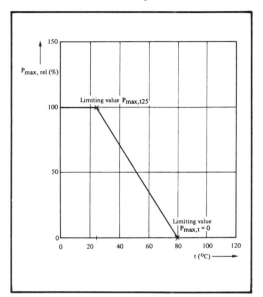

Figure 11.13
*Relative loss power $P_{max,rel}$ as a function
of the ambient temperature t_U for the
luminescence diode TIL 32, in a plastic
package, and as a function of the case
temperature t_G for the luminescence diode
TIL 31 in a metal can*

relative permissible loss power $P_{max,rel}$ as
a function of the temperature t for junction
semiconductors. From the limiting values
$P_{max,t25}$ at $t = 25°C$ and $P_{max,t} = 0$ at
t_{max}, the power reduction factor $D_{P,J-U}$
is calculated for plastic-encapsulated
luminescence diodes, or the power
reduction factor $D_{P,J-U}$ is calculated for
plastic-encapsulated luminescence diodes,
or the power reduction factor $D_{P,J-G}$ for
metal-canned luminescence diodes.

$$D_P = \frac{P_{max,t25}}{t_{max} - t_{25}} \qquad (11.31)$$

The maximum permissible loss power
$P_{max,t25}$ at $t = 25°C$ can be obtained from
the data sheet as follows:

$$P_{max,t25} = U_{F,max,t25} \cdot I_{F,max,t25} \qquad (11.32)$$

The data sheets often contain the current
reduction factor $D_{I,J-U}$ or $D_{I,J-G}$.

$$D_I = \frac{I_{F,max,t25}}{t_{max} - t_{25}} \qquad (11.33)$$

$$D_P = D_I \cdot U_{F,max,t25} \qquad (11.34)$$

The reciprocal of the power reduction
factor D_P gives the mathematical thermal
resistance.

$$R_{th,J-U} = \frac{1}{D_{P,J-U}} \qquad (11.35)$$

$$R_{th,J-G} = \frac{1}{D_{P,J-G}} \qquad (11.36)$$

Example:
The plastic encapsulated luminescence
diode TIL 32 has the following data sheet
values:

$$V_{F,max,t_U25} = 40\ mA \qquad t_{U,max} = 80°C$$

Therefore:

a

$$D_{I,J-U} = \frac{I_{F,max,t_U25}}{t_{U,max} - t_{U25}} = \frac{40\ mA}{80 - 25°C}$$

$$= 0·7272\ mA/°C \qquad (11.37)$$

b

$$D_{P,J-U} = D_{I,J-U} \cdot U_{F,max,t_U25}$$

$$= 0·7272\ mA/°C \cdot 1·6V = 1·1636\ mW/°C \qquad (11.38)$$

216

c

$$P_{max,tU25} = U_{F,max,tU25} \cdot I_{F,max,tU25}$$

$$= 1 \cdot 6 \text{ V} \cdot 40 \text{ mA} = 64 \text{ mW} \qquad (11.39)$$

d

$$R_{th,J-U} = \frac{1}{D_{P,J-U}} = \frac{1\,^{\circ}C}{1 \cdot 1636 \text{ mW}}$$

$$= 859 \cdot 3\,^{\circ}C/W \qquad (11.40)$$

11.2.3
Calculation of loss power for luminescence diodes in plastic packages

The maximum permissible loss power $P_{max,tU}$ at an ambient temperature t_U can be calculated by rearranging the equation (11.27).

$$P_{max,tU} = \frac{t_J - t_U}{R_{th,J-U}} \qquad (11.41)$$

$$P_{max,tU} = D_{P,J-U} \cdot (t_J - t_U) \qquad (11.42)$$

$P_{tU25 - tU}$ is the difference in loss power power between P_{max} at $t_U = 25\,^{\circ}C$ and P_{max} at an ambient temperature of $t_U > 25\,^{\circ}C$.

$$P_{tU25 - tU} = P_{max,tU25} - P_{max,tU}$$

$$\qquad (11.43)$$

$$P_{max,tU} = P_{max,tU25} - \frac{t_U - t_{U25}}{R_{th,J-U}}$$

$$\qquad (11.44)$$

The junction temperature t_J can be calculated with the rearranged equation (11.27).

$$t_J = (R_{th,H-U} \cdot P_{tot}) + t_U \qquad (11.45)$$

$$t_J = \frac{P_{tot}}{D_{P,J-U}} + t_U \qquad (11.46)$$

The heat from diffused luminescence diodes in plastic packages (e.g., TIL 209A), can be led out through the cathode connection, while the heat produced from Si-doped GaAs plastic encapsulated diodes (e.g., TIL 32) can be led off through the anode connection. Here, a valid guide-line value for the thermal resistance $R_{th,J-connection}$ is $300\,^{\circ}C/W$.

11.2.4
Calculation of loss power for luminescence diodes in metal cans with infinitely large heat sinks

The maximum permissible loss power $P_{max,tG}$ at a case temperature t_G is calculated by rearranging the equation (11.25).

$$P_{max,tG} = \frac{t_J - t_G}{R_{th,J-G}} \qquad (11.47)$$

$$P_{max,tG} = D_{P,J-G} \cdot (t_J - t_G) \qquad (11.48)$$

$$P_{max,tG} = P_{max,tG25} - \frac{t_G - t_{G25}}{R_{th,J-G}} \qquad (11.49)$$

The junction temperature t_J is obtained by means of the rearranged equation (11.25):

$$t_J = (R_{th,J-G} \cdot P_{tot}) + t_G \qquad (11.50)$$

$$t_J = \frac{P_{tot}}{D_{P,J-G}} + t_G \qquad (11.51)$$

11.2.5
Calculation of loss power for luminescence diodes in metal cans without heat sinks

The maximum permissible loss power $P_{max,tU}$ at an ambient temperature t_U is calculated by combination of the equations (11.23) and (11.27).

$$P_{max,t_U} = \frac{t_J - t_U}{R_{th,J\text{-}G} + R_{th,G\text{-}U}} \qquad (11.52)$$

The junction temperature t_J is supplied by equation (11.50). The thermal resistance $R_{th,G\text{-}U}$ can be taken from the data sheet. Otherwise the following guide-line values apply:

for TO-18 and TO-46 cans:
$$R_{th,G\text{-}U} = 450^\circ C/W \qquad (11.53)$$

for TO-39 cans:
$$R_{th,G\text{-}U} = 250^\circ C/W \qquad (11.54)$$

A measurement of the case temperature with a temperature probe on uncooled TO-18 and TO-46 metal cans is excessively distorted, since the contact probe draws off too much heat from the metal can.

11.2.6
Calculation of loss power for luminescence diodes in metal cans with heat sinks

The maximum permissible loss power P_{max,t_U} at an ambient temperature t_U is calculated by combination of the equations (11.24) and (11.27):

$$P_{max,t_U}$$
$$= \frac{t_J - t_U}{R_{th,J\text{-}G} + R_{th,G\text{-}K} + R_{th,K\text{-}U}} \qquad (11.55)$$

With GaAs power diodes, the anode connection usually has to be cooled. In the case of Si-doped GaAs power diodes, this connection is the fixing bolt. With insulated mounting with mica washers, thermal resistances up to $R_{th,G\text{-}K} = 1\cdot5^\circ C/W$ are mentioned in the literature. By the use of beryllium washers, hard-anodised aluminium washers, silvered mica washers or aluminium or copper washers coated with $0\cdot1$ mm of Araldite, and by coating the insulting layers on both sides with an effective heat-transfer paste, the thermal resistance can be

reduced to about $R_{th,G\text{-}K} = 0\cdot3^\circ C/W$. For non-insulated mounting, the guide-line value $R_{th,G\text{-}K} = 0\cdot3^\circ C/W$ applies. The thermal resistance $R_{th,K\text{-}U}$ of the heat-sink is to be taken from the heat-sink manufacturer's data sheet.

For luminescence diodes, the following minimum thermal resistances $R_{th,K\text{-}U}$ for the heat-sink are desirable:

Case	P_{max}	$R_{th,K}$
TO 18/TP 46	400 mW	$60^\circ C/W$ (Push-on heat-sink)
TO 5 with bolt	1 W	$10^\circ C/W$
TO 5 with bolt or special }	5 W	$5^\circ C/W$

Several GaAs diodes in an enclosed space heat each other up, so that higher case temperatures must be allowed for. The pill package GaAs diodes TIL 23/TIL 24, used in reading heads, are cooled through an optimally designed circuit board coating (anode connection).

11.2.7
Calculation of the maximum permissible forward current $I_{F,max}$

The maximum permissible diode forward current, $I_{F,max}$, for the calculated loss power $P_{max,t}$ can be determined exactly with the characteristics $I_F = f(U_F)$ and $V_F = f(t)$. For luminescence diodes in metal cans, the case temperature t_G has to be calculated to evaluate the characteristics $V_F = f(t)$ and $\Phi_e = f(t)$.

$$t_G = t_J - (R_{th,J\text{-}G} \cdot P_{tot}) \qquad (11.56)$$

In simplified form, it is sufficient to calculate $I_{F,max,t}$ with the diode forward voltage $V_{F,max,t25}$ at $t = 25^\circ C$, stated in the data sheet. The small error thus produced is an additional safety factor in constant current operation of the diode, since V_F decreases with rising temperature.

218

$$I_{F,max,t} = \frac{P_{max,t}}{U_{F,max,t25}} \qquad (11.57)$$

Example: GaAs diode TIL 31 in TO 18 metal can

Data-sheet values:
Diode forward voltage at $t_G = 25°C$:

$$V_{F,max,t_{G25}} = 1\cdot75 \text{ V}$$

Diode forward current at $t_G = 25°C$:

$$I_{F,max,t_{G25}} = 0\cdot2 \text{ A}$$

Maximum permissible case temperature:

$$t_{G,max} = 80°C$$

Empirical values:
Maximum thermal resistance of the TO 18 can:

$$R_{th,G-U} = 450°C/W$$

Thermal resistance with heat-sink fitted:

$$R_{th,G-K} = 1°C/W$$

The calculation will be carried out:

1
With the data-sheet values at $t_G = 25°C$

2
With an infinitely large heat-sink at $t_G = 50°C$

3
Without heat-sink at $t_U = 50°C$

4
With a push-on heat-sink

$$R_{th,K-U} = 60°C/W \text{ at } t_U = 50°C$$

for the values:

a
Maximum permissible loss power P_{max}

b
Overall thermal resistance R_{th}

c
Power reduction factor D_P

d
Cureent reduction factor D_I

e
Case temperature t_G

f
Maximum permissible diode forward current $I_{F,max}$

1a
$$P_{max,t_{G25}} \quad U_{F,max,t_{G25}} \cdot I_{F,max,t_{G25}}$$

$$= 1\cdot75 \text{ V} . 0\cdot2 \text{ A} = 0\cdot35 \text{ W} \qquad (11.58)$$

1b
$$R_{th,J-G} = \frac{1}{D_{P,J-G}} = \frac{t_{G,max} - t_{G25}}{P_{max,t_{G25}}}$$

$$= \frac{80°C - 25°C}{0\cdot35 \text{ mW}} = 157°C/W \qquad (11.59)$$

1c
$$D_{P,J-G} = \frac{P_{max,t_{G25}}}{t_{G,max} - t_{G25}} = \frac{1}{R_{th,J-G}}$$

$$= \frac{1}{157°C/W} = 6\cdot363 \text{ mW}/°C \qquad (11.60)$$

1d
$$D_{I,J-G} = \frac{I_{F,max,t_{G25}}}{t_{G,max} - t_{G25}} = \frac{200 \text{ mA}}{80°C - 25°C}$$

$$= 3\cdot636 \text{ mA}/°C \qquad (11.61)$$

1e and 1f
See data sheet values.

2a
$$P_{max,t_{G50}} = \frac{t_J - t_G}{R_{th,J-G}} = \frac{80°C - 50°C}{157°C/W}$$

$$= 0\cdot191 \text{ W} \qquad (11.62)$$

2b to 2c
Correspond to 1b to 1c

2e
$$t_G = 50°C = t_U$$

2f
$$I_{F,max,tG50} = \frac{P_{max,tG50}}{U_{F,max,tG25}} = \frac{191 \text{ mW}}{1 \cdot 75 \text{ V}}$$

$$= 109 \text{ mA} \qquad (11.63)$$

3a
$$P_{max,tU50} = \frac{t_J - t_{U50}}{R_{th,J\text{-}G} + R_{th,G\text{-}U}}$$

$$= \frac{80°C - 50°C}{157°C/W + 450°C/W} = 49 \cdot 4 \text{ mW} \qquad (11.64)$$

3b
$$R_{th,J\text{-}U} = R_{th,J\text{-}G} + R_{th,G\text{-}U}$$

$$= 157°C/W + 450°C/W \doteq 607°C/W$$
$$\qquad (11.65)$$

3c
$$D_{P,J\text{-}U} = \frac{1}{R_{th,J\text{-}U}} = \frac{1}{607°C/W}$$

$$= 1 \cdot 747 \text{ mW/}°C \qquad (11.66)$$

3d
$$D_{I,J\text{-}U} = \frac{D_{P,J\text{-}U}}{U_{F,max,tG25}} = \frac{1 \cdot 647 \text{ mW/}°C}{1 \cdot 75 \text{ V}}$$

$$= 0 \cdot 94 \text{ mA/}°C \qquad (11.67)$$

3e
$$t_G = t_J - R_{th,J\text{-}G} \cdot P_{max,tU50}$$

$$= 80°C - 157°C/W \cdot 49 \cdot 4 \text{ mW} = 72 \cdot 3°C$$
$$\qquad (11.68)$$

3f
$$I_{F,max,tU50} = \frac{P_{max,tU50}}{U_{F,max,tG25}} = \frac{49 \cdot 4 \text{ mW}}{1 \cdot 75 \text{ V}}$$

$$= 28 \cdot 23 \text{ mA} \qquad (11.69)$$

4a
$$P_{max,tU50} = \frac{t_J - t_U}{R_{rh,J\text{-}G} + R_{th,G\text{-}K} + R_{th,K\text{-}U}}$$

$$= \frac{(80°C - 50°C) \cdot W}{157°C + 1°C + 60°C} = 137 \cdot 6 \text{ mW}$$
$$\qquad (11.70)$$

4b
$$R_{th,J\text{-}U} = R_{th,J\text{-}G} + R_{th,G\text{-}k} + R_{th,K\text{-}U}$$

$$= 157°C/W + 1°C/W + 60°C/W$$

$$= 218°C/W \qquad (11.71)$$

4c
$$D_{P,J\text{-}U} = \frac{1}{R_{th,J\text{-}U}} = \frac{1}{218°C/W}$$

$$= 4 \cdot 587 \text{ mW/}°C \qquad (11.72)$$

4d
$$D_{I,J\text{-}U} = \frac{D_{P,J\text{-}U}}{U_{F,max,tG25}} = \frac{4 \cdot 587 \text{ mW/}°C}{1 \cdot 75 \text{ V}}$$

$$= 2 \cdot 62 \text{ mA/}°C \qquad (11.73)$$

4e
$$t_G = t_J - R_{th,J\text{-}G} \cdot P_{max,tU50}$$

$$= 80°C - 157°C/W \cdot 137 \cdot 6 \text{ mW} = 58 \cdot 4°C$$
$$\qquad (11.74)$$

4f
$$I_{F,max,tU50} = \frac{P_{max,tU50}}{U_{F,max,tG25}}$$

$$= \frac{137 \cdot 6 \text{ mW}}{1 \cdot 75 \text{ V}} = 78 \cdot 62 \text{ mA} \qquad (11.75)$$

11.3
Radiant power

The radiant power or luminous power produced from a luminescence diode depends on the quantum efficiency Q, the diode forward current I_F and the temperature t_U or t_G. Red GaAsP diodes, yellow and green GaP diodes and IR-emitting GaAs diodes have, to a first approximation, a linear dependence of the radiant power Φ_e on the diode forward current I_F. The linearity improves with small forward currents and low temperatures. Red GaP diodes have a non-linear dependence of the radiant power Φ_e on the diode forward current I_F. With higher diode forward currents, red GaP diodes reach a saturation point. Therefore they are unsuitable for pulse operation.

Figure 11.14 shows the luminous intensity I_V which is at present attainable in mass production, as a function of the forward current I_F for luminescent diodes. The temperature affects the quantum efficiency Q and thus the emitted radiant power Φ_e of a luminescent diode. With rising temperature, the quantum efficiency and radiant power decrease, with falling temperature the reverse is true. In practice, for GaAs diodes, the radiant power is decreased approximately by the factor 0·5 with a temperature rise

of 80°C and it is increased approximately by a factor of 2 with a temperature decrease of 80°C.

Figure 11.15 shows, for comparison, the relative radiation power $\Phi_{e,rel}$ as a function of the temperature t for GaP, GaAs and GaAsP luminescence diodes. It is very often advantageous to show the function Φ_e = f(t) with the diode forward current I_F as a variable. (*Figure 11.16*). For the function Φ_e = f(I_F), as shown in *Figure 11.17*, the temperature t can be selected as the variable.

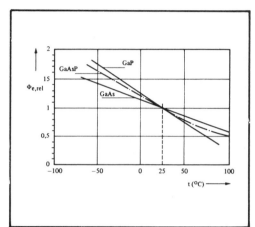

Figure 11.15
Relative radiant power $\Phi_{e,rel}$ as a function of the temperature t for GaP, GaAs and GaAsP luminescence diodes. The forward current is I_F = 20 mA in each case

In the data-sheet, the relative radiant power $\Phi_{e,rel}$ is usually stated for IR-emitting luminescence diodes and the relative luminous intensity $I_{V,rel}$ for light-emitting diodes, as a function of the diode forward current I_F or as a function of the temperature t. Here, the relative radiant power $\Phi_{e,rel}$ or the relative luminous intensity $I_{V,rel}$ relate to the absolute value of Φ_e or I_V for the test conditions stated in the data sheet.

The loss power calculation shown in the previous section 11.2, with the subsequent determination of the maximum permissible

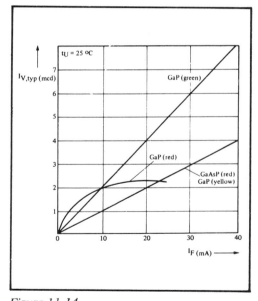

Figure 11.14
Luminous intensity I_V as a function of the forward current I_F for GaP and GaAsP luminescent diodes

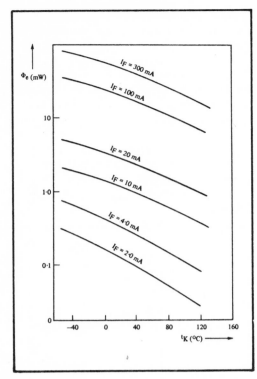

Figure 11.16
Radiant power Φ_e as a function of the heat-sink temperature t_K for the GaAs power diode TIXL 12. The variable is the forward current I_F

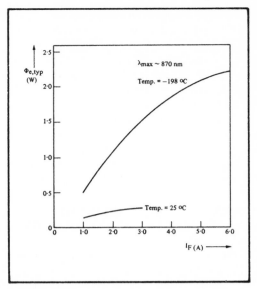

Figure 11.17
Mean radiant power $\Phi_{e,typ}$ as a function of the forward current I_F for the GaAs power diode TIXL 16; the heat-sink temperature t_K is the variable

forward current I_F, is necessary, in order to be able to evaluate the function $\Phi_{e,rel} = f(t)$ and $\Phi_{e,rel} = f(I_F)$, shown in the data sheet.

For the calculated maximum permissible forward current I_F, the relative radiant power $\Phi_{rel(I)}$ can be read from the graph $\Phi_{rel} = f(I_F)$ and the relative radiant power $\Phi_{rel(t)}$ can be read from the graph $\Phi_{rel} = f(t)$ for the calculated temperature t_U or t_G with the parameter I_F stated for the test conditions. *Figures 11.16* and *11.17* show examples of such graphs. The required absolute radiant power Φ_t is the product of the values of $\Phi_{rel(t)}$ and $\Phi_{rel(I)}$ read off and the absolute radiant power Φ_{t25} stated for the test conditions.

$$\Phi_t = \Phi_{rel} (I) \cdot \Phi_{rel(t)} \cdot \Phi_{t25} \qquad (11.76)$$

For the example 4f listed in Section 11.2, the radiant power $\Phi_{e,t}$ will be determined with the data-sheet values.

Data: $I_{F,t_{U50}} = 78 \cdot 62$ mA,

$t_U = 50^{\circ}$C, $t_G = 58 \cdot 4^{\circ}$C,

$\Phi_{e,typ,t_{G25}} = 6$ mW.

In *Figure 11.18,* for a diode forward current of $I_F \approx 78$ mA, a relative radiant power of $\Phi_{rel(I)} = 0 \cdot 75$ is read off. For a case temperature of $t_G = 58^{\circ}$C *Figure 11.19* shows a relative radiant power of $\Phi_{rel(t)} = 0 \cdot 7$. For this working point, the typical emitted radiant power of the TIL 31 is:

$$\Phi_{e,t_{U50}} = \Phi_{e,rel(I)} \cdot \Phi_{e,rel(t)} \cdot \Phi_{e,t_{G25}}$$

$$= 0 \cdot 75 \cdot 0 \cdot 7 \cdot 6 \text{ mW} = 3 \cdot 15 \text{ mW} \qquad (11.77)$$

222

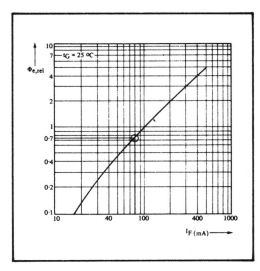

Figure 11.18
Relative radiant power $\Phi_{e,rel}$ as a function
of the forward current I_F for the GaAs
diode TIL 31. The relative radiant power
has been normalised to the value $\Phi_{e,rel}$
= 1 for a forward current of I_F = 100 mA
at $t_G = 25°C$

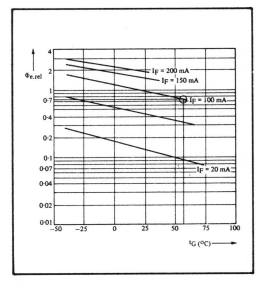

Figure 11.19
Relative radiant power $\Phi_{e,rel}$ as a function
of the case temperature t_G for the GaAs
diode TIL 31, with the forward current
I_F as a variable

Figure 11.20 shows the absolute radiant
power Φ_e as a function of the heat-sink
temperature t_K for the GaAs diode TIXL
16. When using a heat-sink with $R_{th,K-U}$ =
5°C/W and for a maximum ambient
temperature of t_U = 50°C, a diode forward
current of $I_{F,t_{U50}}$ = 990 mA and a case
temperature of t_G = 62·8°C were calculated.

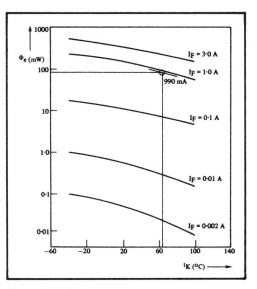

Figure 11.20
Radiant power Φ_e as a function of the heat-
sink temperature t_K for the GaAs power
diode TIXL 16, with the forward current
I_F as a variable

The parameter I_F = 990 mA is entered in
Figure 11.20. For $t_G \approx 63°C$ and I_F = 990
mA, a radiant power of Φ_e = 95 mW is then
read off.

11.4
Radiant efficiency η_e

The radiant efficiency η_e, like the quantum
efficiency Q, depends on the construction
of the luminescence diode and also on the
chosen working point. By considerations of
radiant efficiency, it is possible to select
the optimum working point of a
luminescence diode. The Si-doped GaAs

223

doped-wafer diodes produced by the liquid epitaxy process have the greatest radiant efficiency. It amounts to approximately η_e in 8% at $t_G = 25°C$. At $t_G = -196°C$, it rises about $\eta_e = 25\%$. *Figure 11.21* shows, as an example, the radiant efficiency η_e of the GaAs power diode TIXL 12 as a function of the diode forward current I_F, with the temperature t_G as a variable.

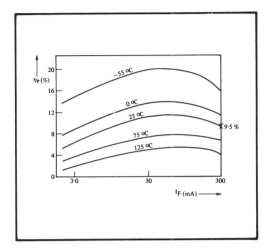

Figure 11.21
Radiant efficiency η_e as a function of the forward current I_F for the GaAs power diode TIXL 12, with the case temperature t_G as a variable

The radiant efficiency η_e of a luminescence diode is calculated in accordance with equation (2.34). For the GaAs diode TIXL 12, it is:

$$\eta_e = \frac{\Phi_e}{P} = \frac{\Phi_{e,min,tG25}}{I_{F,tG25} \cdot U_{F,typ,tG25}}$$
$$= \frac{40 \text{ mW}}{0 \cdot 3 \text{ A} \cdot 1 \cdot 4 \text{ V}} = 9 \cdot 5\% \tag{11.78}$$

This value is entered in *Figure 11.21.*

For the example 4f) listed in Sections 11.2 and 11.3, the radiant efficiency will be calculated. The radiant efficiency according to the data-sheet test conditions serves as a comparative value.

According to example 4f):

$$\eta_e = \frac{\Phi_{e,tU50}}{I_{F,tU50} \cdot U_{F,max,tG25}}$$
$$= \frac{3 \cdot 15 \text{ mW}}{78 \cdot 62 \text{ mA} \cdot 1 \cdot 75 \text{ V}} = 2 \cdot 3\% \tag{11.79}$$

According to the data-sheet test conditions, for the diode TIL 31:

$$\eta_e = \frac{\Phi_{e,tG25}}{I_{F,tG25} \cdot U_{F,max,tG25}}$$
$$= \frac{6 \text{ mW}}{100 \text{ mA} \cdot 1 \cdot 75 \text{ V}} = 3 \cdot 43\% \tag{11.80}$$

11.5
Spectral radiant efficiency

Luminescence diodes are selective emitters. On the other hand, semiconductor laser diodes are decidedly monochromatic emitters. The spectral radiation distribution of a luminescence diode depends on the band spacing of the semiconductor. The maximum wavelength λ_{max} is calculated from the typical band-spacing energy with equation (1.4).

$$\frac{\lambda}{\mu m} = \frac{1 \cdot 24}{\dfrac{W_{Ph}}{eV}} \tag{1.4}$$

Figure 11.22 shows the maximum wavelength for different semiconductor radiating diodes. In addition, the relative spectral sensitivity of the eye, $V(\lambda)$, which is of interest for light-emitting diodes, and the relative spectral sensitivity $s(\lambda)_{rel}$ of silicon photocells, which is of interest for IR-emitting GaAs diodes, are shown as functions of the wavelength λ. The bandwidth $\Delta\lambda_{HP}$ of the selective radiation of a luminescence diode is stated, in a unit of wavelength, between the spectral

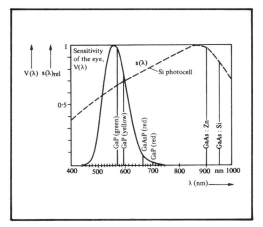

Figure 11.22
Illustration of the various maximum wavelengths λ_{max} for different semiconductor radiation-emitting diodes. In addition, the relative spectral sensitivity of the eye, $V(\lambda)$, which is of interest for light-emitting diodes, and the relative spectral sensitivity $s(\lambda)_{rel}$ of silicon photocells, which is of interest for the IR-emitting GaAs diodes, are stated as functions of the wavelength λ

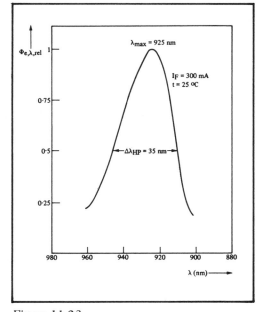

Figure 11.23
Relative spectral radiant power $\Phi_{e,\lambda}$ rel as a function of the wavelength λ for the GaAs power diode TIXL 12

half-power points $HP_{\lambda 1}$ and $HP_{\lambda 2}$ (see Section 10.4). The average bandwidth of luminescence diodes lies between 20 nm and 50 nm.

In *Figure 11.23*, the relative spectral radiant power $\Phi_{e,\lambda,rel}$ is shown as a function of the wavelength λ for the Si-doped GaAs diode TIXL 12, produced by the liquid epitaxy process. The bandwidth $\Delta\lambda_{HP}$ has been specially marked. It is $\Delta\lambda_{HP}$ = 35 nm = 350 A.

Finally, *Figure 11.24* shows $\Phi_{e,rel,\lambda}$ = f(λ) for the diffused red-emitting GaAs diode TIL 209A

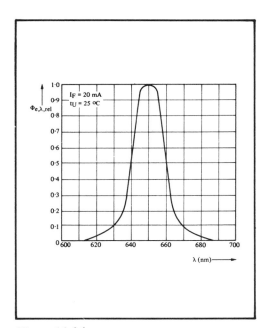

Figure 11.24
Relative spectral radiant power $\Phi_{e,\lambda}$ rel as a function of the wavelength λ for the red-emitting gallium-arsenide-phosphide diode TIL 209A

The spectral radiation distribution of a luminescence diode is temperature-dependent. In the data sheet, the difference between the maximum wavelength λ_{max} at t = 25°C and the maximum wavelength λ_{max} at any given temperature

225

t is stated. As an example, *Figure 11.25* shows the difference in the maximum wavelengths λ_{max} as a function of the temperature t for the GaAs diode TIXL 06. The measurement of radiation wavelengths is carried out with a spectrometer calibrated according to national standards.

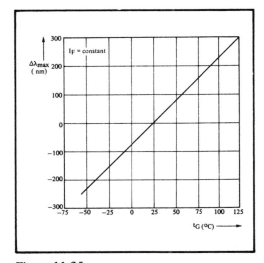

Figure 11.25
Variation of the maximum wavelength $\Delta\lambda_{max}$ as a function of the case temperature t_G for the gallium arsenide diode TIXL 06. The maximum wavelength λ_{max} has been standardised to the value ± 0 nm for a case temperature $t_G = 25°C$

11.6
Electrical Parameters

Two of the most important electrical parameters of luminescence diodes are the diode forward voltage V_F and the diode forward current I_F. The forward voltage is determined by the band spacing of the semiconductor, the diode forward current I_F and the temperature t. Under equal test conditions, therefore, different-coloured light-emitting diodes have different forward voltages. The GaAs diodes have typical forward voltages of $V_F = 1·4$ V (see *Table 11.1*).

Material	Range	Maximum wavelength	$V_{F,typ}$
GaAs	Infra-red	940 nm	1·4 V
GaAsP	Red	650 nm	1·6 V
GaAsP	Orange	610 nm	2·0 V
GaAsP	Yellow	590 nm	3·0 V
GaP	Green	560 nm	3·0 V

Table 11.1
Forward voltages of various luminescence diodes

The diode forward voltage V_F is shown as a function of the diode forward current I_F in *Figure 11.26* for the GaAs diode TIXL 16 and in *Figure 11.27* for the GaAs diodes TIXL 12 and TIXL 13, at a constant temperature t = 25°C. From this V-I characteristic, the dynamic impedance r_S can be determined in simplified form for the normal working range of a luminescence diode through the differential quotient $\Delta V_F/\Delta I_F$. With a straight characteristic section, as in the *Figures 11.26* and *11.27*, the dynamic impedance r_S can be stated directly. The dynamic impedance of GaAs diodes lies approximately between $r_S = 0·1\Omega$ and 2Ω. Modulation final stages for luminescence diodes are to be matched to the dynamic impedance r_S.

The diode forward voltage of luminescence diodes is temperature dependent. The temperature coefficient is stated by the differential quotient dV_F/dt,

i.e.:

$$TK = \frac{dU_F}{dt} \tag{11.81}$$

GaAs diodes have smaller temperature coefficients than silicon components. For the normal working range of a luminescence diode, the temperature coefficient can be calculated in simplified form through the differential quotient $\Delta V/\Delta t$. For GaAs diodes, temperature coefficients of

226

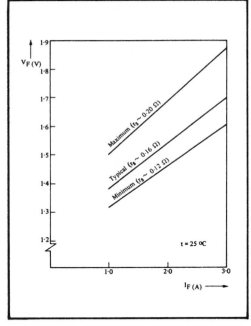

Figure 11.26
Forward voltage V_F as a function of the
forward current I_F for the GaAs power
diode TIXL 16

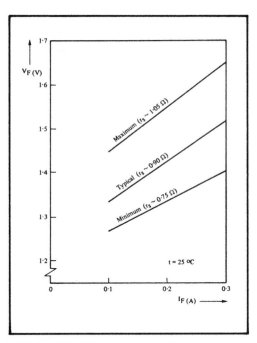

Figure 11.27
Forward voltage V_F as a function of the
forward current I_F for the GaAs power
diodes TIXL 12 and TIXL 13

TK = –1·2 mW/°C to TK = –1·5mV/°C
are obtained. *Figure 11.28* shows the
function V_F = f(t) for the TIXL 16 and
Figure 11.29 shows this function for the
TIXL 12 and TIXL 13. The forward current
I_F is the variable.

For the exact determination of the working
point of a luminescence diode, the function
I_F = f(V_F), which is often stated, is usually
sufficient. In *Figure 11.30*, the function
I_F = f(V_F) is shown for comparison,
for various semiconductor luminescence
diodes. As a further example, *Figure 11.31*
shows the diode forward current I_F of the
red-emitting GaAsP diode type TIL 209A as
a function of the diode forward voltage V_F.

As well as optoelectronic applications,
these light-emitting diodes can be used
advantageously as reference diodes for
low voltages, since they have a steep slope
$\Delta I_F/\Delta V_F$ and a low temperature

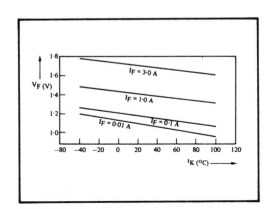

Figure 11.28
Forward voltage V_F as a function of the heat-
sink temperature t_K for the GaAs power
diode TIXL 16; the forward current I_F is the
variable

coefficient of about TK = 2 mV/°C. *Figure
11.32* shows the function I_F = f(V_F) for
the GaAs diode TIL 23 with the
temperature t_G as a variable.

227

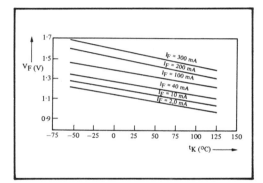

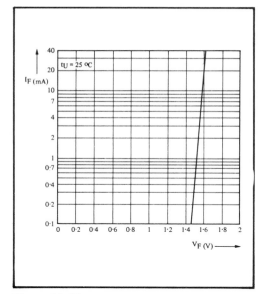

Figure 11.29
Forward voltage V_F as a function of heat-sink temperature t_K for the GaAs power diodes TIXL 12 and TIXL 13. The forward current I_F is the variable

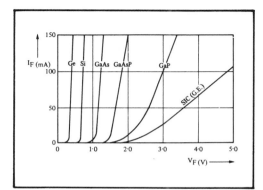

Figure 11.30
Forward current I_F as a function of the forward voltage V_F for semiconductor diodes of various materials

The temperature coefficient TK can be calculated in simplified form through the differential quotient $\Delta V/\Delta t$, its value is $1{\cdot}33$ mV/$^\circ$C.

11.7
Pulse operation

In pulse operation, luminescence diodes are controlled by a pulsed current I_F. The peak pulse current amplitude depends on the construction of the diode and on the cooling. A certain power- and current-

Figure 11.31
Forward current I_F as a function of the forward voltage V_F for the red-emitting GaAsP diode TIL 209A

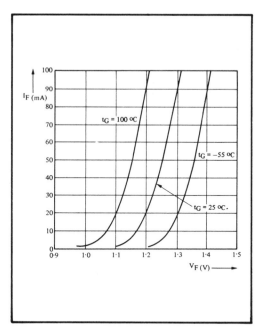

Figure 11.32
Forward current I_F as a function of the forward voltage V_F for the GaAs diodes TIL 23 and TIL 24, with the case temperature t_G as a variable

228

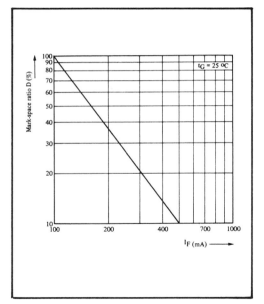

Figure 11.33
Mark-space ratio D in % as a function
of the forward current I_F for the GaAs
diodes TIL 23 and TIL 24

density must not be exceeded in the wafer
and in the bond connection wire. Further,
the peak pulse current depends on the
mark-space ratio D and the pulse width
t_p. *Figure 11.33* shows the mark-space
ratio D and *Figure 11.34* the pulse-width
t_p as functions of the diode forward current
I_F for the GaAs diodes TIL 23 and TIL 24
at $t_G = 25°C$.

If such functions do not appear in the
data sheet, these illustrations can serve
as an approximate solution. For this, the
abscissa value of $I_F = 100$ mA is to be
replaced in each case by the maximum
permissible diode forward current $I_{F,t25}$

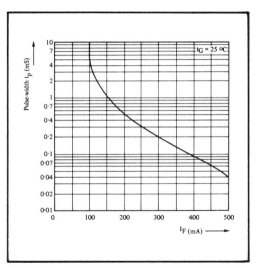

Figure 11.34
Pulse width t_p is ms as a function of
the forward current for the GaAs diodes
TIL 23 and TIL 24

at $t = 25°C$ for the corresponding GaAs
diode. The other abscissa values are corrected
with the ratio $I_{F,t25}/100$ mA.

Example: TIL 24: $I_{F,max,tG25} = 100$ mA

With $I_{F,tG25} = 500$ mA, the maximum
permissible pulse duration is $t_p = 40$ μs
and the maximum permissible mark-space
ratio D=10%.

Example: TIL 31: $I_{F,max,tG25} = 200$ mA

With $I_{F,tG25} = 1$ A, the maximum
permissible pulse width is $t_p = 40$ μ
and the maximum permissible mark-space
ratio D=10%.

12
Radiation measurements

12.1
General Considerations

Every optical radiation has a given spectral distribution. This is taken into account through the spectral densities of the radiometric parameters. Every radiometric parameter can be related to a differential range of the wavelength λ or the frequency ν. The spectral radiant power is linked with each of these spectral parameters. Therefore, for the sake of simplicity, the spectral distribution of an actual radiation source is related to the basic parameter, radiant power. Integration of the spectral radiant power over the whole wavelength range gives the radiant power (see Section 2.3.4).

$$\Phi_e = \int_0^\infty \Phi_{e,\lambda} \qquad (12.1)$$

In general, every optical radiation is evaluated by a photodetector. But also, every photodetector evaluates the optical radiation falling upon it in accordance with its spectal sensitivity. In the field of photometry, this evaluation is adequately illustrated in Chapter 5 and also in DIN standards 1301/1304/5031/5033. The visible radiation from light-sources for illumination purposes or from displays for the representation of information is evaluated by the human eye as a "photodetector". It is related to the specified sensitivity curve of the eye, $V(\lambda)$.

$$\Phi_V = \int_{\lambda_1}^{\lambda_2} \Phi_{e,\lambda} \cdot V(\lambda).d\lambda \qquad (12.2)$$

Luminous fluxes are measured with photodetectors which are specially corrected to the spectral sensitivity of the eye. In comparison with measurements in the other optical wavelength ranges, photometry has two advantages: Every visible radiation is related only to an obligatory spectral sensitivity of a photodetector or of the eye. The part of the radiation which can be evaluated is measured with simple measuring instruments in photometric parameters.

The calculation and measurement of an optical radiation in the wider sense takes place in accordance with similar considerations to those in photometry. Firstly an optical radiation is measured as a whole, secondly the part of the radiation evaluated by a photodetector is determined selectively. (See Chapter 9).

12.2
Measurement of colour temperature of standard light A

The measurement of the sensitivity of Si photodetectors is at present carried out with the radiation known as standard light A. This kind of radiation is defined by the relative spectral radiation distribution S_λ of the Planck radiation between 320 nm and 780 nm at a colour temperatur $T_f = 2856$ K (see DIN 5033, Part 7). The radiation functions S_λ in the IR range which is of interest with an emitter temperature $T_V = 2856$ K are not taken into account.

The scientific lamps (Wi-series lamps) which have previously been operated with a colour temperature $T_f = 2856$ K show different radiation functions in the IR range. This is due to the different absorptions of the glass and the different emitter temperatures between the filament fixing points and the central part of the filament. The newly-developed filament lamps Wi 40 and Wi 41 (Figure 12.1) therefore have self-supporting filaments.

For exact radiation measurements in the IR range, the radiation function S_λ of the

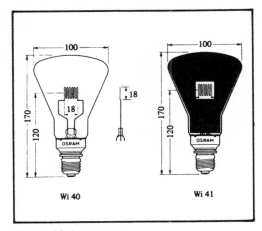

Figure 12.1
Osram scientific lamps

filament lamp used and the difference factor in relation to the equivalent function S_λ of a Planck radiator emitting at T_V = 2856 K should be known. A recommended Wi lamp to DIN standards and the statement of exactly calculated conversion factors for the wavelength range of interest would, for example, be a solution.

The Wi lamp is to be operated in the previous way at T_f = 2856 K. The colour-temperature T_f can be measured, for example, with filament pyrometers, the principle of which is illustrated in *Figure 12.2*.

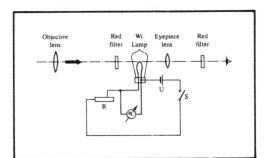

Figure 12.2
Principle of the filament pyrometer

The filament of the built-in lamp is observed with the eye through the eyepiece lens. The interposed red filter, with the spectral sensitivity of the eye, V_λ, gives the desired narrow spectral range. An image of the Wi filament to be measured is formed by the objective lens in the plane of the filament of the built-in lamp.

Equality of radiance is set by varying the lamp current or lamp voltage, so that the filament cap of the built-in lamp no longer stands out from the incandescent material to be measured, but almost disappears. The voltmeter in parallel with the built-in filament lamp is calibrated with the colour temperature.

Furthermore, the colour temperature of the Wi lamp can be adjusted by a second method, in accordance with the lamp manufacturer's data sheets, by means of the lamp current or the lamp voltage. For this, calibrated lamps of types Wi 40 or Wi 41 are needed. The lamp manufacturer's conditions must be observed. The Wi lamp will be operated vertically, with the cap downward. The internal glass frame, if any, is the side away from the detector. The screw ring is the anode. The Wi 40 lamp used then had the following values: V_L = 32.0 V (D.C.); I_L = 5.716 A.

Since discrepancies can occur through contact resistances between the lampholder and base when the lamp voltage is adjusted, it is most convenient to work with the current. It should then be possible to adjust the value of the stated lamp current exactly, to rather better than the third decimal place. Therefore, only finely adjustable and well stabilised mains supplies are suitable for the power supply.

Often, no precision ammeter is available to measure the lamp current. In such cases, the lamp current is determined through the voltage drop in a precision series resistance. For simplicity, the A-V-Ω Multizet S, in conjunction with the associated shunt M955-A1 for current measurements up to 30 A, is used as the series resistance. The Multizet is to be switched to the 100 mV range. With a lamp current of 5.716 A, the digital voltmeter shows a voltage of 19.0533 mV.

234

During such current and voltage measurements, the lamp leads must be connected to the current terminals of the shunt and the digital voltmeter to the voltage terminals of the Multizet. One measurement lamp Wi 40 should only be used as a calibration radiation source. Further, long-term measurements should be carried out with a second measurement lamp. The lamp Wi 40 should only be used as a calibration radiation source. Further, long-term measurements should be carried out with a second measurement lamp. The lamp Wi 40 can be operated for approximately 50 - 100 hours with the manufacturer's stated data without recalibration. Since this measurement lamp is in any case mounted on an optical bench with a home-made adapter, an expensively-made precision lampholder can be dispensed with. The current supply leads and the measurement leads for voltage measurement are soldered directly to the lamp. This arrangement also has the advantage, without great expense, that the current and voltage values recorded by the lamp manufacturer can be compared one with the other with two digital voltmeters. The tolerances of simple measurement resistors become visible and can thus be taken into account.

12.3
Measurement of radiant power with thermal photodetectors

The standard light A radiation has a great spectral width, corresponding to the Planck radiation distribution. Such wide-band radiant powers are also measured with wide-band, radiation-sensitive receivers. Since radiation detectors, which are based on the principle of the internal or external photo-effect, are, almost without exception, selective detectors, only the thermal types of detector are suitable for use as wide-band photodetectors. Thermal photodetectors are almost black bodies of small dimensions. They utilise the effect, that an absorbed incident radiant energy heats up their bodies. The absorbing detector surface is blackened,

so that the degree of absorption is as great as possible, in accordance with Kirchhoff's law (equation 4.39).

In general, heat increases the vibrations of the crystal lattice of the absorbing substance. These vibrations are called phonons. Corresponding to the rise in temperature of the receiver, they cause a change in its electrical resistance, its contact potential or the product of pressure x volume of an enclosed quantity of gas in pneumatic detectors. For the greater part, the efficiency of thermal detectors is low.

Certain measuring instruments, which utilise the change in electrical resistance with a rise in temperature, are called bolometers. They are used for temperature and radiation measurement. For radiation measurement, a metal strip (e.g., platinum), coated with lamp-black, is used as a bolometer.

The principle of a bolometer circuit is shown in *Figure 12.3*. When not irradiated, the Wheatstone bridge is balanced. Through irradiation, the change in resistance of the bolometer (R 1) unbalances the bridge, so that a current flows through the galvanometer. The current value is proportional to the temperature rise and thus to the absorbed radiant power. Changes in the

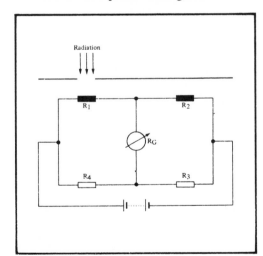

Figure 12.3
Principle of a bolometer

ambient temperature do not falsify the measured value, since the bridge is temperature-compensated by R 2. The resistance R 2 consists of an identical, but darkened bolometer. The battery must deliver a highly constant voltage. The newer semiconductor bolometers, equipped with thermistors, show an exponential change in resistance as a function of temperature.

Another detection principle which is often used is based on the change in the contact potential of a thermocouple. A thermocouple consists of a small soldered or welded joint between two metal wires with different thermoelectric potentials. As is shown in *Figure 12.4* a constantan wire

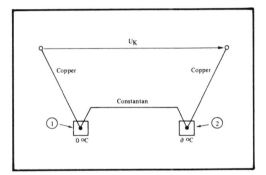

Figure 12.4
Principle of construction of a thermocouple, composed of a constantan wire and the wires of the copper leads

forms a thermocouple with the copper measurement leads. For contact measurement, junction 2 is brought into good thermal contact with the substance to be measured. If, during this measurement, junction 1 is kept at a constant temperature of $0°C$ in melting ice, then the contact voltage is a measure of the temperature difference in relation to $0°C$.

Figure 12.5 shows a thermocouple made from the dissimilar metal wires M 1 and M 2. The measurement leads are not used to form a thermocouple. In the same way, they must not form a new thermocouple at the connection points A 1 and A 2. The metal wires M 1 and M 2 are connected by the

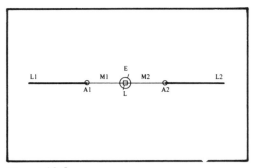

Figure 12.5
Thermocouple with separate measurement leads.

junction L, which is in good thermal contact with the blackened receiver surface E. Incident radiation is absorbed by the black receiver surface E. The receiver E is heated up, so that the temperature of the thermocouple junction rises. To a first approximation, the thermocouple voltage varies in proportion to the rise in temperature and thus to the absorbed radiant power. The thermoelectric voltages of a single thermocouple are very small. Therefore, a number of thermocouples are often connected in series to form a thermopile. The thermoelectric voltage is thus multiplied according to the number of thermocouples. The greater mass of the thermopile gives longer response times, in comparison with thermocouples.

Figure 12.6 shows a thermopile from the American firm of Eppley Laboratory Inc.

The thermopile is usually located in an evacuated housing, so that the heat losses through convection by the surrounding air are reduced. A blackened receiver foil, with the measurement junctions, is fitted behind the quartz window. The reverse side of the receiving foil is made reflective, so that its absorption and sensitivity can be neglected. Without irradiation, the measurement junctions are kept at the same temperature as the reference junctions. The reference junctions are screened from incident radiation. If radiation falls through the quartz window onto the blackened foil, the temperature of

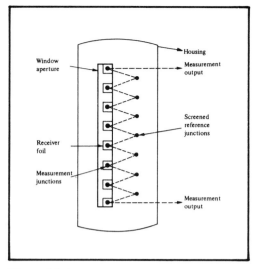

Figure 12.6
Thermopile as produced by Eppley Lab. Inc.

the measurement junctions rises in comparison with the reference junctions. Changes due to the ambient temperature are compensated for by the reference junctions.

Radiation measuring instruments with thermocouples and thermopiles sometimes have calibration curves or calibration tables for correction of the measured values. Such radiation measuring instruments measure the incident radiant power. They are usually calibrated in μW, mW or watts. The irradi-

ance is determined by dividing the measured value by the effective sensitive receiving surface area.

12.4
Measurement of radiant power of the standard light A radiation with the thermopile

Radiation measurements are advantageously carried out in a separate laboratory room. The daylight entering through the window will be shut out with a closely-woven blackout curtain. The front of the curtain is rubber-coated. Any further room lighting will be switched off. The undesired stray radiation from the actual radiation source must not fall on the measuring receiver through reflections from the room walls, the work-bench surfaces and instrument surfaces. The room walls should therefore be painted matt black. The bench top and instrument surfaces will be covered with matt black material or cardboard. Measurement errors mainly occur with thermal photodetectors, if they are subjected to draughts of air. Therefore, the windows and doors are to be well closed and fans and air-conditioning installations switched off.

As shown in *Figure 12.7*, the Wi lamp, the shutter and the thermopile are mounted on an optical bench at equal heights in the

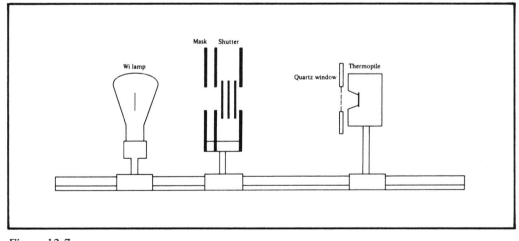

Figure 12.7
Measurement set-up on the optical bench for measurements of radiant power.

237

optical axis. The Wi lamp is operated with standard light A, as stated previously. The shutter is arranged approximately 10 cm away from the filament. The shutter aperture should be somewhat larger than the maximum dimensions of the filament spiral or the window aperture of the Wi lamp. The whole shutter must completely screen the thermopile, like a mask, against the lamp radiation outside the beam of radiation needed for measurement, since the longer-wavelength IR radiation would heat up the quartz window of the thermopile unnecessarily. The matt black back of the shutter is towards the receiver. Its reflective front is directed towards the radiation source. The incident stray radiation from the filament lamp is very well reflected. The shutter can therefore not cause interference as a secondary radiation source. For the same reason, the lamp Wi 41 is also more suitable than the Wi 40. The radiation from the Wi 41 only emerges through a window in the lamp bulb. The remaining internal surface of the lamp bulb is coated with an opaque film. This so-called skirt has the disadvantage, however, that it also acts as a thermal radiator ($T \approx 500$ K).

For measurements of radiant power, Texas Instruments often use a silver-bismuth thermopile from Eppley Laboratory Inc. This is supplied already calibrated. This is supplied already calibrated. This firm can carry out this calibration in accordance with the rules of the National Bureau of Standards, USA, or of the National Physical Laboratory, England.

The quartz window of the thermopile must be clean. The response time of the Eppley thermopile is about 30 s. The measurement begins with the shutter closed. The value due to the residual stray radiation is noted as the dark value. Following this, the shutter is opened. The distance of the thermopile from the Wi lamp is varied, until the desired radiant power is indicated. This measured value is called the exposure value. With the shutter closed, the dark value is checked once more. The true value of the incident

effective radiant power is calculated from the exposure value, corrected with the Eppley calibration curve, minus the corrected dark value. The actual value does not yet correspond exactly to the desired radiant power. The measurement is repeated, until the actual and the desired radiant power values agree. The actual purpose of this radiant power measurement is the subsequent calibration of a solar cell and the determination of the photocurrent sensitivity of Si photodiodes and phototransistors.

12.5
Measurement of irradiance of a standard light A radiation with Si photodetectors

Irradiance measurements are necessary to determine the photocurrent or the photosensitivity of a photodetector. The photosensitivity of Si photodiodes and Si phototransistors is related to given irradiance values. Sensitive phototransistors are measured with low irradiances such as 1 mW/cm^2, 2 mW/cm^2, 5 mW/cm^2 and 9 mW/cm^2, and less sensitive phototransistors and photodiodes with higher irradiances such as 20 mW/cm^2.

From the preceding measurement of radiant power with the thermopile, the irradiance is the quotient of the radiant power and the effective receiving area. With the thermopile, the required irradiance, corresponding to the data in the data-sheet of the photodiode or phototransistor typs, is determined exactly at the relevant distance from the radiation source. The photocurrent sensitivity of the corresponding photodiode or phototransistor is measured at the same position originally occupied by the thermopile.

For a series of measurements in the test department, such radiation measurements are very time-consuming. In addition, the large requirement for calibrated measurement detectors demands very inexpensive equipment designs. In the test department, Texas Instruments use calibrated Si solar cells as measurement detectors. The spectral

238

sensitivity of Si solar cells and the spectral emission of the standard light A radiation source can be regarded as constant. The proportion of the radiation which can be evaluated by the Si solar cell also remains constant. Therefore, the actual irradiance can be determined from the relative measurement with an Si solar cell.

The radiation measurements in the same arrangement with Si solar cells are less troublesome, since the longer-wavelength radiation from the filament lamp is not evaluated. Also, the Si solar cells are unaffected by their previous history, so that the photocurrent sensitivity has not been impaired by previous measurements with high irradiances. The value of the short-circuit current of a calibrated solar cell can be reproduced exactly with the same irradiance. The short-circuit is measured with the same low-resistance microammeter. The voltage drop on the microammeter should only be 10 mV. If large-area solar cells ($A > 1$ cm^2) are measured with irradiances above 5 mW/cm^2, then a voltage drop on the microammeter up to 100 mV can be permitted. It is convenient, almost to short-circuit the solar cell with a low resistance (e.g., 10 Ω). The short-circuit current is measured through the voltage drop across this resistance, with a digital voltmeter.

The calibration of the solar cell only needs to be checked infrequently, since it has high long-term constancy. The calibration itself is carried out with the thermopile, by measuring the irradiance for various distances from the filament of the standard light A radiation source. The thermopile is then replaced by the solar cell in the same position for each measurement point.

For each measurement point, the short-circuit current of the solar cell and the irradiance, determined with the thermopile, are plotted in a graph. As an example, *Figure 12.8* shows the short-circuit current of a solar cell as a function of the irradiance. If the radiation source can still be regarded as a point source from each measurement

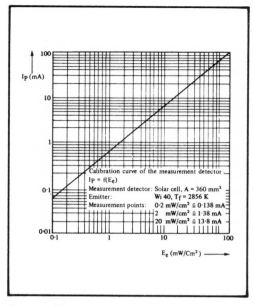

Figure 12.8
Short-circuit current Ip of a Si solar cell as a function of the irradiance E_e. If the short-circuit current Ip of the solar cell is not a linear function of the irradiance E_e, then the solar cell is defective or a measurement error has occurred

point and the solar cell has a strictly linear relationship between its short-circuit current Ip and the irradiance E_e, then the values of the measured short-circuit current Ip = f(E_e) produce a straight line. Further measurement points can also be calculated with the inverse-square law.

Si photodiodes or phototransistors could also be considered as calibrated detectors. Of course, phototransistors can only be calibrated for fixed measurement points, since their non-linearity I_C = f(E_e) causes severe disturbance. Photodiodes are more suitable. They have a linear relationship for Ip = f(E_e), at least over a wide range. However, because their sensitive areas are mostly very small, these photodetectors require very careful adjustment on the optical bench. The wafer should be centred in the component within close tolerances. Built-in lenses are undesirable, in order to avoid possible misalignment effects.

Components with flat glass windows should be used, since glass windows can be cleaned without damage, in constrast to plastic windows. Furthermore, components with metal cases are preferable for this purpose, in order to obtain a low thermal resistance between the junction and the case. The wafer of a photodiode, which is heated by the long-wavelength radiation from the lamp, is cooled better. Therefore measurement errors due to the temperature rise of the wafer can be neglected with short measurement times.

Calibrated photodiodes or phototransistors are only used more advantageously for measurement and calculation of the tolerances of the photocurrent for prototypes in the laboratory. They do not form a good substitute for a radiometer. If necessary, a radiometer can be built with a commercially-available Si solar cell. As previously described, the voltage drop across the short-circuiting resistance of $R = 10 \; \Omega$ is measured with a digital voltmeter. This arrangement is quite suitable for relative measurements. For absolute measurements and in case of simple requirements, the Si solar cell can be calibrated with a calibrated photodiode. Regarding the previously-mentioned standard light A source as a point source, a calibration curve $I_p = f(E_e)$ can be prepared for the Si solar cell used with the inverse-square law. In this connection, it is explicitly pointed out, that all commercially available radiation measuring instruments with silicon photodetectors, themselves have to be calibrated to the standard light A radiation with the methods described.

12.6
Measurement of a luminescence diode radiation with Si photodetectors

12.6.1
General Measurement Problems

Radiation measurements can be carried out with commercially available measuring instruments and also with simple measuring detectors. The commercial measuring instruments mostly contain a PIN photodiode as a measurement detector, while commercially available Si solar cells or, if necessary, Si photodiodes are suitable for use as simple measurement detectors. Silicon photodetectors are used as measurement detectors for the following reasons:

a
The selectively emitted radiation from current luminescence diodes falls in the spectral sensitivity range of Si photodetectors.

b
Si photodetectors are very sensitive, so that very small radiant powers can be measured.

c
Si photodetectors have high long-term constancy and are not dependent on their previous history.

d
Practical radiation measurements with Si photodetectors are simpler to carry out, as compared with those with thermal photodetectors.

The Si photodetectors of commercial measuring instruments are obtainable with attached subtraction filters.

These spectrally-corrected Si photodetectors are divided into three groups (see *Figure 12.9*):

1
For measurement of radiation in radiometric values within the wavelength range from 450 nm to 950 nm with tolerances of approximately $\pm 5\%$ (see curve a). With such photodetectors, the measurement of Si-doped GaAs diodes is difficult, since the measuring detector no longer covers 100% of the spectral radiation distribution of the GaAs diode. With an exact spectral

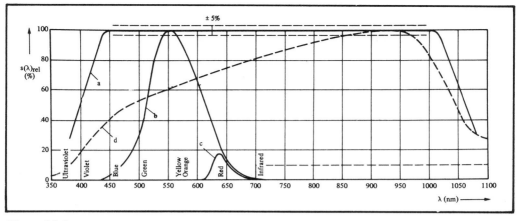

Figure 12.9
Spectral sensitivities of various corrected Si photodetectors

sensitivity distribution curve and the exact spectral radiation distribution of the GaAs diode, a correction factor can be determined graphically.

2

For measurement of mixed radiation in photometric units within the photometric sensitivity curve of the eye, $V(\lambda)$, corresponding to curve b. This mainly means use as a Luxmeter.

3

For measurement of monochromatic or selective radiation in photometric units for a small spectral range within the photopic sensitivity curve of the eye, $V(\lambda)$. With these photodetectors, for example, the selective radiation of luminescence diodes is measured. Curve c shows the spectrally corrected sensitivity for measurement of red-emitting GaAsP diodes. For comparison, curve d reproduces the relative sensitivity $s(\lambda)$ of an Si PIN diode without a filter.

The measurement of medium radiant powers can be carried out with a neutral, calibrated grey filter placed before the Si photodetector. Very high radiant powers (> 1 W) are determined with an optical beam switch connected before the Si photodetector, since grey filters heat up with high radiant

powers. A beam splitter, with a defined division ratio, for example, is an optical beam switch.

A measuring photodetector with a sufficiently large photosensitive area A_E always measures the radiant power Φ_e or the luminous power Φ_V of an incident, narrowly-focussed, parallel beam or radiation, e.g., from laser sources. From radiation sources arranged as a point radiator, a photodetector only ever measures the irradiance E_e or the illuminance E_V. The radiant intensity or luminous intensity of a point source of radiation in the normal direction is calculated with the equation (3.25) (see Section 3.3), while with the measurement arrangements which are usual in practice, $\cos\varphi_E$ can be made $= 1$.

$$I = \frac{E.r^2}{\Omega_o} \tag{12.3}$$

The radiance or luminance in the normal direction is calculated according to equation (3.19).

$$L = \frac{I}{A_S} \tag{3.19}$$

In most cases, the radiance or luminance can only be determined with very wide tol-

241

erances, because the effective emitting area A_S of luminescence diodes is not exactly known. Only in the case of high-quality GaAs power diodes is the emitting area A_S known exactly.

Luminescence diode arrays for illumination or irradiation purposes often have to be dealt with as surface emitters. To characterise surface emitters, the irradiance values are stated for planes at various distances.

12.6.2
Measurements of relative spectral sensitivity with a monochromator

The relative spectral sensitivity $s(\lambda)_{rel}$ of an Si photodetector can be determined with a monochromator. A monochromator can select, from the wide-band emission spectrum of a Wi lamp at, for example, $T_f = 2856$ K, a narrow-band radiation in the wavelength interval $\Delta\lambda$.

In principle, a simple monochromator consists of the inlet slit, a prism, lenses or mirrors for parallel alignment or focussing or a radiation and the outlet slit. The incident radiation through the inlet slit is split up through the wavelength-dependent refraction of the prism. The arrangement of the outlet slit permits selection of the radiation within very small wavelength ranges.

Texas Instruments often use the double monochromator, Model 99, from Perkin Elmer. *Figure 2.10* shows, schematically, the optical path of the incoming radiation:

Inlet slit S 1, parabolic mirror M 1, prism PR, Littrow mirror M 2, prism PR, parabolic mirror M 1, mirror M 3, divided mirror M 4, chopper CH, mirror M 3, parabolic mirror M 1, prism PR, Littrow mirror M 2, prism PR, parabolic mirror M 1, mirror M 5 and outlet slit S 2.

The chopper CH interrupts the beam for its second passage through the monochromator.

The monochromatic beam of the desired wavelength interval $\Delta\lambda$ can be determined exactly by measurement through its chopper frequency, in comparison with the beam which emerges after the first pass with an unwanted wavelength range.

The relative spectral sensitivity $s(\lambda)_{rel}$ of a Silicon photodetector is determined in two steps. First the spectral radiation distribution N_λ of the Wi lamp is measured at $T_f = 2056$ K. With the radiation falling through the inlet slit of the monochromator and the thermopile arranged at the outlet slit, the spectral radiation distribution N_λ is recorded for each wavelength interval $\Delta\lambda$. The thermopile used must have a flat spectral sensitivity curve between 300 nm and 1300 nm.

Following this, in the same measurement arrangement and with the same radiation N_λ, the spectral sensitivity $s(N)_\lambda$ of the Si photodetector used is measured for each wavelength interval $\Delta\lambda$. The ratio $s(N)_\lambda$ to N_λ for each wavelength interval $\Delta\lambda$ gives the relative spectral sensitivity $s(\lambda)_{rel}$ of the Si photodetector.

The measured relative spectral sensitivity values for an Si solar cell are listed in *Table 12.1*.

Figure 12.11 shows the function $s(\lambda)_{rel} = f(\lambda)$ of the Si solar cell used, derived from these values.

12.6.3
Irradiance measurement for luminescence diode radiation with Si photodetectors

A luminescence diode radiation can only be evaluated by measurement with an Si photodetector, if the absolute spectral sensitivity distribution $s(\lambda) = f(\lambda)$ is known. However, among commercially available Si photodetectors, this is only stated in the data sheet for the large-area photodiodes and for solar cells. For radiation measurements, a large-area photodetector is usually desired.

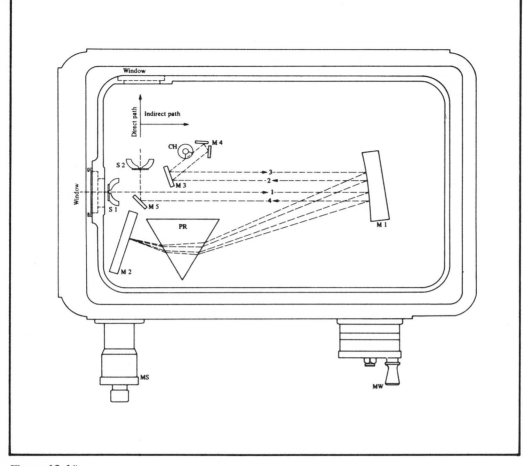

Figure 12.10
Construction principle of the double monochromator, model 99, from Perkin-Elmer

When the requirements are not severe, the data-sheet characteristic $s(\lambda) = f(\lambda)$ is also adequate for measurement purposes. The selective radiant power of a luminescence diode, falling on the photodetector, can be read off from the function $s(\lambda) = f(\lambda)$ with the measured photocurrent I_P of an Si photodiode or the measured short-circuit current I_P of a solar cell and with the known maximum wavelength λ_{max} of the selective luminescence radiation. If the measurement detector has a photosensitive area of $A_E = 1\ cm^2$, the incident radiant power is equal to the irradiance for 1 cm² reference area.

As a rule, the data sheets on small-area Si photodiodes contain the relative spectral sensitivity distribution $s(\lambda)_{rel} = f(\lambda)$. In the same way, only the relative spectral sensitivity distribution $s(\lambda) = f(\lambda)$ is obtained initially if, for exact spectral measurements of large-area Si photodiodes or Si solar cells, the spectral sensitivity distribution is determined with a monochromator (see previous Section 12.6.2). In principle, the relative spectral sensitivity distribution curve of a Si photodetector can be converted into the absolute spectral sensitivity $s(\lambda)$ with one absolute spectral sensitivity measurement at a wavelength measure-

243

Column 1	Column 2	Column 3	Column 4	Column 5
λ	$s(\lambda)_{rel}$	$s(\lambda)$	$Q(\lambda)$	$s(\lambda)_{th}$
nm		A/W	%	A/W
1·100	12·8	0·07	7·9	0·886
1·050	31·9	0·175	20·7	0·845
1·000	61·1	0·335	41·5	0·806
950	85·4	0·468	61·1	0·766
900	97·6	0·535	72·5	0·737
850	100·0	0·548	80·0	0·684
800	97·9	0·537	83·2	0·644
750	92·9	0·509	84·2	0·604
700	85·2	0·467	82·7	0·564
650	76·6	0·420	80·1	0·524
600	66·9	0·367	76·5	0·483
550	56·0	0·305	69·2	0·443
500	44·3	0·243	60·2	0·403
450	31·7	0·174	47·9	0·363
400	18·2	0·100	31·0	0·322

Table 12.1
Spectral measured values of an Si solar cell

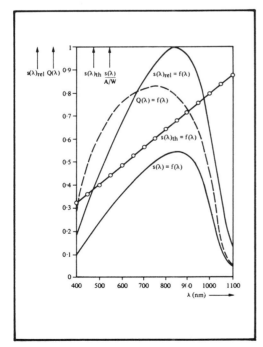

Figure 12.11
Spectral characteristics of a solar cell

ment λ_{Meas}. In simplified form, it is sufficient to measure the absolute sensitivity with a spontaneously emitting GaAs diode with known radiant intensity I_e and maximum wavelength λ_{max}. Somewhat more accurate measured results are to be expected, if the spectral radiation distribution of the GaAs diode used is measured with a monochromator and the part of the radiation evaluated by the Si photodetector is then determined graphically.

It is advantageous to measure the absolute spectral sensitivity $s(\lambda)$ of the Si photodetector at several different wavelengths. Texas Instruments measure the absolute spectral sensitivity $s(\lambda)$ of Si photodetectors with an argon laser at $\lambda = 514$ nm, with a HeNe laser at $\lambda = 632.8$ nm and with a spontaneously emitting GaAs diode at $\lambda = 910$ nm. The absolute spectral sensitivity values of the Si solar cell used are listed in Table 12.1, column 3. Figure 12.11 shows the function $s(\lambda) = f(\lambda)$ plotted.

12.7

Measurement of the total radiant power of a luminescence diode

The total emitted radiant power of a luminescence diode can be determined with an integrating radiant power meter. The Ulbricht spherical photometer, for measurement of total luminous flux, is well known from illumination engineering (see *Figure 12.12*). The spherical photometer belongs to the category of integral photometers, since it integrates all partial luminous fluxes. The light emitted from the measurement lamp is scattered and reflected on the matt white interior wall of the sphere, so that every surface element has the same remitted luminance. A small aperture in the sphere permits the measurement, with a sensitive photometer, of the remitted luminance of the aperture area. Direct radiation from the light source onto the aperture is prevented by a screen B. For the measurement of the radiant power of IR-emitting GaAs diodes, spherical photometers are made from small aluminium hollow spheres.

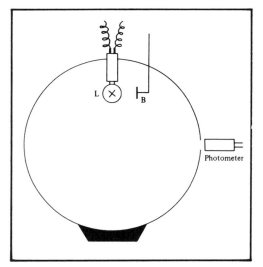

Figure 12.12
Construction principle of a spherical photometer

For the serial measurement of luminescence diodes, another measurement method has become established (see *Figure 12.13*).

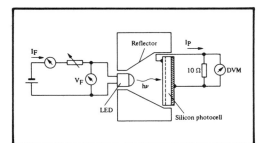

Figure 12.13
Measurement of the total radiant power of a luminescence diode with an Si solar cell

A luminescence diode is arranged opposite a large-area Si solar cell. In addition, this diode is located in a special reflector, so that the total emitted radiation falls on the Si solar cell. With the measured short-circuit current Ip and the absolute spectral sensitivity value (A/W) read off from the characteristic $s(\lambda) = f(\lambda)$ at the maximum wavelength λ_{max} of the luminescence radiation, the radiant power can be calculated in accordance with the rearranged equation (9.25).

$$\Phi_e = \frac{I_P}{s(\lambda)} \tag{12.4}$$

The quantum efficiency $Q(\lambda)$ of the luminescence diode, which is also of interest, can be calculated according to the equation (11.5):

$$Q(\lambda)_D = \frac{I_P}{I_F \cdot Q(\lambda)_{SC}} \tag{11.5}$$

The quantum efficiency $Q(\lambda)_{SC}$ of the solar cell, which is necessary for this, can be determined by means of equations (9.9) and (9.4):

$$Q(\lambda)_{SC} = \frac{s(\lambda)}{s(\lambda)_{th}} \tag{9.9}$$

$$Q(\lambda)_{SC} = \frac{I_P}{\Phi_{e,\lambda} \cdot \lambda} \cdot \frac{1 \cdot 24 \text{ mW}}{10^6 \text{ A}} \tag{9.4}$$

To complete the spectral parameters of the solar cell used, the values for $s(\lambda)_{th}$ and $Q(\lambda)_{SC}$ are therefore entered in the *Table 12.1* and shown in *Figure 12.11* as characteristic curves.

245

13
Optoelectronic couplers

Optoelectronic couplers or source/detector systems are used in many varied forms in almost all branches of electronics. For example, the so-called twilight switches are simple examples of such systems. Here, the natural daylight serves as the source radiation. The actual twilight switch is the detector. Optoelectronic rangefinders, for example, are more complicated source/detector systems. Spontaneously emitting GaAs diodes or lasers are used as the radiation source.

The small couplers described in Section 10.3 relate to very short source-detector distances, which lie below the critical photometric distance. Those for short, medium or long ranges relate to source-detector distances above the photometric critical distance. For the radiation calculation of couplers or systems, the fundamental principles worked out in Part I are needed, with the parameters of optoelectronic semiconductor components described in Part II and with practical radiation measurements.

13.1
Direct Couplers

By the term "direct couplers", it is understood that the photodetector is irradiated directly by the radiation source. In most practical applications, the normal to the source coincides with the normal to the detector. This means, that the photodetector receives the radiation in its major direction of reception and this is emitted in the major direction of the source. *Figure 13.1* shows the principle of direct couplers. The irradiance E_e falling on the photodetector can be calculated for non-parallel source emission by equation (3.25), where $\cos\varphi_E$ is made = 1.

$$E_e = \frac{I_e}{r^2} \cdot \Omega_o \qquad (3.25)$$

If, as shown in *Figure 13.2*, the source S is aligned at an angle φ_S to the detector E, then its radiant intensity $I_{e,\varphi}$ can be calculated with the value of $I_{e,o}$ in the normal to the source and the function $f(\varphi_S)$ stated in the data sheet. The angle φ_S is formed between the normal to the source and the

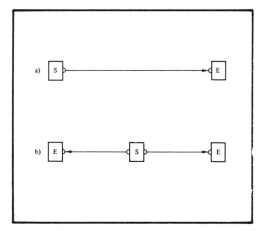

Figure 13.1
Principle of a direct coupler:
a) Single coupler
b) Double coupler

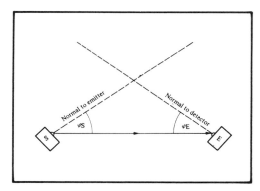

Figure 13.2
Direct coupler with a source turned through the angle φ_S and a detector turned through the angle φ_E.

direction from the source to the detector.

$$I_{e,\varphi} = I_{e,o} \cdot f(\varphi_S) \qquad (10.12)$$

The simplest method of determining the radiation falling on the photodetector from a parallel beam of radiation from the source is by measurement.

The photocurrent Ip of a photodetector working in the linear part of the characteristic Ip = $f(E_e)$ can be calculated through the rearranged equation (9.25).

$$I_P = s \cdot E_e \qquad (13.1)$$

For photodetectors working non-linearly, the amplified photocurrent I_M or I_C is read off the characteristic curve I = $f(E_e)$ in the data sheet.

If, as in *Figure 13.2*, the detector E is aligned at an angle φ_E to the source, then its sensitivity s_φ can be calculated with the value of s_0 in the normal to the detector and the function $f(\varphi_E)$ stated in the data sheet. The angle φ_E is formed between the normal to the detector and the direction from the detector to the source.

$$s_\varphi = s_0 \cdot f(\varphi_E) \qquad (13.2)$$

13.2
Reflection Optoelectronic Couplers

Reflection couplers differ from direct systems in that the radiation from the source falls on the detector via a beam-deflecting device. *Figure 13.3* shows the principle of a reflection coupler. To deflect the beam from the source, mirrors, triple reflectors, diffuse reflecting surfaces or any desired materials are used. The reflectivity φ of commercially available reflectors can be taken from the data sheet, that of known substances from tables or characteristic curves. The radiance returned diffusely from a reflector was calculated in Section 2.3.3. Diffuse reflection was described in Section 6.4.

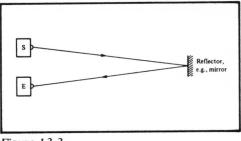

Figure 13.3
Principle of a reflection coupler

A beam of rays from a source can also be reflected onto the photodetector through several mirrors in series. Mirrors reflect an incident ray in accordance with the law of reflection. (See Section 6.4). They must be adjusted exactly. Their sensitivity to vibration causes serious problems.

Instead of direct couplers, reflection couplers can also be used by the back-reflection method. *Figure 13.4* shows the

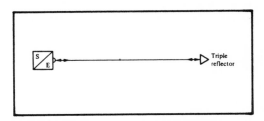

Figure 13.4
Principle of a back-reflection (autocollimation) coupler

principle of this variation, known as the back-reflection or autocollimation coupler. The source and the detector are mounted in one housing. Triple reflectors (back-reflectors) are used instead of mirrors to reverse the beam. A triple reflector throws the radiation back to its point of origin. In comparison with a mirror, it can be mis-aligned by a few degrees. The triple reflector is an exact corner of a cube, made of a transparent medium, e.g., glass. The back-reflection takes place almost without losses through total reflection. *Figures 13.5* and *13.6* show the principle of a triple reflector.

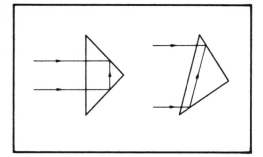

Figure 13.5
Principle of back-reflection with triple
reflectors, in two-dimensional
representation

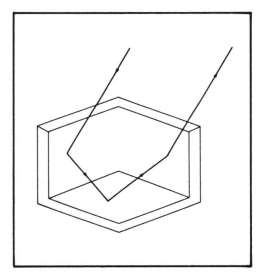

Figure 13.6
Principle of back-reflection with a corner of
a cube, to be considered as reflective, in three-
dimensional representation

A further important optical component in
back-reflection couplers is the optical
switch. It is needed to separate the emitted
and received radiation, since these both
occupy the same or almost the same
radiation space ("light tube"). The optical
switch then prevents an optical short-circuit,
so that the source does not irradiate the
photodetector directly. Radiation losses
always occur at an optical switch
(*Figure 13.7*).

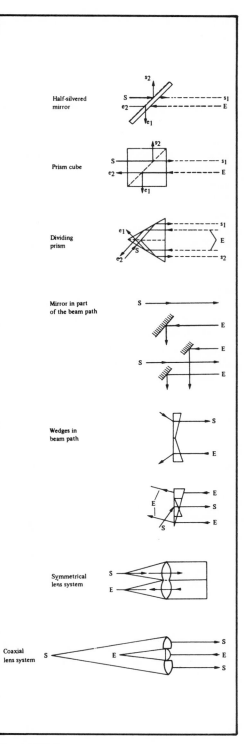

Figure 13.7
Principle of optical switches

251

13.3
Couplers with non-stationary source emission

Reflection couplers on the back-reflection principle are also operated with non-stationary source emission. This means that the beam from the source is usually moved periodically over a given deflection range. *Figure 13.8* shows the principle of a back-reflection coupler with non-stationary source emission. The focussed radiation from the source strikes the triple reflector via the rotating mirror and the parabolic mirror. The radiation thrown back from the triple reflector passes in the opposite direction along the same beam path to the optical switch and then to the photodetector. In this case, the optical switch is a partially-transparent mirror. The rotating mirror guides the emitted beam periodically over the preset deflection range (moving beam principle). A light beam is produced, and can be used, for example, for accident-prevention on machines or for measuring the length of materials.

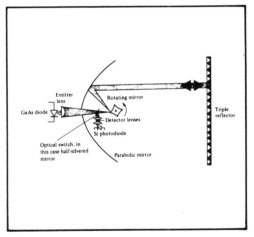

Figure 13.8
Principle of a back-reflection coupler with non-stationary radiation

13.4
Simple examples of couplers

Optoelectronic components are generally selected according to the operating conditions needed, the most suitable constructional form for the particular application, the quality and quantity supplied and the purchase price.

The radiant power and radiant intensity of a luminescence diode, and the sensitivity of a photodetector are determined according to radiometric considerations and calculations.

In Chapter 3 and Section 10.1, the solid-angle and radiant intensity calculations of luminescence diodes with approximately spherical radiation distributions were described. For this, the relationship

$$\Omega_K = 2\pi(1-\cos\varphi) \cdot \Omega_o \qquad (3.4)$$

applies. With the minimum radiant power $\Phi_{e,min}$ of a luminescence diode, stated in the data sheet, the radiant intensity $I_{e,o,min}$ is calculated.

$$I_{e,o,min} = \frac{\Phi_{e,min}}{\Omega_K} \qquad (10.16)$$

The radiant intensity $I_{e,o}$ of luminescence diodes with narrow emission peaks is determined more advantageously by means of an irradiance measurement in the normal to the source and with equation (2.40). The photometric limiting distance of the measurement detector from the luminescence diode is to be maintained as a minimum.

$$E_e = \frac{\Phi_e}{A_E} \qquad (2.40)$$

The radiant intensity tolerances of luminescence diodes can be determined roughly, equally well with the tolerance calculation described in Section 10.2 or by batch measurements. In *Table 13.1*, the radiant intensity values of various GaAs diodes are listed.

Photodetectors show different sensitivities

| Type | Radiant intensity I_e in mW/sr | |
	Calculated min. values	Typical measured values
TIL 23		1
TIL 24		2
TIL 31		25
TIL 32		0.5
TIL 06	0·21	
TIXL 12	11	15
TIXL 13	5·5	7
TIXL 14	16·5	22
TIXL 15	8·3	11
TIXL 16	43	46
SL 1183	1300	1600
SL 1278	1	3
TIXL 26	0·26	
TIXL 27	3·9	5

Table 13.1
Radiant intensity table for GaAs diodes
(For test conditions, see data sheets)

according to the type of incident radiation. The sensitivity of Si photodetectors is stated, in accordance with the data-sheet test conditions, for the tungsten filament lamp radiation with a colour temperature of T_F = 2856 K or 2870 K (2856 K = Standard light A). The data-sheet characteristic curves $I_P = f(E_e)$ and $I_E = f(U_{CE})$, with the variable E_e, can only be evaluated for this type of radiation down to a minimum irradiance of approx. E_e = 1 mW/cm².

For photodiodes and photocells, the stated absolute sensitivity s or the characteristic curve $I_P = f(E_e)$ can usually be extrapolated linearly to smaller sensitivity values (see Section 9.4). The absolute sensitivity s_M of a phototransistor is calculated approximately with the corresponding typical characteristic curve $B_{rel} = f(I_C)$ (See Section 9.6).

The sensitivity s for different radiation to that stated in the data sheet, e.g., radiation Z, is calculated for photodiodes and phototransistors with the relevant actinic value $a_e(Z)$ and, for phototransistors, also

with the corresponding typical characteristic $B_{rel} = f(I_C)$ (See Section 9.8).

For exact tolerance calculations, the characteristic $B_{rel} = f(I_C)$ and the actinic value a_e of the radiation used are to be determined for the photodetector selected.

For practical radiation calculations, the typical sensitivity $s(LIR)_{930nm}$ for the luminescence radiation LIR_{930nm} of a Si-doped GaAs diode has been determined with various Si photodetectors. *Table 13.2* shows the typical photocurrents or collector currents for $s(LIR)_{930nm}$ and for various irradiances E_e. The numerical value or I_C at E_e = 1 mW/cm² corresponds, for example, to the typical sensitivity $s(LIR)_{930nm}$ in

$$\frac{\mu A}{mW/cm^2}$$

The expected typical collector current I_C of a detector in an opto coupler will be calculated with the values from *Tables 13.1* and *13.2*.

Example:
A coupler is to be designed approximately with the GaAs diode TIL 31 and the phototransistor TIL 81, for a distance r = 6 cm. From *Table 13.1*, we obtain $I_{e,typ,TIL31}$ = 25 mW/sr. The irradiance falling on the phototransistor amounts to:

$$E_e = \frac{I_e}{r^2} \cdot \Omega_o = \frac{25 \text{ mW.sr}}{sr.6.6.cm^2}$$

$$= 0.694 \text{ mW/cm}^2 \qquad (13.3)$$

According to *Table 13.2*, the interpolated collector current is

$I_{C,TIL81}$ = 4.43 mA.

In an accurately adjusted coupler, with a pair, TIL 31 and TIL 81, selected according to the data, this value can be reproduced fairly accurately.

Type	$\dfrac{I_{C(LIR)}}{\mu A} = s(LIR)_{typ}.E_e$	Irradiance E_e		
		1 mWcm^{-2}	0.5 mWcm^{-2}	0.1 mWcm^{-2}
H 62	Photodiode	44 μA	22 μA	4.4 μA
TIL 64	Phototransistor	77 μA	36 μA	6 μA
TIL 67	Phototransistor	1300 μA	645 μA	120 μA
TIL 78	Phototransistor	1030 μA	505 μA	90 μA
TIL 81	Phototransistor	6218 μA	3235 μA	610 μA
LS 613	Phototransistor	258 μA	116 μA	20 μA
LS 614	Phototransistor	342 μA	157 μA	28 μA

Table 13.2
Typical photocurrents or collector currents for s(LIR)930nm of Si photodiodes and Si phototransistors.

13.5
Opto couplers with lenses

The distance to be spanned by an optical coupler is primarily dependent on the optical aids used. Suitable reflectors, lenses or lens combinations in the beam path of the source and the detector cause a considerable increase in the radiant power falling on the photodetector.

It is only a secondary factor that the radiation source must have a sufficiently high radiant power, in order to produce an adequate irradiance for the photodetector with the optical aids selected.

Lastly, the photo-current sensitivity of the detector must be mentioned. It can be compensated for within wide limits by the subsequent amplifier stage.

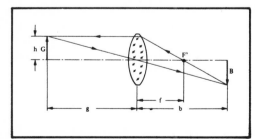

Figure 13.9
Formation of image of an object by a double-convex lens

Figure 13.9 shows an object, an image of which is formed by a double-convex lens. The focal length f of the lens permits the calculation of the most important relationships with relation to the position and size of an object and the position and size of the image.

The general equation for a lens reads:

$$\frac{1}{f} = \frac{1}{b} + \frac{1}{g} \qquad (13.4)$$

The magnification ratio will be calculated:

$$V = \frac{B}{G} = \frac{b}{g} \qquad (13.5)$$

In these: f = focal length, b = distance to image, g = distance to object, h = height of incident ray, B = image size, F = focus on image side, G = size of object and V = magnification ratio.

For the formation of a useful image, single lenses are only suitable with very small aperture ratios. The aperture ratio is the ratio of the diameter d of the inlet aperture to the focal length f of the lens. The diameter d of the inlet aperture is determined either by the diameter d of the beam of radiation or by the diameter d of an aperture mask which limits the beam (see

254

Figure 13.12). The full aperture corresponds to the diameter d of the lens (see *Figure 13.13*). The quality of the image from single lenses is only sufficient for demands of low sensitivity, even with a very small aperture ratio.

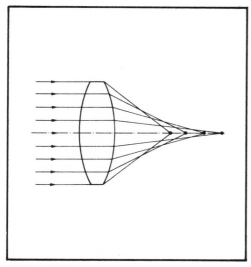

Figure 13.10
Occurrence of spherical aberration

With a larger image height h of the incident radiation, the image points move on the optical axis away from the focus F' (see *Figures 13.9* and *13.10*). This aperture error is called spherical aberration; it can be reduced to a certain degree by the shape of the lens, if the refraction of the rays is divided as equally as possible between the two lens surfaces. The lenses corrected in this way only have an aberration minimum for a single direction of incidence of the radiation and only for a single magnification ratio.

With a plano-convex lens, the refractions are most favourably distributed, if rays parallel to the axis fall on the convex side of the lens (see *Figure 13.11*).

Ideally, the emitted radiation of a coupler should form an image of the radiation source at infinity. In this way, a beam of rays parallel to the axis is obtained from a point

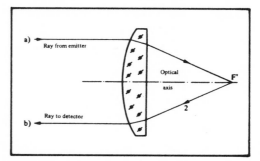

Figure 13.11
Parallel radiation through a plano-convex lens; a) for an incoming received ray, b) for an outgoing emitted ray.

source. Similarly, in the ideal case, an image of a radiation source at an infinite distance should be formed on the photodetector.

The radiant power of the image of the source on the photodetector is firstly proportional to the effective receiving lens area A_E, which is limited by the aperture or by the cross-section of the beam of rays from the source, and secondly it is inversely proportional to the square of the image distance b. The effective receiving lens area A_E is proportional to the square of the diameter d of the inlet aperture. With a source at an infinite distance, the image distance b is equal to the focal length f.

For this case, the radiant power Φ_e of the image of the source falling on the photodetector is proportional to $(d/f)^2$. The fraction d/f represents the relative aperture or the aperture ratio. The aperture ratio d/f is expressed by twice the tangent of the half aperture angle α. The half aperture angle α is called the aperture. The reciprocal of the aperture ratio d/f gives the stop number of the lens:

$$2 \tan \frac{\alpha}{2} = \frac{d}{f} \quad (13.6) \qquad n = \frac{f}{d} \quad (13.7)$$

Figure 13.12 shows the relative aperture of a lens, fixed by an aperture stop. In *Figure 13.13*, the relative aperture of a lens with full stop setting is shown. Finally,

255

Figure 13.14 shows lenses with equal relative aperture at full stop setting.

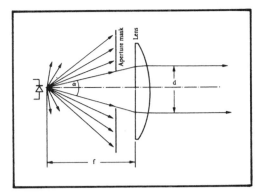

Figure 13.12
Determination of the relative aperture with an aperture stop; the angle α is the aperture angle

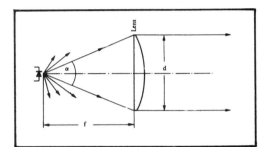

Figure 13.13
Determination of the relative aperture at full stop setting

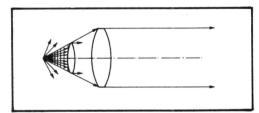

Figure 13.14
Lenses with equal relative aperture at full stop setting

In practice, if requirements are not very severe, both the production of the parallel beam and its focussing on the photodetector are carried out with only one lens in each case.

The front lens to suit an optoelectronic component depends in each case on the specification and on the aperture angle Θ_{HP} of the emission or reception peak. When selecting a lens, the aperture ratio relates at first to the full aperture of the lens. Narrow peaks have a low ray divergence, so that planoconvex lenses with small aperture ratio are conveniently selected.

If an image of the source is to be formed, relatively accurately, on the photodetector, irrespective of its emission peak, then planoconvex lenses with small aperture ratio are also needed. Spherical maxima have a wide divergence, so that aspherical condenser lenses are used to advantage. A large aperture ratio of the aspherical lens is to be selected, in order to cover a large part of the spherical peak. The whole spherical maxima can be covered with suitable concave mirrors, but not with lenses. Since an exactly parallel beam of radiation cannot be produced with a single lens, the lens for optoelectronic components with flat windows is to be adjusted to form an image of the wafer on the distant object plane (see *Figure 13.15*)

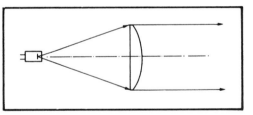

Figure 13.15
The front lens for an optoelectronic component with a flat window is adjusted to form an image of the wafer on the distant object plane

For an optoelectronic component with built-in lens, the attached lens is either to be adjusted to form an image of the wafer or of the edge of the lens on the distant object plane. (See *Figures 13.16* and *13.17*.)

In general, large aperture ratios·d/f produce a beam which also has a large diameter. Opto couplers designed in this way are uncritical in their construction and can be

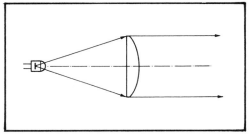

Figure 13.16
Here, the front lens for an optoelectronic component with built-in lens is adjusted to form an image of the wafer on the distant object plane

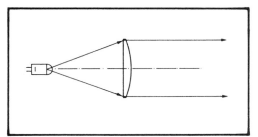

Figure 13.17
Here, the front lens for an optoelectronic component with built-in lens is adjusted to form an image of the edge of the lens on the distant object plane

S ◐ ‖ ➤ r ———————————➤ ◑ ‖ E

Source: GaAs diode TIL 31	Dis-tance	Detector: Phototransistor TIL 81	Collector current of the TIL 81
Parameters: $t_G = 25°C$; $I_F = 100$ mA		$s(PA)_{2870K} = 14.7$ mA/5 mW/cm² $\qquad V_{CE} = 5V$	
Main peak: $I_E = 21$ mW/sr		$E_{e,TIL31} = 0·1$ mW/cm² $\quad 0·5$ mW/cm² $\quad 1$ mW/cm²	
Total emission: $\Phi_e = 3·72$ mW		$I_C =$ $\quad$ 0·72 mA $\quad\quad$ 2·9 mA $\quad\quad$ 8·3 mA	

Source lens	$\dfrac{r}{m}$	Receiver lens	$I_{C,TIL81}$
Dimensions in mm		Dimensions in mm	mA
d = 42 f = 50 $\quad$ Double convex	1	d = 31.5 $\quad$ Aspherical f = 25 $\quad$ condenser lens	1.2
d = 22.4 $\quad$ Aspherical f = 18 $\quad$ condenser lens	1	d = 22·4 $\quad$ Aspherical f = 18 $\quad$ condenser lens	0.066
d = 22.4 $\quad$ Aspherical f = 18 $\quad$ condenser lens	1	d = 31·5 $\quad$ Aspherical f = 25 $\quad$ condenser lens	0.165
d = 31.5 $\quad$ A erical f = 25 $\quad$ condenser lens	1	d = 31·5 $\quad$ Aspherical f = 25 $\quad$ condenser lens	0.190
d = 22.4 $\quad$ Aspherical f = 18 $\quad$ condenser lens	1.20	d = 22·4 $\quad$ Aspherical f = 18 $\quad$ condenser lens	0.043
d = 42 f = 50 $\quad$ Planoconvex	2.50	d = 42 f = 50 $\quad$ Planoconvex	0.200
d = 31.5 $\quad$ Aspherical f = 25 $\quad$ condenser lens	3.00	d = 31·5 $\quad$ Aspherical f = 25 $\quad$ condenser lens	0.025
d = 62 f = 190 $\quad$ Double convex	3.00	d = 64 f = 150 $\quad$ Double convex	2.2
d = 62 f = 190 $\quad$ Double convex	3.50	d = 31·5 $\quad$ Aspherical f = 25 $\quad$ condenser lens	0.450
d = 62 f = 190 $\quad$ Double convex	5.25	d = 64 f = 150 $\quad$ Double convex	8.0
d = 62 f = 190 $\quad$ Double conves	11.0	d = 64 f = 150 $\quad$ Double convex	3.0
The forward current $I_F = 100$ mA was raised to $I_F = 200$ mA (permissible at $t_G = 25°C$)	11.0		6.5

Table 13.3
Examples of direct opto couplers with a calibrated GaAs diode TIL 31 and a calibrated Si phototransistor TIL 81

257

very easily adjusted. Small aperture ratios d/f also give a beam of small diameter. Depending on their cost, these couplers are sensitive to vibration and need precision adjusting devices.

The examples listed in *Tables 13.3* and *13.4* were constructed with lenses from Spindler and Hoyer. This firm is one of the few which supplies individual lenses, even in small batches. The well-proven GaAs diode TIL 31 was chosen as the source. With a housing temperature $t_G = 25°C$ and a diode forward current $I_F = 100$ mA, a radiant intensity $I_{e,o} = 21$ mW/sr in the normal direction and a total radiated power $\Phi_e = 3.72$ mW were measured on the specimen. As the detector, the phototransistor TIL 81, which matches the TIL 31, was used. The housing temperature was $t_G = 25°C$ and the collector-emitter voltage $V_{CE} = 5$ V. The specimen used had the following sensitivities:

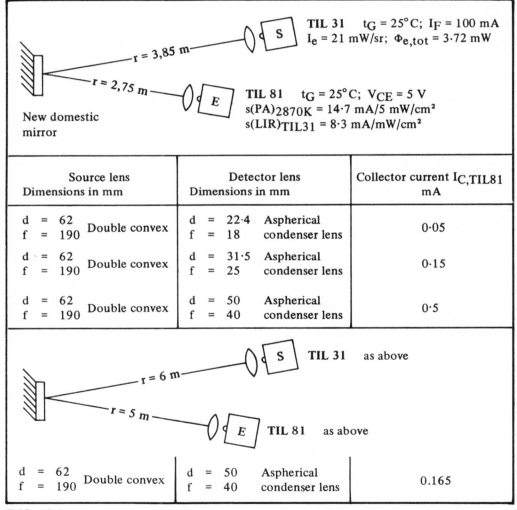

TIL 31 $t_G = 25°C$; $I_F = 100$ mA
$I_e = 21$ mW/sr; $\Phi_{e,tot} = 3.72$ mW

New domestic mirror

TIL 81 $t_G = 25°C$; $V_{CE} = 5$ V
$s(PA)_{2870K} = 14.7$ mA/5 mW/cm²
$s(LIR)_{TIL31} = 8.3$ mA/mW/cm²

Source lens Dimensions in mm	Detector lens Dimensions in mm	Collector current $I_{C,TIL81}$ mA
d = 62 f = 190 Double convex	d = 22·4 Aspherical f = 18 condenser lens	0·05
d = 62 f = 190 Double convex	d = 31·5 Aspherical f = 25 condenser lens	0·15
d = 62 f = 190 Double convex	d = 50 Aspherical f = 40 condenser lens	0·5

TIL 31   as above

TIL 81   as above

| d = 62
f = 190 Double convex | d = 50 Aspherical
f = 40 condenser lens | 0.165 |

Table 13.4
Examples of reflection opto couplers with a calibrated GaAs diode TIL 31 and a calibrated Si phototransistor TIL 81

$$s(PA)_{2870K} = \frac{14.7 \text{ mA}}{5 \text{ mW/cm}^2} \quad \text{and}$$

$$s(LIR)_{TIL31} = \frac{0.72 \text{ mA}}{0.1 \text{ mW/cm}^2} \quad \text{and}$$

$$\frac{2.9 \text{ mA}}{0.5 \text{ mW/cm}^2} \quad \text{or} \quad \frac{8.3 \text{ mA}}{1 \text{ mW/cm}^2}$$

The radiant power and the sensitivity values of the specimen used lie below the typical values for these components.

13.6
Opto couplers with unmodulated optical radiation

Unmodulated opto couplers are still relatively new at the present time. In the consumer sector, in particular, they are used more and more frequently. For example, they are used in tape recorders for automatic tape-switching. In comparison with modulated optical opto couplers, they have a considerable price advantage, since they need considerably fewer components, both on the transmitter and the receiver side. At the source end, the radiation source is either connected to a constant-current source or, in very low-priced versions, to a simple power supply. At the receiving end, a DC-coupled amplifier is usually driven through a photodiode, a phototransistor or a photoresistor. Such detector circuits have one important disadvantage, since they cannot distinguish between the actual radiation from the transmitter and radiation from external sources. Very sensitive detectors respond both to background radiation and to reflected ambient radiation. For example, during interruptions of an unmodulated optical coupler, light-coloured material can reflect the radiation from the surroundings onto the detector. No response signal can be given. For an adequate contrast-current ratio, the desired useful photocurrent $I_{P(useful)}$ should be greater, by a factor of 10, or better still by a factor of 50, than the photocurrent $I_{P(interference)}$ pro-duced by the unwanted maximum stray radiation from outside sources. It therefore follows that: *Unmodulated optical couplers should have a powerful source and a relatively insensitive detector.*

The source will be arranged in a position with the darkest possible background, so that the detector only records the useful radiation and none or only little of the background radiation.

By skilful mechanical installation, the photosensitive sensors can be protected from background or ambient radiation. Tubes are mainly used to screen this unwanted radiation. The interior walls of these tubes can reflect residual ambient radiation onto the sensor. Therefore, the interior walls are sprayed with matt-black photographic lacquer or synthetic lacquer. Matt black surfaces absorb a very large proportion of the optical radiation. The consumer industry often uses black plastic tubes made by injection moulding. The plastic used should be particularly tested for its transmission in the near IR range. For example, black coloured polystyrene has a very good transmission in the near IR range.

Further protection from background and ambient radiation is offered by very narrowly focussed reception characteristics of the photodetector. If the sensitivity is highly dependent on the angle of incidence of the rays, this has the effect that background radiation is only received in the major direction. Detection of ambient or background radiation from other angles is thus effectively suppressed. The radiometric characteristics of the photosensitive sensor are greatly narrowed with optical reflectors or lenses. With correct arrangement of these optical aids, the apparent photosensitive sensor area is magnified several times. The incident irradiance is thus also increased by a similar factor. With very closely focussed detector characteristics, the interfering background radiation can sometimes be shaded by the actual source structure.

Opto couplers are subjected to various environmental effects. They are protected from mechanical stresses, such as shock, impact or vibration, by very rigid housings in cast iron or glass-fibre-reinforced plastic. The housings are given an appropriate paint coating against atmospheric oxidation or corrosive substances such as acids and alkalies. The electrical components and wiring are protected against moisture and condensation by a special, thin sprayed-on film of lacquer. Regeneratable desiccant cartridges absorb the moisture in the housing. Water-cooled housings give protection against great heat. Objective lenses are protected against the accumulation of dust by dust-protection tubes (*Figure 13.18*). Between the built-in baffles, the dust collects in so-called dust chambers. In addition, the lens can be cleaned with compressed air. The flow of air passes from the lens, through the dust-protection tube, to the exterior. The accumulated dust is carried away to the open air.

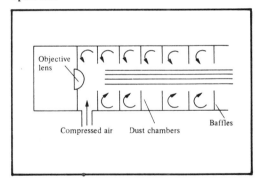

Figure 13.18
Dust-protection tube

Misting of the lenses and front windows is prevented by a heating system. Electrical heating wires can be cast into the front windows. In tropical designs, all electrical components, including the wiring and the chassis, are given a coating of a fungicide lacquer, as a protection against mildew. All necessary openings, e.g., for air circulation and cooling, must be closed by fine-meshed tropical screens. In general, the housings should be closed and sealed against dust, moisture, water splashes, oil, insects and small animals (in the tropics). By suitable construction, opto couplers can be safely used in explosion-hazard areas.

13.7
Opto couplers with modulated optical radiation

The criteria listed in Section 13.6 also apply substantially for couplers with modulated radiation. The transmitter of the coupler contains a stage to modulate the forward current of the luminescence diode. The modulated radiation produced falls on a photodetector. The receiver amplifier is usually RC-coupled. Either amplitude, frequency or pulse modulation can be selected. Pulse modulation has the advantage, that the luminescence diode can be loaded with considerably higher pulse currents, as compared with continuous-wave operation. At the receiver end, a more favourable signal/noise is obtained.

The relatively large spectral sensitivity range of junction photodetectors partly coincides with the spectral emission distribution of possible interference radiation sources. Non-selective coupler receivers therefore respond to modulated interference radiation. Filament lamps on the A.C. mains have an emission modulated at 100 Hz with relatively few harmonics. Gas discharge lamps, e.g., fluorescent lamps on the A.C. mains, have an emission modulated with 100 Hz and also with relatively strong harmonics up to 3 kHz and weaker harmonics up to 10 kHz. A remedy is provided by optical filters, placed in front of the photodetectors. Band-passes with interference filters would be an elegant but expensive solution. But the spectral sensitivity distribution of the photodetector can also be narrowed with inexpensive black glasses or edge filters. Filters cause the signal-to-noise ratio to be affected by the reflection and attenuation losses.

An improvement in the ratio of useful signal to interference and also in the

260

signal/noise ratio is achieved, if the bandwidth of the receiver is narrowed and a modulation frequency $f_{mod} > 10$ kHz is selected.

With suitable circuit design, coupler receivers with photodiodes or photocells can still be used safely, even if the interference irradiance is several order of magnitude greater than the modulated useful irradiance. It must be remembered, that the interference photocurrent $I_{P(int)}$ produced in a photodiode causes a high current noise and thus worsens the signal/noise ratio. Photodiodes with small dark currents are used in sensitive coupler receivers. This results in a low current noise and a low NEP value. When phototransistors are used, both the useful irradiance and the interference irradiance are to be measured exactly. In principle, the working point of the phototransistor should be determined by means of the useful irradiance with the base current $I_{P(useful)}$ (photocurrent of the collector-base photodiode). The useful irradiance must have a minimum value, so that the phototransistor still reliably amplifies the incident useful radiation in complete darkness (see data-sheet graph $I_C = f(E_e)$).

The phototransistor is shielded from high interference irradiance levels, since the phototransistor, if driven into the saturation range, can neither detect nor amplify the incident useful modulated radiation.

14
Operation of luminescence diodes with direct current

Operation of Luminescence Diodes with Direct Current

When designing circuits with luminescence diodes, account must first be taken of the fact that these have a very low differential internal resistance of only a few Ohms. In addition, the tolerances on the forward voltage, V_F, due to scatter between devices, and its temperature-dependence must also be taken into account. For these reasons, these diodes should only be fed from circuits, which have a high internal resistance. In the simplest case this is achieved by selecting a correspondingly high supply voltage V_b and setting the desired diode current with the series resistance R_V (*Figure 14.1a*). A more elegant method, however, is to feed the diode from a constant-current source (*Figure 14.1b*). The diode variables previously mentioned can then be disregarded.

14.1
Operation through series resistances

A voltage source, e.g., a battery, in series with a resistance is a simple current source. In this case, however, fluctuations in the working voltage cause a variation of the forward current I_F and thus of the radiant power of the diode. *Figure 14.2* shows the effect of the series resistance R_V on the forward characteristic of luminescence diodes. High working voltages and thus large series resistances cause relatively smaller variations of the radiant power in case of voltage fluctuations.

In practice, R_V is determined by the pre-determined working voltage in the equipment. *Figure 14.3* shows two circuits with series resistances for two different working voltages.

14.2
Operation from constant-current sources

It is more advantageous, to operate luminescence diodes with a constant-current source. In this case, fluctuations in the working voltage have no effect on the forward current I_F and thus on the radiant power of the luminescence diode. Simple constant-current sources can be built both

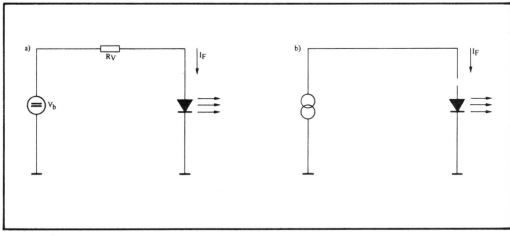

Figure 14.1
Principle of operation of luminescence diodes, a) through series resistance, b) from constant current source.

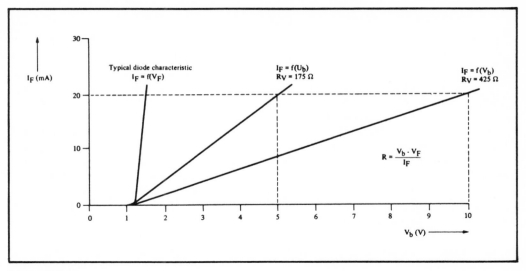

Figure 14.2
Typical forward characteristic of luminescence diodes with and without series resistance

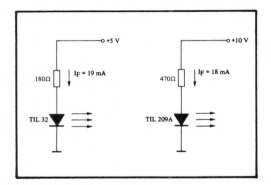

Figure 14.3
Operation of luminescence diodes through series resistance

with bipolar transistors and with field-effect transistors.

Figure 14.4 shows the output characteristics $I_{DS} = f(V_{DS})$ of an N-channel junction FET, with the voltage between gate and source V_{GS} as a variable. In the left-hand part of the graph, the resistance range, the output current I_{DS} is strongly dependent on the voltage V_{DS} applied between drain and source. In the right-hand part of the graph, the current saturation range, the output current I_{DS} only varies very slightly as a

function of the applied voltage V_{DS}. With circuits of this kind, care is to be taken, that the transistor is operated in this range under all conditions. For the circuits shown in *Figure 14.5*, field-effect transistors with a slope of 5 to 20 mA/V and a pinch-off voltage of about 5 to 7 V are needed. With operating currents of $I_F = 5$ to 40 mA, which are necessary for low-power luminescence diodes, the necessary gate bias voltage V_{GS} is then 0 to 5 V. In both circuits in *Figure 14.5*, the necessary gate bias voltage V_{GS} is produced automatically through the resistance in the source lead. The required diode current can be adjusted exactly with the 250Ω potentiometer. The necessary working voltage for these circuits is determined in accordance with the following scheme:

Gate bias voltage V_{GS}	0 to 5 V
Drain-source voltage V_{DS}	> 4 V
Forward voltage of the diode V_F	1·5 V
Minimum working voltage V_b	10·5 V

Because of the voltage drop across the drain-source path, which is sometimes very high,

266

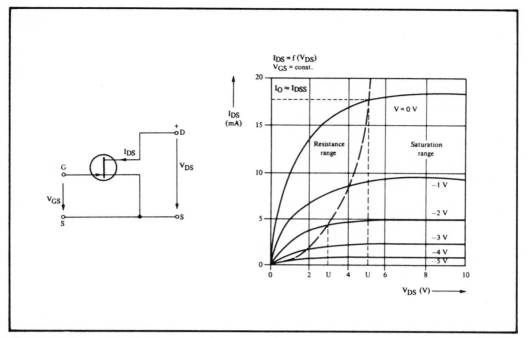

Figure 14.4
Output characteristic curves of a self-conductive N-channel FET

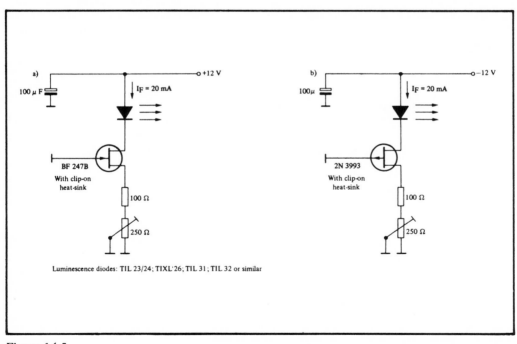

Figure 14.5
Operation of luminescence diodes with FET constant sources, a) with N-channel FET,
b) with P-channel FET

the loss power in the transistor becomes very large, so that these circuits can generally only be used up to currents of approx. 40 mA.

Constant-current sources can also be constructed with bipolar transistors. In these, however, a separate bias voltage, which is stabilised – as shown in *Figure 14.6* – with

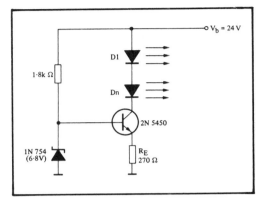

Figure 14.6
Series connection of luminescence diodes with a constant-current source.

a Zener diode, must be supplied to the base of the transistor. At the same time, with all circuits of this kind, several luminescence

diodes can be connected in series. The diode current is calculated in accordance with the formula:

$$I_F = I_C \approx I_E$$

$$= \frac{V_Z - V_{BE}}{R_E} = \frac{6.8\ V - 0.7\ V}{270\ \Omega} = 22{\cdot}6\ mA$$

When designing these circuits, care must be taken, that the transistor is not operated in the saturation range ($V_{CE} < V_{BE}$).

The maximum possibly number, n, of diodes in the collector lead is then calculated in accordance with the formula:

$$V_b \geqslant n.V_F + V_{CEmin} + V_E$$

$$n \leqslant \frac{V_b - V_{CEmin} - V_E}{U_F} = \frac{24\ V - 0{\cdot}7\ V - 6{\cdot}1\ V}{1{\cdot}6\ V}$$

$$n \leqslant 10$$

Further, simple current sources can be constructed with two transistors (*Figure 14.7*). In this case, the current is again determined by the emitter resistance R_E. The transistor

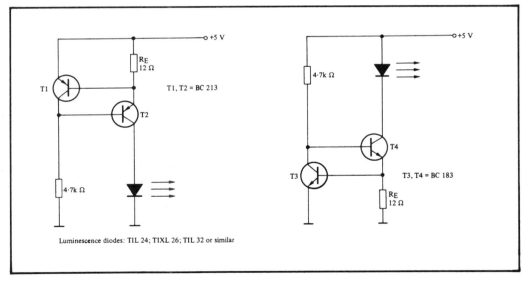

Figure 14.7
Constant-current source with transistors

T 1, the base-emitter voltage V_{BE} of which serves at the same time as the reference voltage, measures the voltage drop on the emitter resistance and then drives the transistor T 2. In this case, the diode current is calculated in accordance with the formula:

$$I_F \approx I_{E2} = \frac{V_{BE1}}{R_E} = \frac{0.65\ V}{12\ \Omega} = 54\ mA$$

If luminescence diodes are operated in equipments in which large fluctuations of the working voltage are to be expected, it is advisable to stabilise the supply voltage for the diodes (*Figure 14.8*). In this case, the

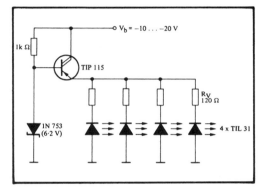

Figure 14.8
Operation of luminescence diodes from a constant-voltage source; the supply voltage for the luminescence diodes is stabilized with a Zener diode

diodes are operated in parallel. In order to ensure accurate current sharing, each diode has its own series resistance. The diode current I_F is determined by the emitter voltage of the transistor and the series resistance R_V and is calculated with the formula:

$$I_F = \frac{V_Z - V_{BE} - V_F}{R_V}$$

$$= \frac{6.2\ V - 1.5\ V - 1.6\ V}{120\ \Omega}$$

$$I_F = 25.8\ mA$$

Since, in this circuit, all the anodes of the luminescence diodes are at ground potential, they can, if required, be mounted on a heatsink, without the need for special measures for insulation.

14.3
Drive by logic circuits

In digital systems, luminescence diodes often have to be switched on and off by digital signals. In this case, the circuits must be so designed, that digital signals can carry out the desired functions directly. The circuit in *Figure 14.9* can be driven directly from TTL circuits. As well as the maintenance of the required diode current I_F, it must also be ensured, that definite currents flow in the diode with the levels supplied from TTL circuits. The diode is therefore connected in the emitter circuit. Therefore, a voltage of at least $V_b = V_F + V_{BE} = 1.6\ V + 0.7\ V = 2.3\ V$ must be present at the base of the transistor, so that a current flows through the diode. The corresponding input voltage before the diodes D 1 and D 2 is then $V_I = V_b - V_D = 2.3\ V - 0.7\ V = 1.6\ V$. Since, with TTL circuits, $V_{ILmax} \leqslant 0.8\ V$ and $V_{IHmin} \geqslant 2.9\ V$, the circuit is TTL-compatible at this point.

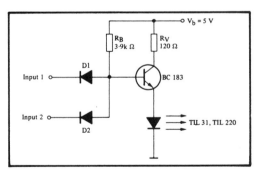

Figure 14.9
TTL-compatible control circuit for luminescence diodes

The diode current is calculated by the formula

$$I_F = I_C + I_B$$

269

Since $I_C \geqslant I_B$, the calculation can be simplified:

$$I_F = \frac{V_b - V_{CEsat} - V_F}{R_V}$$

$$= \frac{5\ V - 0{\cdot}3\ V - 1{\cdot}6\ V}{180\ \Omega}$$

$$I_F = 20{\cdot}6\ mA$$

In order to drive the transistor as far as possible into the saturation range, the calculation is only based on a current gain $h_{FE} \geqslant 30$. Thus R_B $3{\cdot}9\ k\Omega$. The input current I_{IL} of the circuit thus lies below 1 mA; this corresponds to a fan-in = 1.

In the same way, circuits which are compatible with High-Level-Logic families can be built with discrete components. *Figure 14.10* shows a circuit, which is designed to drive circuits of the HLL family "300". Since large fluctuations of working voltage are permitted here (V_b = 10.5 to 16.5 V), it is not advisable to adjust the diode current I_F through a series resistance. In *Figure 14.10*, therefore, a circuit similar to that in *Figure 14.7* is used.

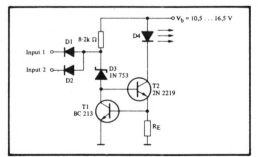

Figure 14.10
High-Level-Logic gate to drive luminescence diodes

The Zener diode D 3 in the input circuit matches the threshold voltage V_{Th} at the input to the corresponding values of the logic family and is calculated at:

$$V_{Th} = V_{RE} + V_{BE2} - V_{D3} - V_{D1}/2$$

$$= 0.7\ V + 0.7\ V + 6.2\ V - 0.7\ V = 6.9\ V$$

The maximum possible diode current is determined by the loss power in the transistor T 2 (P_{Vmax} = 0.8 W) and is calculated at:

$$I_{Fmax} = \frac{P_{Vmax}}{V_{bmax} - V_{RE} - V_F}$$

$$= \frac{0.8\ W}{16{\cdot}5\ V - 0{\cdot}7\ V - 1{\cdot}6\ V} = 56\ mA$$

($t_U = 25°C$)

The emitter resistance R_E is then:

$$R_E = \frac{V_{BE1}}{I_F} = \frac{0{\cdot}7\ V}{56\ mA} = 12.5\ \Omega$$

In the same way, luminescence diodes can be driven directly from TTL circuits. The types SN 7416N and SN 7417N, which can deliver an output current of I_{OL} = 40 mA, are particularly suitable for this. In this case, the current is determined once again by a series resistance (*Figure 14.11*). This is calculated in accordance with the formula:

$$I_F = \frac{V_{CC} - V_{OL} - V_F}{R_V} = \frac{5\ V - 0{\cdot}7\ V - 1{\cdot}6\ V}{R_V} = \frac{2{\cdot}7\ V}{R_V}$$

$$R_V = \frac{2{\cdot}7\ V}{I_F}$$

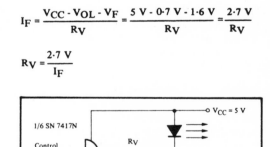

Figure 14.11
Driving of luminescence diodes by TTL circuits

In principle it is also possible, to connect the luminescence diode between the integrated circuit output and ground, if the circuit in question has an inverting (totempole)

270

output *(Figure 14.12)*. The current is then determined by the internal organisation of the integrated circuit.

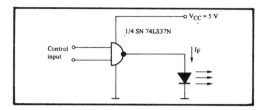

Figure 14.12
Driving of luminescence diodes by TTL circuits

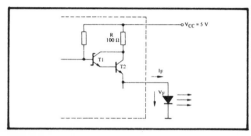

Figure 14.13
Circuit diagram for determination of the diode current when a gate of type SN 74LS37N is used.

Figure 14.13 shows the part of the circuit of the gate SN 74LS37N which determines the output current. The current through the luminescence diode is now calculated from the formula:

$$I_F = \frac{V_{CC} - V_{CEsat1} - V_{BE2} - V_F}{R}$$

$$= \frac{5\ V - 0.3\ V - 0.7\ V - 1.8\ V}{100\ \Omega} = 24\ mA$$

However, two points must be noted with this circuit: firstly the tolerance of the resistance R is ± 30%, so that reproducible values can only be achieved with difficulty. Secondly, the maximum permissible power dissipation of the I.C. (P_{Vmax} = 60 mW for a 14-pin packager) must be observed.

Considerably higher currents can be achieved with the integrated interface circuits of the SN 75400 series. The maximum permissible output current I_{OL} for SN 75450 - 454 is about 300 mA per output, so that when both outputs are connected in parallel, a diode current of 600 mA can be achieved *(Figure 14.14)*. Of course, a separate resistor has to be connected in each collector lead, in order to achieve exact current sharing. The necessary series resistances are then calculated according to the formula:

$$R_V = \frac{V_{CC} - V_{OL} - V_F}{I_{OL}} = \frac{5\ V - 0.7\ V - 1.6\ V}{300\ mA} = 9\ \Omega$$

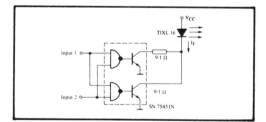

Figure 14.14
Driving of luminescence diodes using interface circuits

15
Photodetector circuits

15.1
Principle of operation

Photodetectors are used to measure and evaluate radiation. The simplest optoelectronic couplers contains any radiation source as the emitter and a sufficiently sensitive photodetector for the direct control of a switching stage. *Figure 15.1* shows

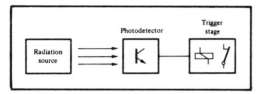

Figure 15.1
Simplest opto coupler for unmodulated radiation

the schematic diagram of a very simple coupler. The photodetector makes a Yes-No statement:

Yes = Sufficient radiation
No = Insufficient or no radiation

The measurement of grey steps or contrast current ratios is not possible by the above principle.

Figure 15.2 shows the block diagram of a coupler for more severe requirements, where the photodetector circuit has been considerably extended. The photodetector is selected for a desired sensitivity range. The amplifier which follows is designed for a defined voltage or current gain. The threshold switch ensures definite switching points. In many applications, a large hysteresis is chosen, so that in operation at the limit of sensitivity, the modulated ambient (interference) radiation from filament and fluorescent lamps does not cause triggering in time with the modulation. The

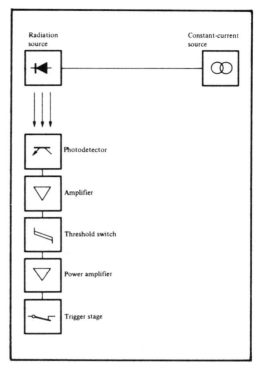

Figure 15.2
Principle of a coupler for unmodulated radiation

power amplifier control the switching stage, which consists, in the simplest case, of a switching transistor or − if electrical isolation is necessary at the output − of an optocoupler or a relay.

Figure 15.3 shows the principle of measurement of radiant power. For this, photodetectors with a linear relationship between the output signal and the incident radiant power are needed.

The output signal of the photodetector is fed to a DC voltage amplifier. The amplified DC voltage then controls a measuring instrument, on which the received radiant power can be read.

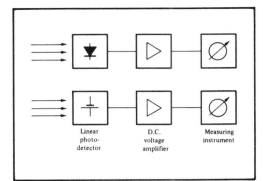

Figure 15.3
Principle of analogue measurement of optical radiations

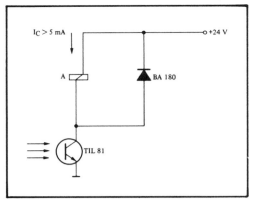

Figure 15.4
Simple relay control with the sensitive phototransistor TIL 81

15.2
Detector circuits for two-pole junction photodetectors

Two-pole junction photodetectors are understood to mean photocells, photodiodes and phototransistors without base connection.

15.2.1
Direct relay control with phototransistors

Sensitive phototransistors already deliver relatively high collector currents. With adequate irradiances, low-current relays can be controlled directly. The phototransistor TIL 81 has the following typical parameters:

$$I_C = 20 \text{ mA at } E_{e,2856K} = 5 \text{ mW/cm}^2$$

These values are sufficient, as is shown in *Figure 15.4*, for a direct relay control with this phototransistor. The light-current relay is reliably turned on with an irradiance of $E_e = 5$ mW/cm², if the response current of the relay is below 5 mA. The diode BA 180 protects the phototransistor from induced voltages when the phototransistor is turned off.

15.2.2
Photodarlington circuits

Phototransistors can be extended with subsequent bipolar transistors to form Darlington circuits. Such circuits are more sensitive in proportion to the current gain B (= h_{FE}) of the subsequent transistor. The total current gain B_D of the Darlington configuration is:

$$B_D = B_{Phototr.} \cdot B_{bip.Tr.}$$

On this basis, simple but sensitive relay controls can be produced. As an example, *Figure 15.5* shows an NPN phototransistor with a subsequent NPN transistor as an NPN photodarlington in a collector or emitter-follower connection. *Figure 15.6* shows a complementary circuit with an NPN phototransistor and a subsequent PNP transistor as an NPN photodarlington. In both circuits, the relay pulls in when the phototransistor is irradiated.

In *Figure 15.7*, the circuit of a simple Luxmeter with a photodarlington can be seen. With average illuminances and the design data stated, the Luxmeter has a sufficiently linear characteristic $I_e = f(E_e)$. The circuit can be calibrated by comparison with another Luxmeter.

276

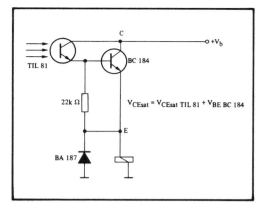

Figure 15.5
NPN Photodarlington as an emitter follower

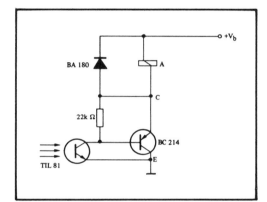

Figure 15.6
NPN Photodarlington in a complementary
circuit

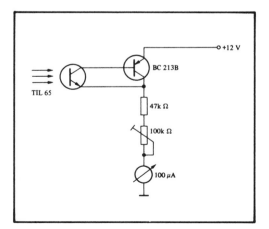

Figure 15.7
Simple Luxmeter

15.2.3
Thyristor and Triac control with phototransistors

In the same way, thyristors can be driven by phototransistors. *Figure 15.8* shows the theoretical circuit for this. If radiation falls on the phototransistor with sufficient intensity, a current flows in the gate and fires the thyristor. The resistance R 1 and the capacitor C 1 prevent the firing of the thyristor by rapidly rising leakage currents and interference voltages. In addition, the firing threshold of the circuit can be varied with R 1.

Figure 15.9 shows a thyristor equivalent circuit driven by a phototransistor. The collector of the PNP transistor drives the base of the NPN transistor and conversely the collector of the NPN transistor drives the base of the PNP transistor. Through the back-to-back coupling of the two transistors, a bistable circuit is obtained. The phototransistor initiates the trigger process through the base of the NPN transistor (cathode gate). Thyristors and thyristor equivalent circuits fire unintentionally below their firing threshold, if the ignition current or the anode voltage rise very rapidly (rate effect). A remedy is provided by the resistance R 1 and the capacitor C 1 at the cathode gate and the capacitor C 2 at the anode gate.

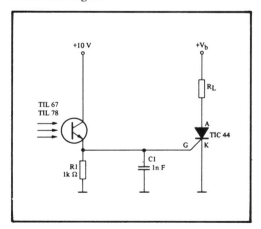

Figure 15.8
Thyristor control with a phototransistor

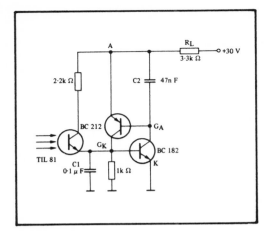

Figure 15.9
Control of a thyristor equivalent circuit formed from two transistors

Very sensitive thyristor drive circuits can be constructed with a photodarlington arrangement. In *Figure 15.10*, the thyristor fires when the phototransistor is irradiated, while in *Figure 15.11* the thyristor is fired when the phototransistor is darkened. Furthermore, as well as thyristors, triacs can also be controlled. In this case, the phototransistor has to be adapted for operation on an alternating voltage. *Figure 15.12* shows a simple circuit for A.C. operation. A bridge rectifier is connected in series with the phototransistor, in order to ensure that the voltage on the transistor has the correct polarity. In *Figure 15.13*, the characteristics $I_C = f(V_{CE})$ of this "A.C. phototransistor" are shown. The family of characteristics, which are already known, in the 1st quadrant, is repeated in the 3rd quadrant.

A further circuit for operation on alternating voltages is shown in *Figure 15.14*. During a positive half-wave, the phototransistor T 2 is short-circuited by the diode D 2 connected in parallel with it. The phototransistor T 1 receives its collector voltage with the correct polarity, fed through the diode D 2. Conversely, during a negative half-wave, transistor T 1 is short-circuited by diode D 1 and the phototransistor T 2 receives its collector voltage through the diode D 1.

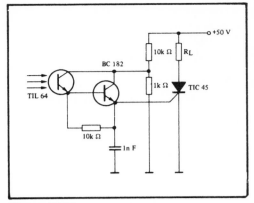

Figure 15.10
Thyristor control with an NPN phototransistor-Darlington

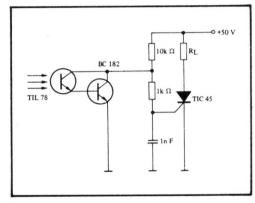

Figure 15.11
Thyristor control with an NPN phototransistor-Darlington

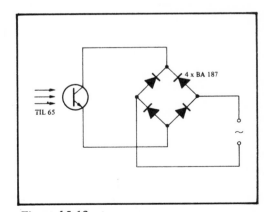

Figure 15.12
Simple phototransistor circuit for A.C. operation

278

This circuit serves for the brightness control or regulation of a filament lamp by phase-angle control. Here, the triac, in series with the filament lamp which acts as the load, forms the main circuit. The gate of the triac is driven through the trigger diode from the RC phase-shifter network. When the control voltage on R 1 and C 1 reaches the breakdown voltage of the trigger diode, this turns on and fires the triac. The triac is turned off again when the AC voltage passes through zero.

The A.C. phototransistor" is in parallel with the phase-shifter capacitor C 1. When the phototransistors T 1 and T 2 are darkened, the conduction angle in the triac is determined by the time-constant R 1 x C 1. If radiation falls on the phototransistors T 1 and T 2, part of the current through R 1 is diverted through the transistors and thus the charging of the capacitor is slowed down, so that the breakdown voltage of the trigger diode is not reached until later. Thus the conduction angle in the triac is reduced and the brightness of the filament lamp decreases. Conversely, with low irradiances, the conduction angle in the triac increases, as does the brightness of the lamp.

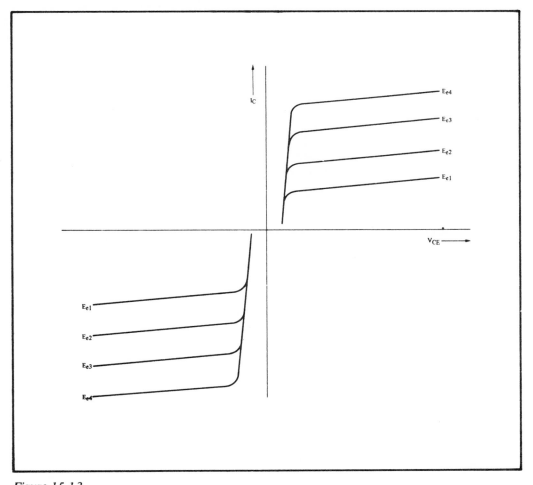

Figure 15.13
Characteristic curves $I_C = f(V_{CE})$ of a so-called A.C. phototransistor, with the irradiance E_e as variable

279

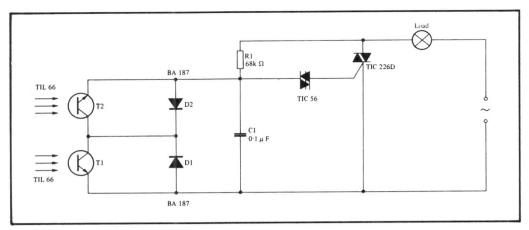

Figure 15.14
Phototransistor circuit for A.C. operation with two diodes and two NPN phototransistors for lighting control with filament lamps

15.2.4
Driving of transistor- and operational-amplifiers with phototransistors, photodiodes and photocells

Two-pole radiation-sensitive semiconductors can control a D.C. amplifier by various principles.

Since a phototransistor or a photodiode behaves like a constant current source, variations in the working voltage cause little or no variation in the output current.

The output current of a phototransistor or a photodiode either controls an amplifier directly, or the amplifier is controlled by the voltage produced on the working resistance of the photodetector. If the output impedance of the photodetector is high in comparison with the input impedance of the amplifier, the term "current control" is used. In the converse case, the term "voltage control" is used.

Figure 15.15 shows the current control of transistor amplifiers with NPN phototran-

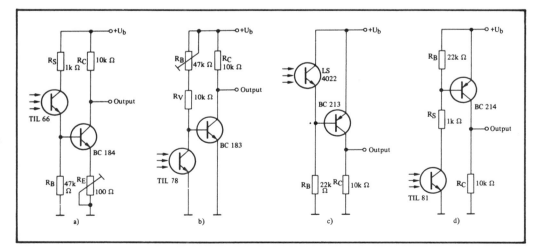

Figure 15.15
Current control of transistor amplifiers with NPN phototransistors

280

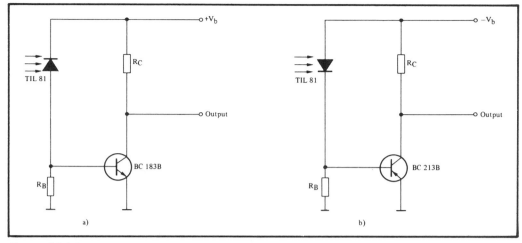

Figure 15.16
Current control of transistor amplifiers with photodiodes

sistors. The collector current of the photo-transistor controls the base of an amplifier transistor, which is operated in common emitter connection. Interfering leakage currents are led off through the base resistance R_B. In addition, the sensitivity and in some cases the switching time of the phototransistor can be varied with a variable resistance in the base circuit. Scatter between individual components can sometimes be compensated for by a feedback resistor R_E in the emitter circuit. If very high irradiances fall on the phototransistor, a protective resistor is to be provided to limit the loss power. If radiation falls on the phototransistor in circuits a) or c), the output voltage decreases in each case, while in the circuits b) and d) it increases.

Figure 15.16 shows the current control of transistor amplifiers with photodiodes, while the latter, together with the transistor, form an NPN phototransistor in circuit a) and a PNP phototransistor in circuit b).

In comparison with phototransistors, photodiodes have smaller photocurrent tolerances, since the effect of the current gain B is eliminated. Also they have a linear characteristic $I_P = f(E_e)$. In this, I_P is the photo-current and E_e the incident irradiance. If one of these two characteristics is needed,

the use of photodiodes is recommended, despite the greater expenditure on amplification.

Figure 15.17 shows the voltage control of transistor amplifiers with NPN phototransistors. The collector current of the photo-transistor produces a voltage, which controls the subsequent amplifier transistor, on the working resistance R_A. The amplifier transistor is operated as in emitter-follower. Scatter between individual phototransistors is compensated for by a variable working resistance R_A. With this, the switching times of the phototransistors are affected considerably more than with the resistor R_B in *Figure 15.15*.

In *Figure 15.18*, the voltage control of an operational amplifier with a photodiode is shown. The inverting input receives its input voltage through the feed-back resistance R 1 from the output of the operational amplifier. If radiation falls on the photodiode TIL 81, a current flows through its working resistance R_A. There is a positive voltage drop across the working resistance R_A, and this drives the non-inverting input of the operational amplifier. The voltage gain is determined by the resistance ratio R 1/R 2. The offset voltage can be balanced with the resistance R 3. The circuit is

281

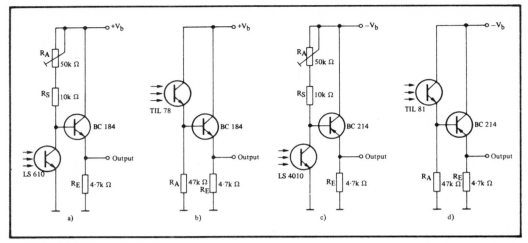

Figure 15.17
Voltage control of transistor amplifiers with NPN phototransistors

excellently suited for sensitive Luxmeters. It can equally well be used in optoelectronic couplers or as a detector for modulated radiation. The upper frequency limit is determined firstly by the operational amplifier and secondly by the junction capacity of the photodiode and its working resistance R_A.

Power matching of a photodetector to the following circuit is mainly applied with solar cells for power generation. Basically,

a photosensor represents a generator. It can work in the no-load, the matched or the short-circuited mode. In the no-load mode, the sensor must not be loaded, so that an amplifier with a high input impedance or a high-resistance measuring instrument must be connected to it.

Figure 15.19 shows the no-load voltage $V_{P,L}$ as a function of the irradiance E_e for a photosensor. The characteristic curve has an exponential form, so that saturation

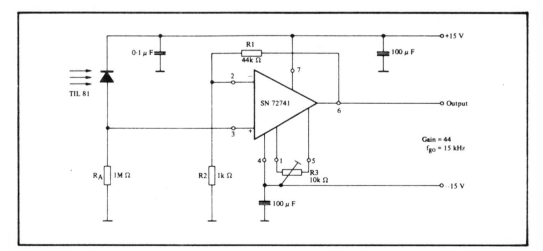

Figure 15.18
Photodiode D.C. amplifier

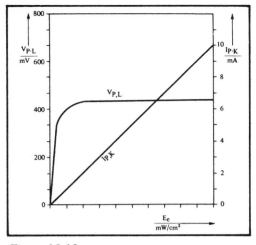

Figure 15.19
No-load photo-voltage $V_{P,L}$ and short-circuit photocurrent $I_{P,K}$ as functions of the irradiance E_e for a photocell

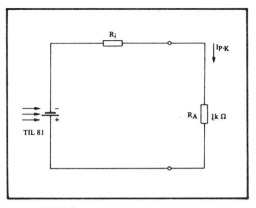

Figure 15.20
Short-circuited operation of the photo-sensor TIL 81

occurs at high irradiances. The no-load operation of a photosensor has little practical importance.

In *Figure 15.20*, the short-circuited operation of a small-area photosensor is represented. In this case, the internal resistance R_i of the sensor is large in proportion to the load resistance R_A. With large-area photosensors, the load resistances R_A have values of approx. 0.1 Ω

to 100 Ω, while all small-area photosensors can be used with load resistors of over 1 kΩ. In contrast to the no-load voltage $V_{P,L}$, which has an exponential characteristic, there is a strictly linear relationship between the short-circuit photocurrent $I_{P,K}$ and the irradiance E_e. This can also be seen from *Figure 15.19*. Therefore, photo-sensors are usually operated in the short-circuit mode practice.

Figure 15.21 shows an operational amplifier driven by two TIL 81 photosensors, connected in opposing parallel. The two input resistors $R_A/2$ of the amplifier form,

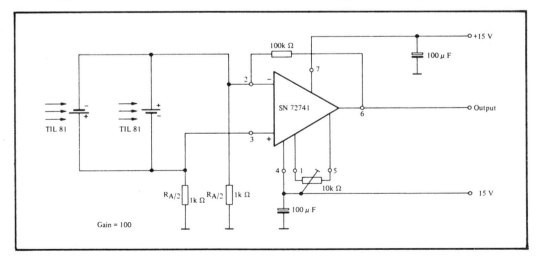

Figure 15.21
D.C. amplifier with photosensors connected in opposing parallel

283

together, the load resistance of the photo-sensors. If an equal radiant power falls on both sensors, then the voltage at the amplifier input, and thus also at its output, is zero. If the two devices are irradiated unequally, the difference signal is amplified, by a factor of 100. This circuit is used in measuring instruments, for example, for relative measurement of the photocurrent sensitivity (with one photosensor as reference and a second photosensor as the test device), as a direction-dependent detector for couplers and, where requirements are not severe, as a distance/voltage transducer for follower controls.

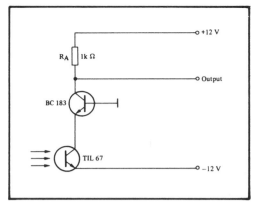

Figure 15.23
Phototransistor cascade circuit with
improved dynamic characteristics

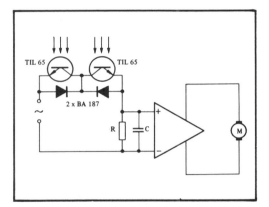

Figure 15.22
Equivalent circuit of an A.C. phototransistor
for position control

Figure 15.22 shows the principle of a follower or position control with an "A.C. phototransistor" circuit. With equal irradiance on the phototransistors, the capacitor C is charged up during both A.C. voltage half sine waves through the transistors, with the same charge but opposite polarity. The resultant voltage is therefore approximately zero. With unequal irradiance on the phototransistors, the difference signal is amplified, to energise a motor, for example.

Figure 15.23 shows the principle of a phototransistor cascade circuit to improve the dynamic range. In this, the working resistance of the phototransistor has been

replaced by a common base connected NPN transistor. The input impedance r_{BE} of the base circuit is very low and is approximately:

$$r_{BE} \approx \frac{V_T}{I_C} \approx \frac{26 \text{ mV}}{1 \text{ mA}} \approx 26 \ \Omega$$

V_T is the thermal voltage. In accordance with the equation (10.62)

$$f_g = \frac{1}{2 \ \pi . R_A . C}$$

a high limiting frequency is achieved, since the working resistance of the phototransistor is only about 1/100th of the original R_A value. This very low resistance has the effect, that the voltage gain and thus the dynamic collector-base capacitance (Miller capacitance) remains small. The voltage gain V_U of the whole cascade circuit corresponds to that with an emitter-connected phototransistor.

Figure 15.24 shows a further cascade circuit with a photodarlington transistor. It also serves to improve the dynamic characteristics. The resistance R_B has been inserted between the base and emitter of the transistor T 2, in order to divert the leakage currents of the transistors.

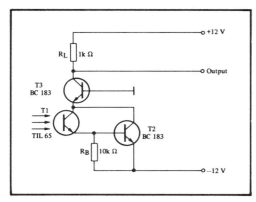

Figure 15.24
Circuit to improve the dynamic charac-
teristics of a photodarlington transistor

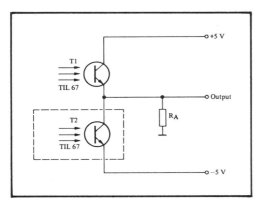

Figure 15.25
Temperature compensation for photo-
transistors without base connection, with a
selected pair of phototransistors

15.2.5
Control of multivibrators with phototransistors

Figure 15.26 shows the control of a
Schmitt trigger by a phototransistor. If a
sufficiently high irradiance falls on the
phototransistor TIL 65, the transistor T 1
becomes conductive. Thus the base current
and also the collector current of the transistor T 2 is reduced. As a result, the
potential at the emitter rises and the
current in the transistor T 1 increases
further, until the circuit suddenly trips
over, so that the output is switched off.
This circuit can also be constructed with
NPN transistors. It is used as a simple
optoelectronic detector or "twilight
switch"

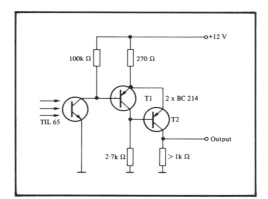

Figure 15.26
Control of a PNP transistor trigger circuit
with an NPN phototransistor

In *Figure 15.25*, a method for temperature
compensation of two-pole junction semiconductor devices is shown. For this,
matched pairs of transistors are used. With
rising temperature, the leakage current of
the phototransistor T 1 is compensated for
by the leakage current of the (non-
irradiated) phototransistor T 2, which
flows in the same direction. Instead of
phototransistors, photodiodes can also be
used.

In *Figure 15.27*, a somewhat more complicated circuit for a photodetector is
shown. The output signal of the phototransistor T 1 is amplified further in the
subsequent Darlington stage and fed
through a decoupling amplifier T 4 to the
input transistor T 5 of the Schmitt trigger.
The output transistor T 6 of the Schmitt
trigger controls the relay A directly. When
the phototransistor is irradiated, the relay
is pulled in. The sensitivity of this circuit
can be varied with the variable resistor R 1.

285

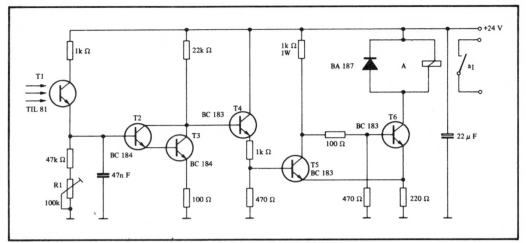

Figure 15.27
Optoelectronic detector

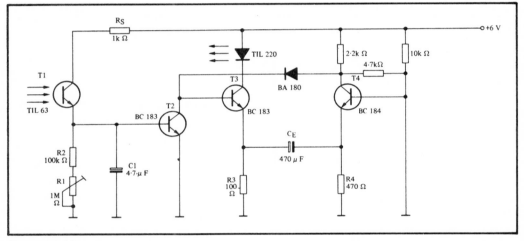

Figure 15.28
Twilight switch with flasher

Figure 15.28 shows a twilight switch combined with a flasher. With the onset of dusk, the collector current of the phototransistor T 1 is no longer sufficient to keep the transistor turned on. The transistor T 2 blocks, whereupon the base voltage and also the emitter voltage on T 3 rise. This voltage change also acts, through the capacitor C_E, on the emitter of T 4, so that the current in this transistor decreases. The collector voltage of this transistor is fed back through the diode BA 180, so that the circuit suddenly trips over to the new state.

The capacitor C_E is now discharged through the resistor R 4, until the transistor T 4 again becomes conductive and thus blocks the transistor T 3. This process is repeated periodically, so that the luminescence diode TIL 220 flashes.

In *Figure 15.29*, a more sensitive Schmitt trigger, with the operational amplifier SN 72741, is shown. In this application, the operational amplifier only needs one supply voltage.

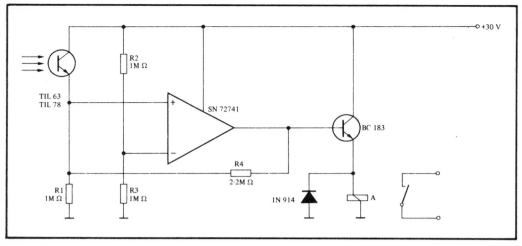

Figure 15.29
More sensitive Photo-Schmitt-Trigger with the operational amplifier SN 72741

The inverting input of the amplifier is set to half the supply voltage by a symmetrical voltage divider R 2, R 3. The non-inverting input of the amplifier is driven by the voltage produced on the load resistance R_A of the phototransistor. For Schmitt trigger operation, the operational amplifier has a feed-back, through the resistance R 4, from the output to the non-inverting input.

When the phototransistor TIL 78 is irradiated, the potential at the non-inverting

input rises. When it reaches half the supply voltage, the Schmitt trigger switches its output from the LOW state (approx. 0 V) to the HIGH state (approx. V_b).

The hysteresis of the Schmitt trigger is determined by the ratio R_4/R_A, while R_A is given a value according to the required sensitivity. The sensitivity of this circuit increases with decreasing supply voltage, since at the same time the threshold voltages of the Schmitt trigger decrease and thus a

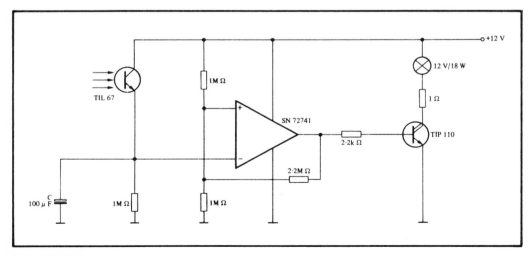

Figure 15.30
Parking-light switch with the phototransistor TIL 67

287

smaller current is necessary through R_A. This circuit is used as an optoelectronic detector or twilight switch. *Figure 15.30* shows an extended circuit for use as a parking light switch for motor vehicles. The capacitor C slows down the response of the circuit, so that it does not operate with short-term variations in the light.

15.3
Detector circuits with three-pole photo-transistors

Three-pole phototransistors are understood to mean phototransistors in which the base connection is brought out and used.

15.3.1
Operating modes of three-pole photo-transistors

Phototransistors with base connections can work in various operating modes.

15.3.1.1
Photocell and photodiode operation of the phototransistor TIL 81

Both the collector-base junction and the base-emitter junction can be operated either as a photocell or as a photodiode. For the base-emitter junction, these two modes are of no practical importance, since firstly their photosensitive area is very small and secondly the high emitter doping does not permit a high reverse voltage in photodiode operation (max. 5 V). Also, the area of the collector-base junction is about ten times as large as that of the base-emitter junction. Therefore, the former is generally used as the photodetector for photocell and photodiode operation.

When an NPN phototransistor is operated in the photocell mode, a negative potential is obtained at the collector and a positive potential at the base.

In the diode mode, the collector-base diode is biassed in the reverse direction. Negative potential is applied to the base and positive potential to the collector.

The maximum reverse voltage V_{CBO} of the collector-base diode of the phototransistor TIL 81 is 50 V. *Figure 15.31* shows the typical current-voltage characteristics for a silicon P-N junction. The 3rd quadrant applies for photodiode operation with the function $I_P = f(V_R)$. The incident irradiance E_e is the variable. For a constant incident irradiance $E_{e,1}$, $E_{e,2}$ . . . , the working point A of the photodiode can be determined by drawing in the working resistance R_A. The working resistance $R_{A,3}$, for example, has a higher value than $R_{A,4}$. The 4th quadrant shows photocell operation. The points of intersection of the individual characteristic curves with the negative ordinate show the short-circuit current $I_{P,K}$ for the relevant incident irradiance E_e. The values of the short-circuit currents $I_{P,K}$ almost correspond to the currents I_P in the diode mode. Here, the dark current in diode operation has been neglected. The points of intersection of the individual curves with the positive abscissa characterise the no-load voltage $V_{P,L}$ for the incident irradiance E_e in each case. For a constant incident irradiance $E_{e,1}$, $E_{e,2}$ . . . , the working point A of the photocell is determined by drawing in the working resistance R_A. Here, the working resistance $R_{A,1}$ has a higher value than $R_{A,2}$. The 1st quadrant shows the typical forward characteristic of a Si photodiode. During exposure to radiation, the forward voltage V_F increases slightly. This forward voltage V_F corresponds to the no-load voltage $V_{P,L}$ in photocell operation.

15.3.1.2
Advantages of the phototransistor TIL 81 in photodiode and photocell operation

The universally applicable junction photo-detector TIL 81 is built into a modified TO 18 metal case. It can therefore be fixed

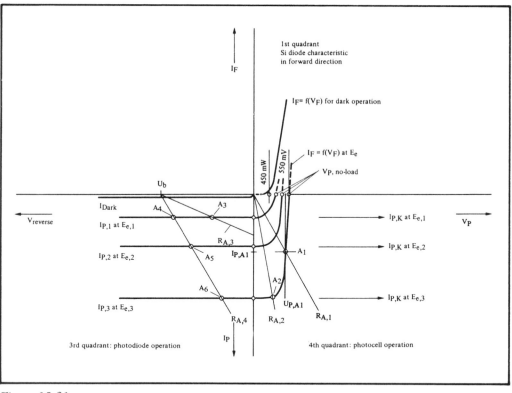

Figure 15.31
Current-voltage characteristics of a silicon photosensitive P-N junction

mechanically very simply by a press fit or clip, in guide tubes or transistor plug bases, or by mounting on a printed circuit. In comparison with unencapsulated or large-area photocells and photodiodes, firstly mechanical fixing and adjustment is simpler, and furthermore, an additional lens or objective can be fitted more conveniently, in accordance with the laws of optics, to the small lens area of the TIL 81.

The absolute sensitivity of this phototransistor in photodiode or photocell operation is

$$S_{2856K} = \frac{170\,\mu A}{25 mW/cm^2}$$

This value can only be used for comparison of the sensitivity of exactly equivalent photodiodes and photocells.

A comparison of the quality of different photodiodes and photocells can be made with the values NEP (Noise Equivalent Power) and specific detectivity described in Chapter 9. In most cases, these values are not stated for simple junction photodetectors. Here, approximate calculations or measurements of the signal-noise ratio can help. The noise level of a junction photodetector is very dependent on its dark current, so that as a simplification, the dark current serves as a further comparative criterion for different Si photodiodes and photocells.

Solar cells made by the epitaxial technique cannot be used as sensitive detectors, since they have very high dark currents and also, in the diode mode, have very low breakdown voltages. Also, large-area photocells or photodiodes made by the epitaxial or planar

289

technique are less suitable for the detection of modulated radiation, because of their high junction capacitance. Furthermore, the high sensitivity of a large-area cell, as compared with the TIL 81 is the photocell mode, is no advantage, since in short-circuited operation, the load resistance of a large-area photocell can only have 1/100th to 1/10th of the value for the small-area phototransistor TIL 81.

In photodiode operation, the data-sheet of the TIL 81 guarantees a maximum dark current of $I_{Dmax} = 10$ nA at $t_U = 25°C$. Typical dark currents lie two to three orders of magnitude lower, i.e., in the picoampere range. Therefore, this phototransistor shows a low intrinsic noise level in photocell and diode operation, so that it can be used as a detector for very weak incident radiations. As an example, let us mention an IR telephony unit with the TIL 81 in the photocell mode as detector and with the low-power GaAs diode TIL 31 as the transmitter. With these two, a good signal/noise ratio is still attained at distances greater than 10 m. With the use of additional lenses, distances of several hundred metres can be spanned. Without external optical components, the irradiance falling on the TIL 81, under the conditions

for the TIL 31: $\Phi_e = 5$ mW, $I_e = 20$ mW/sr; source-detector distance r = 10 m, has the following value:

$$E_e = \frac{I_e}{r^2} \cdot \Omega_0 = \frac{20 \text{ mW} \cdot \text{sr}}{\text{sr} \cdot 100 \text{ m}^2}$$

$$= \frac{20 \cdot 10^6 \text{ nW}}{1 \cdot 10^6 \text{ cm}^2} = 20 \frac{\text{nW}}{\text{cm}^2}$$

The lower limit of detection of radiation for the TIL 81 is still considerably lower. For the reception of HF-modulated radiation, the TIL 81 is operated more favourably as a photodiode. The applied reverse voltage reduces the junction capacitance. The collector-base capacitance of the TIL 81 in photocell operation is approx. $C_{CB} = 40$ pF, and in photodiode operation with a reverse voltage of 15 V it is about $C_{CB} = 12$ pF.

Both in photocell and photodiode operation, the phototransistor TIL 81 shows little scatter between individual specimens, while in transistor operation the scatter is determined by the differing D.C. gain B.

Photocell and photodiode operation of the phototransistor TIL 81 is clearly identified by circuit symbols (see *Figure 15.32*).

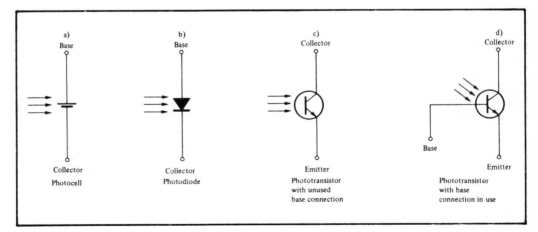

Figure 15.32
Circuit symbols for the Si junction photodetector TIL 81 in various operating modes

15.3.1.3
Phototransistor operation of the TIL 81

Figure 15.32 shows, under c), the circuit symbol for the TIL 81 with the base connection unused and, under d), that with the base connection in use. Phototransistors with base connection, in emitter-connection, and phototransistors without base connection are used as detectors for unmodulated radiation.

Phototransistors with an external base connection can be driven independently of any incident radiation. Among other uses, they are used as reference transistors in differential amplifiers. The external base

connection also permits the elimination of side-effects, for example, such as reduction of the collector-base dark current, improvement of the limiting frequency and control of the working point. The phototransistor TIL 81 with a base connection can be used very economically in multivibrator circuits, since only one additional transistor is needed.

Reception of modulated radiation with phototransistors requires stable working-point conditions. In common emitter-connection, the working point is mainly determined by the incident irradiance. For this, *Figure 15.33* shows the modulation and working-point conditions of the photo-

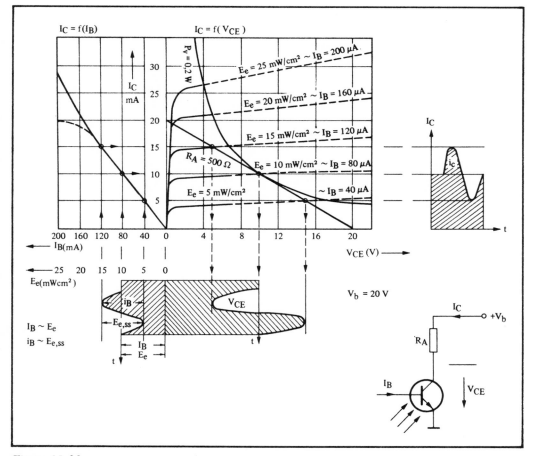

Figure 15.33
Modulation and characteristics of a TIL 81 phototransistor

transistor TIL 81. If a very weak modulated radiation signal and a very intense interference radiation fall simultaneously on the detector, the interference radiation drives the phototransistor into voltage saturation. The modulated radiation signal will not be amplified. If a very intense modulated radiation falls on the phototransistor, the useful radiation drives the phototransistor into voltage saturation; the output signal will be distorted. If, on the other hand, only a very weak modulated radiation falls on the phototransistor, then the photo-current produced by the collector-base diode is insufficient to drive the transistor.

A remedy is provided by a defined external irradiation or an additional base control current, to fix the working point. In practice, a strong interference radiation is usually to be expected, so that the working point has to be shifted according to the interference intensity.

Phototransistors in common emitter connection or without base connection are therefore usually unsuitable for reception of modulated radiation.

The emitter follower offers an alternative,

if a fixed potential from a high-impedance voltage source is applied, as shown in *Figure 15.34*, to the base of the phototransistor.

The working point of the transistor in the dark condition is determined by the base voltage divider R 1, R 2, which forms a voltage source of 3.7 V with an internal resistance $R_O = 50$ kΩ. The collector current is then

$$I_C \approx I_E = \frac{V_B - V_{RE}}{R_E} = \frac{3 \cdot 7 \text{ V} - 0 \cdot 7}{10 \text{ k}\Omega}$$

$$= 0 \cdot 1 \text{ mA}$$

The input impedance r_i of the emitter follower is obtained, with h_{FE} TIL 81 $\geqslant 100$, as

$$r_i \approx R_E.h_{FE} = 10 \text{ k}\Omega.100 = 1 \text{ M}\Omega$$

Thus the load resistance of the photodiode is primarily determined by the resistance R_O, through which the useful and the interference photocurrents i_p and I_p flow. Since this resistance is relatively low and the voltage gain of the subsequent emitter follower is $V_u \approx 1$, even high interference

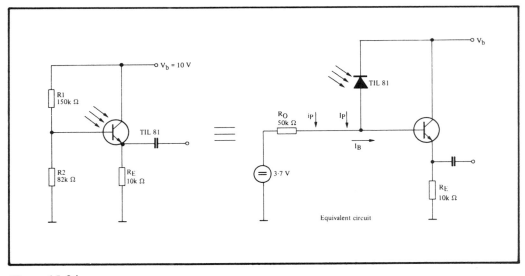

Figure 15.34
Phototransistor used with an emitter follower for reception of modulated radiation

292

currents Ip are unable to overload the whole circuit. The output impedance of the circuit is calculated from the formula

$$r_O = \frac{V_T}{I_C} + \frac{R_O}{h_{FE}} = \frac{26 \text{ mV}}{0.1 \text{ mA}} + \frac{50 \text{ k}\Omega}{100}$$

$$= 760 \ \Omega$$

Since the voltage gain of the circuit is $V_u \leqslant 1$, the effect of the Miller capacitance is also slight. The upper frequency limit is determined by the collector-base capacitance C_{CB} and the resistance R_O.

15.3.2
Driving of amplifiers with three-pole phototransistors

Figure 15.35 shows a cascade circuit to improve the dynamic characteristics of the phototransistor TIL 81. In addition to the circuit in *Figure 15.23*, the base of the phototransistor T 1 is driven through the feed-back resistors R 2, R 3 from the collector of the NPN transistor. The D.C. feed-back through R 2, R 3 has the effect that the dark current of the phototransistor is now hardly affected by its current gain. The quiescent current of the circuit is set with the resistor R 3.

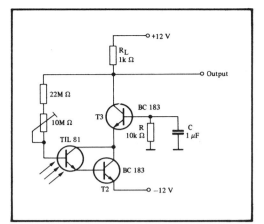
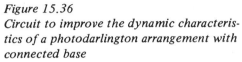

Figure 15.36
Circuit to improve the dynamic characteristics of a photodarlington arrangement with connected base

The photodarlington cascade in *Figure 15.36* is of similar construction to the circuits in *Figure 15.24* and *Figure 15.25*. This circuit is very sensitive. With high irradiances, the photodarlington transistor, consisting of T 1 and T 2, and also the base-connected NPN transistor T 3, go into saturation. For current limitation, a series resistance R is connected in the base lead of T 3 and the base is capacitatively earthed. With rising temperature, the collector-base leakage current I_{CBO} increases. It acts as the base

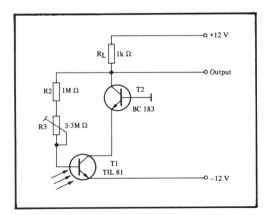

Figure 15.35
Circuit for improvement of the dynamic characteristics of a phototransistor with base connection

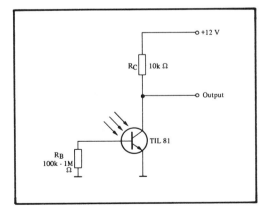

Figure 15.37
Temperature compensation of the leakage current I_{CBO} for a phototransistor by the base resistance R_B

293

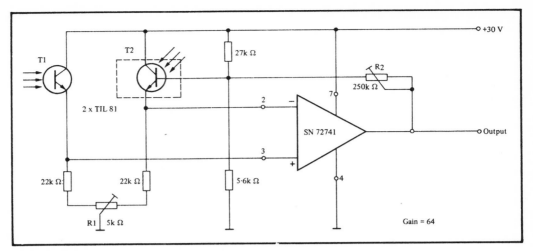

Figure 15.38
Phototransistor D.C. amplifier with temperature compensation

control current and is further amplified, according to the D.C. gain B of the phototransistor. This effect is particularly undesirable with low irradiances.

The dark current is reduced, by connecting a base resistor R_B, which diverts the collector-base leakage current to ground, between base and ground (*Figure 15.37*). At the same time, the sensitivity of the phototransistor is reduced, since part of the photocurrent I_P of the collector-base photodiode is also diverted.

Figure 15.38 shows the temperature compensation of phototransistors in a difference stage. It contains the phototransistor T 1 as photodetector and an obscured phototransistor T 2 as reference transistor. The temperature response of the two phototransistors has no effect, since the subsequent operational amplifier is only driven by a difference signal. Scatter between specimens of the phototransistors and of the operational amplifier are compensated by the potentiometer R 1. The desired gain is set with the resistance R 2. This circuit is used as a scanning amplifier and as the detector in optoelectronic couplers.

15.3.3
Phototrigger and photomultivibrator circuits

Phototrigger circuits can be constructed very economically with a phototransistor and a bipolar transistor. *Figure 15.39* shows a photothyristor circuit with the NPN phototransistor TIL 81 and the PNP transistor BC 212. The collectors of the two transistors each drive the base of the

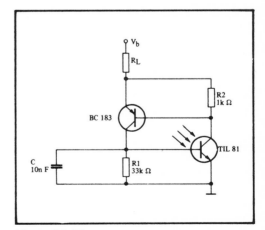

Figure 15.39
Photothyristor, simulated with NPN phototransistor and PNP transistor

294

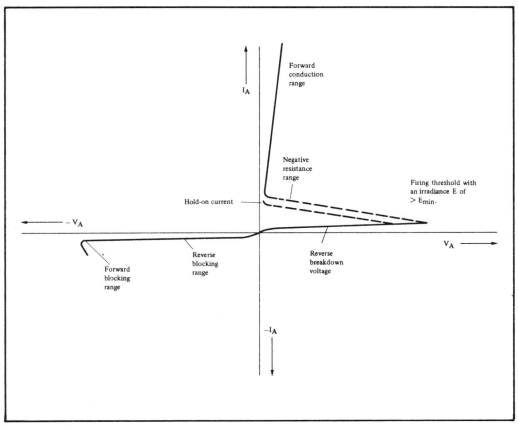

Figure 15.40
Theoretical characteristic of a photothyristor

other transistor. With sufficiently high irradiance, the phototransistor turns on the transistor T 2. Through the feed-back to the base of the phototransistor, the latter remains conductive, even if the irradiance decreases again. The response sensitivity of the thyristor is primarily determined by the base resistance R 1, which diverts part of the photocurrent, while the capacitor C and the resistance R 2 prevent undesired firing of the thyristor in case of rapid changes in the anode voltage. The thyristor can be extinguished again by switching off the supply voltage.

The characteristic in *Figure 15.40* shows, in the 1st quadrant, the characteristic of the thyristor in the forward range and, in the 3rd quadrant, the reverse characteristic.

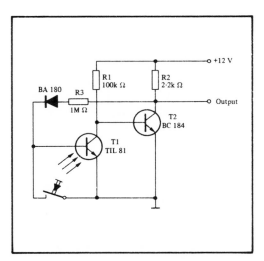

Figure 15.41
Phototrigger circuit with reset button

295

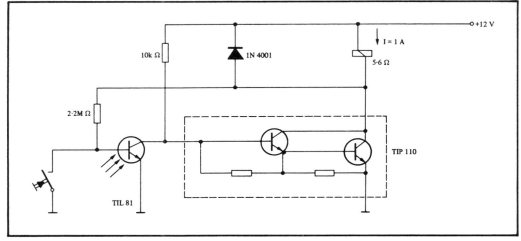

Figure 15.42
Phototrigger with reset button and NPN-Darlington Power stage

This circuit is used as a simple detector for A.C. operation, in firing circuits for power thyristors and for the storage of non-recurring optical events.

As a further example, *Figure 15.41* shows a phototrigger circuit with a reset button. In the dark condition, the transistor T 2 is driven through R 1 and is conductive. If radiation falls on the phototransistor T 1, the base control current of T 2 is reduced by the amount of the collector current of T 1. The collector voltage of T 2 rises, so that the base of the phototransistor TIL 81 is driven additionally through R 3 and R 2. The circuit trips to the new state, which is maintained, because of the feedback, even if the phototransistor is no longer irradiated. The diode BA 180 prevents a part of the photocurrent from flowing away through the feed-back resistance R 3. The original operating condition can be re-established by the reset button.

In *Figure 15.42*, the transistor T 2 has been replaced by the NPN-Darlington transistor TIP 110.

The circuit in *Figure 15.43* is constructed as a phototrigger in a complementary circuit. If the phototransistor TIL 81 is irradiated, both transistors T 1 and T 2

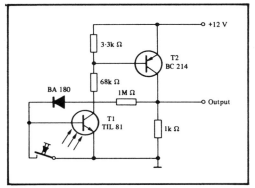

Figure 15.43
Phototrigger in complementary circuit; in the quiescent state (without radiation), both transistors are non conducting

change to the conductive state, which is maintained until the reset button is operated. In the circuit in *Figure 15.44*, T 2 has been replaced by the PNP-Darlington transistor TIP 115 in order to be able to drive high-power loads. The applications of such photo-trigger circuits include optical threshold switches, twilight switches, sensors in alarm installations etc.

A further two-state circuit is the mono-stable multivibrator shown in *Figure 15.45*. The output transistor T 2 drives a counter relay. In the steady-state condition, radi-

296

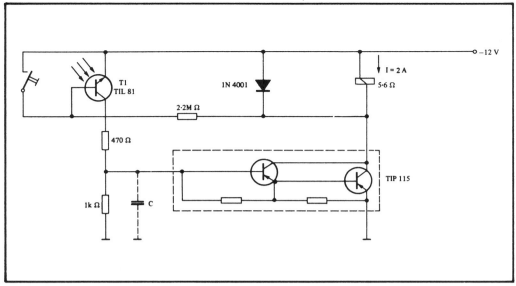

Figure 15.44
Phototrigger with PNP-Darlington power stage

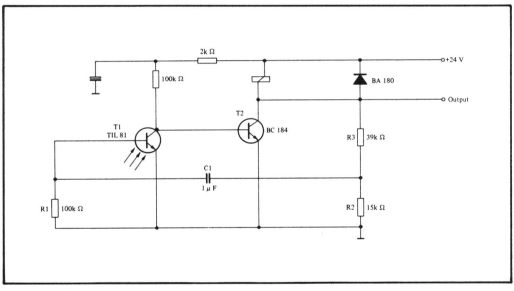

Figure 15.45
Phototrigger circuit, as monostable multivibrator with a counter relay

ation falls on the phototransistor T 1. The output transistor is blocked and the counter relay is de-energised. If the radiation falling on the phototransistor TIL 81 is briefly interrupted, T 2 becomes conductive. The collector voltage on T 2 decreases.

The base of the phototransistor receives a negative potential through the feed-back capacitor C 1, so that T 1 changes to the non conducting condition and T 2 further into the conductive state. The capacitor is now discharged through the resistors R

297

and R 2 ∥ R 3 and, if exposed to light, through the photocurrent of the transistor T 1. The drop-out time of the relay therefore depends on the irradiance falling on the phototransistor, while high irradiances give short drop-out times.

Figure 15.46 shows a timing switch with the N-channel junction FET Type BC 264A. The circuit also works as a monostable multivibrator. If radiation falls with sufficient intensity on the phototransistor TIL 81, the phototransistor switches to the conductive state and the FET is turned off. After the discharge of the capacitor C 1 through R 1 and R 2, the FET becomes conductive again, irrespective of the radiation falling on the phototransistor. The high input impedance of the field-effect transistor does not load the timing circuit, so that the time constants can be constructed with very high resistances and small capacitor values.

Finally, in *Figure 15.47*, the circuit of a sensitive photo-Schmitt-trigger with a low dark current is shown. With very low irradiances and high operating temperatures, the not inconsiderable residual current of the collector-base photodiode is to be diverted through a base resistor. The resultant loss of sensitivity is compensated by the variable resistor R 1. A Schmitt trigger (T 3, T 4) is driven through the decoupling amplifier T 2. The circuit is used in sensitive optoelectronic couplers.

15.4
Simple couplers with filament lamps and two-pole phototransistors

1
Construction with few, inexpensive components.

2
Little mechanical adjustment work.

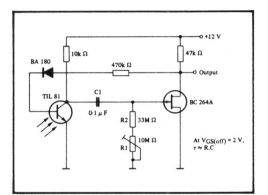

Figure 15.46
Photo-time-switch with FET

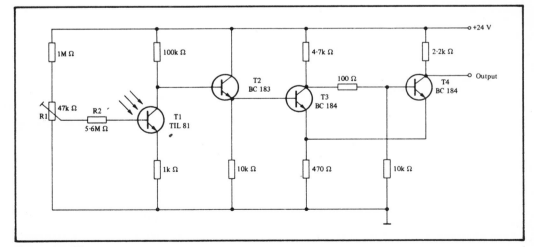

Figure 15.47
Photo-Schmitt-trigger with low dark current

3

The couplers are to be designed optically and electrically with very high safety factors.

4

A command signal is initiated by a power switch directly at the detector output

Simple couplers only make a Yes-No decision. They are used to control various functions, such as vane-wheel sensing in anemometers, sensing and monitoring the rotational speed of rotating parts, cam sensing, film frame alignment in film projectors and end-of-tape switching in audio and video tape recorders.

Such simple opto couplers need a powerful radiation source. Thus the amplifier cost at the receiving end is kept low and a good signal/interference ratio is achieved against stray radiation. A filament lamp satisfies these conditions. The type of filament is mainly determined by the distance from source to detector. The filament lamps used can include types with and without lenses, selected and non-selected miniature lamps, dial lamps, torch bulbs, telephone indicator lamps and small and large special filament lamps for optical applications. Allowance must be made for their sensitivity to vibration, temperature-dependence and wide tolerances on radiant intensity. Colour temperature data is only obtainable for the optical filament lamps.

The distribution temperature which is actually needed is seldom stated. The filament of the radiation source used should be as straight and evenly wound as possible. Where possible, the lamp will be operated with D.C., since A.C. would cause undesired modulation of the radiation. To improve reliability, the lamps will be operated at about 10 to 20% below their rated voltage. Depending on the type and whether the lamp is a gas-filled or vacuum type, the colour temperature then falls to about 2000° K. The spectral radiant power distribution (emission function) of the lamp is

then still just adequately matched to the spectral sensitivity distribution of silicon junction detectors. However, the very large proportion of thermal radiation can cause an unwanted heating of the photodetector. Therefore, an IR filter (e.g., RGN 9 or RG 38) may sometimes have to be placed in front of the photodetector.

As a sensor for simple couplers, for example, the two-pole, plastic-encapsulated phototransistor TIL 78, which has a typical aperture angle of $\pm\ 20°$, is particularly suitable. With an irradiance of $E_{e,2856K}$ = 20 mW/cm^2, it delivers photocurrents of at least 1 mA and typically 7 mA. This high sensitivity for a plastic phototransistor also makes it possible to keep the expenditure on amplification low. Taking account of the stated tolerances, the distances stated for the couplers described below contain a high safety factor. With exact adjustment, greater distances can be spanned.

As a first example, *Figure 15.48* shows a simple coupler for r = 10 mm with a dial-light lamp and a photo-Darlington transistor. When the phototransistor TIL 78 is irradiated, the relay pulls in.

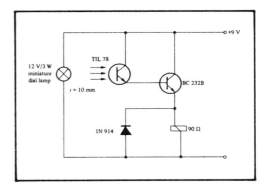

Figure 15.48
Opto coupler for r = 10 mm

The coupler in *Figure 15.49* is designed for r = 15 mm. The phototransistor is followed by a voltage amplifier, which drives the collector-connected power stage. The relay drops out, when the radiation beam is interrupted.

299

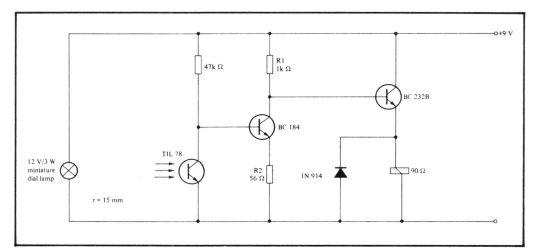

Figure 15.49
Opto coupler for r = 15 mm

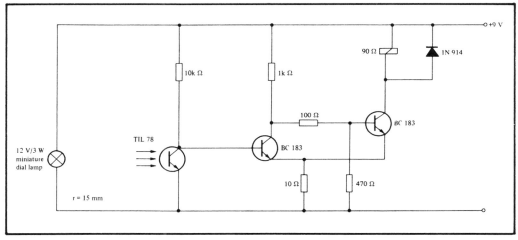

Figure 15.50
Opto coupler for r = 15 mm with threshold switch (Schmitt trigger)

In the coupler in *Figure 15.50*, also for r = 15 mm, the phototransistor TIL 78 drives a Schmitt trigger. On irradiation, the relay pulls in.

In *Figure 15.51*, an opto coupler for r = 30 mm is shown. The phototransistor drives a sensitive two-transistor amplifier. On irradiation, the relay pulls in.

Finally, *Figure 15.52* shows a coupler for an alarm installation. If the radiation falling on the phototransistor is interrupted or the contact mat is trodden on, the relay pulls in. The radiation source is a pocket torch with an IR filter RG 9 attached. The detector is mounted at the focus of a flashlight reflector. An IR filter RG 38 is also fitted in front of the phototransistor TIL 65. With this arrangement, distances of r ⩾ 5 m can be spanned. Despite the IR filter, the sensitive detector of the coupler must be shielded against stray radiation.

300

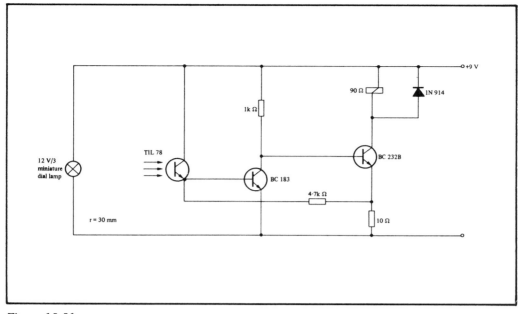

Figure 15.51
Opto coupler for r = 30 mm

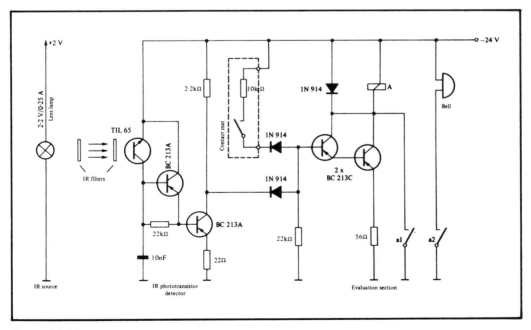

Figure 15.52
Simple opto coupler with contact mat for an alarm installation

15.5
Logic circuits with phototransistors

Logic circuits can also be constructed with phototransistors. They are used for the logical evaluation of one or more radiation sources directly at the measurement point. High potential differences can exist between the radiation sources and the photodetector logic. The logic levels are characterised as follows:

Optical logic level
L (low) = Phototransistor not irradiated
H (high) = Phototransistor irradiated

Electrical logic level
L (low) = Voltage almost zero
H (high) = Almost supply voltage

Figures 15.53 to *15.56* show the four basic logic circuits.

Figure 15.53:
AND gate consisting of two TIL 78 phototransistors in series.

Figure 15.54:
NAND gate consisting of three LS 611 phototransistors in series.

Figure 15.55:
OR gate consisting of three LS 613 phototransistors in parallel.

Figure 15.56:
NOR gate consisting of three LS 612 phototransistors in parallel.

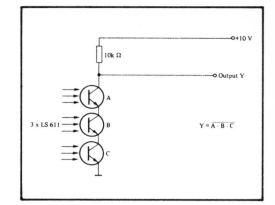

Figure 15.54
NAND gate with phototransistors (3 inputs)

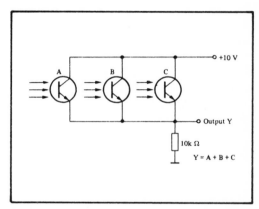

Figure 15.55
OR gate with phototransistors (3 inputs)

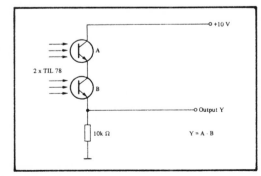

Figure 15.53
AND gate with phototransistors (2 inputs)

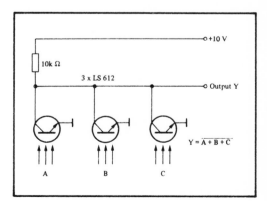

Figure 15.56
NOR gate with phototransistors (3 inputs)

302

A further logic circuit is illustrated in *Figure 15.57*. Here, the output switches to L-level, if both phototransistors are irradiated. Thus the circuit is a NAND gate. The diode in the emitter circuit and the base diverter resistance on transistor T 3 prevent the transistor from being turned on by residual currents of the phototransistors.

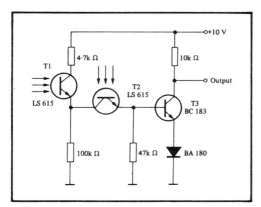

Figure 15.57
Logic circuit with two phototransistors and an NPN transistor; it represents a NAND gate

The circuit in *Figure 15.58* delivers an H-signal at its output, if the phototransistor T 2 is irradiated and the phototransistor T 1 is not irradiated.

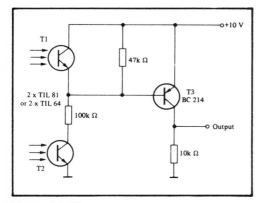

Figure 15.58
Logic circuit with two phototransistors and a PNP transistor; this is an OR gate

Figure 15.59 shows a circuit to give L-level at the output, when the phototransistor T 1 is irradiated and phototransistor T 2 is not irradiated.

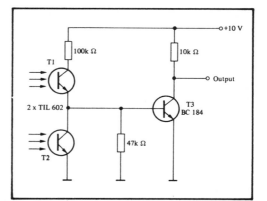

Figure 15.59
Logic circuit with two phototransistors and an NPN transistor, it fulfils the NOR function

In the same way, Exclusive-OR gates can be constructed with phototransistors and additional active components (*Figure 15.60*). If neither of the two phototransistors are irradiated, the transistors T 3 and T 4 are non conducting because their bases and emitters are almost at the potential of the supply voltage. If only the phototransistor T 1 is now irradiated, the transistor T 3 becomes conductive, because its emitter potential falls

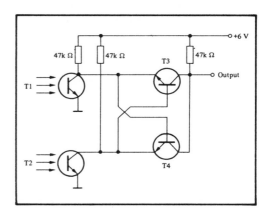

Figure 15.60
Exclusive NOR gate with phototransistors

almost to 0V and it receives a base current supply through the collector resistance of the phototransistor T 2. If the phototransistor T 2 now also turns on, then both transistors T 3 and T 4 are again non conducting, because both their bases and their emitters are connected to 0 V through the phototransistors.

15.6
Photodetector circuits to drive TTL integrating circuits

In industrial control installations and in punched tape and card readers, Si junction photodetectors are used to drive TTL integrated circuits. TTL-compatible photodetectors can drive a TTL circuit directly. Si phototransistors are TTL-compatible, if the collector current produced by the incident radiation is at least as great as the necessary input current of the TTL circuits. For this, they must have a sufficiently small saturation voltage $V_{CEsat} \leqslant 0.4$ V.

Input currents and voltages of the SN54/74 Series Schmitt triggers

Non-TTL-compatible photodetectors need a matching or interface circuit for the corresponding TTL circuit. Also, TTL circuits only work reliably, if the control signals have a rise and fall time of t_r, $t_f \leqslant 200$ n secs. The output signals from phototransistors do not satisfy this requirement. Indeed, the optical turn-on and turn-off of a radiation falling on the phototransistor usually takes several milliseconds or longer. Therefore it is necessary, to speed up the edges of the signals from the photodetector with a

Schmitt trigger. in order to prevent oscillation of the subsequent TTL circuits.

The input levels and currents of the most commonly used TTL Schmitt triggers are summarised in the following table.

In this:
V_{T+} = Positive response threshold,
V_{T-} = Negative response threshold,
V_{IH} = H-level input voltage,
V_{IL} = L-level input voltage,
I_{IH} = H-level input current,
I_{IL} = L-level input current.

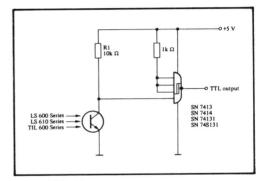

Figure 15.61
Direct control of TTL Schmitt triggers with a phototransistor

Figure 15.61 shows a phototransistor interface circuit to drive TTL Schmitt trigger circuits. With a collector voltage of $V_{CE} = V_{T-min}$, the irradiance falling on the phototransistor should produce a minimum collector current of $I_{CE} = I_{IL}$. Under these conditions, the TTL output will be safely switched to the logical H level. The maximum value of the resistance R 1 is calculated as follows:

Type	V_{T+max} (V)	V_{T-min} (V)	V_{IH} (V)	at	I_{IH} (mA)	V_{IL} (V)	at	I_{IL} (mA)
'13	2·0	0·6	2·4		0·04	0·4		−1·6
'14	2·0	0·6	2·4		0·04	0·4		−1·2
'132	2·0	0·6	2·4		0·04	0·4		−1·2
'S132	1·9	1·1	2·4		0·05	0·5		−2·0

$$R_{1\,max} = \frac{V_b - V_{T+max}}{I_{CE} - I_{IL}}$$

Unused TTL inputs are connected together and through a pull-up resistance (1 kΩ) to the supply voltage. This an H-level is ensured at these inputs. Because of their small mechanical dimensions, the LS 600, LS 610 and TIL 600 phototransistors are particularly suitable for punched tape and card readers.

Figure 15.62 shows a drive circuit with inverted output function. When the phototransistor is irradiated, an L-level is obtained at the output, if a voltage of at least V_{T+max} is present at the Schmitt trigger input.

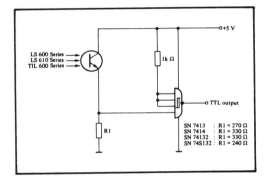

Figure 15.62
Direct control of TTL Schmitt triggers with a phototransistor

The necessary collector current I_{Cmin} of the phototransistor is calculated as:

$$I_{C,min} = \frac{V_b - V_{IL}}{R_1} + I_{IH}$$

The resistance R 1 is obtained as

$$R_1 = \frac{V_{T-min}}{I_{IL}}$$

If the radiation falling on the phototransistor is weak, a matching amplifier is necessary between the phototransistor and the Schmitt trigger. For this, *Figure 15.63* shows a phototransistor interface circuit and *Figure 15.64* shows a photodiode interface circuit to drive TTL circuits.

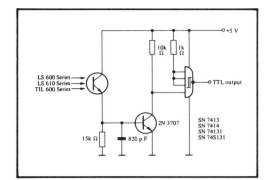

Figure 15.63
Phototransistor interface circuit for driving TTL Schmitt triggers

Punched-tape and punched-card reading heads for a large number of reading channels can be constructed more economically, if the signals from the phototransistors or diodes are led directly, through any necessary interface circuits, to the TTL gates. In this form, however, the output signals would not be usable, since the gate oscillates during the turn-on and turn-off of the phototransistor, because the signal edges are too slow (*Figure 15.65*). In order still to be able to evaluate the read signals, an additional marker channel, for which the drive perforation track can be used in the case of punched tape, is needed. This signal must then be fed through a Schmitt trigger, which delivers a satisfactory TTL signal at its output. With this strobe or interrogation signal, the read amplifiers are then only conducting at a time when definite logic levels can be relied on at the input of the amplifier.

Figure 15.66 shows the corresponding circuit. Since the SN 75450 circuit already contains the necessary transistors, this circuit can be constructed very economically.

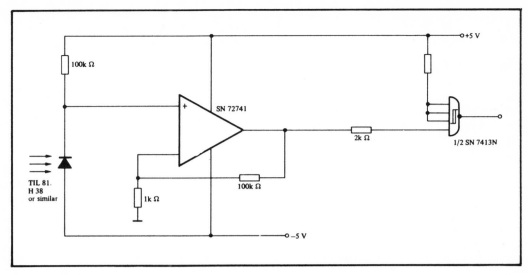

Figure 15.64
Photodiode interface circuit with operational amplifier

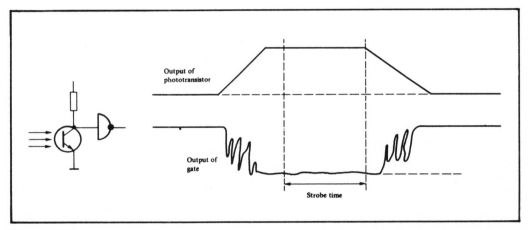

Figure 15.65
Oscillation of a TTL gate when driven with signals with insufficiently fast rise and fall times

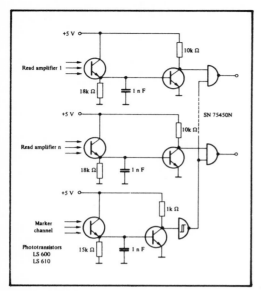

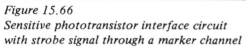

Figure 15.66
Sensitive phototransistor interface circuit
with strobe signal through a marker channel

16
Modulated transmitters with luminescence diodes

16
Modulated transmitters with luminescence diodes

The radiation from a luminescence diode can be modulated. Basically, both sine-wave and pulse modulation are possible. If a luminescence diode is to be controlled with a sinusoidal alternating current, then its working point has to be set with a bias current I_F.

The necessary thermal calculation for luminescence diodes was described in Section 11.2. In pulse operation, luminescence diodes need no additional bias current (see also Section 11.7).

16.1
The simplest modulator circuits for luminescence diodes

Figure 16.1 shows the simplest modulator circuit for luminescence diodes. The GaAs diode TIL 32 is controlled with a sinusoidal alternating current. The Si diode connected in parallel protects the luminescence diode from high reverse voltages. Because of the non-linear forward characteristic of the GaAs diode, the signal emitted will be distorted.

Figures *16.2* and *16.3* show two further theoretical circuits for modulcators. Through a switch S, the current through the luminescence diode is increased in *Figure 16.2*, or short-circuited in *Figure 16.3*, in time with the clock frequency.

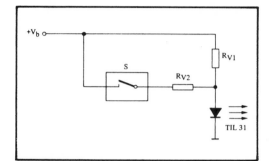

Figure 16.2
Simple modulator circuit with periodical keying of the GaAs diode by the switch S

If the contact S in *Figure 16.2* is replaced by a phototransistor (e.g., TIL 81) opposite the luminescence diode, and a moving punched tape is inserted between the luminescence diode TIL 31 and phototransistor TIL 81, to produce a timing frequency, then the coupler shown in *Figure 16.4* is obtained. Here, an optical

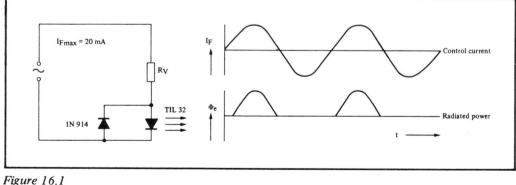

Figure 16.1
Control of a luminescence diode with alternating current

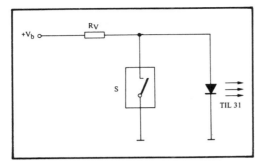

Figure 16.3
Simple modulator circuit with periodical short-circuting of the GaAs diode with the switch S

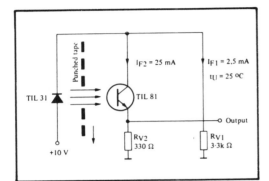

Figure 16.4
Opto coupler with optically-closed feed-back loop

feed-back loop is used to key the luminescence diode. If the phototransistor is obscured, the forward current through the luminescence diode TIL 31 is determined by the series resistance R_{V1}.

$$I_{F1} = \frac{V_b - V_{F,TIL31}}{R} = \frac{10\ V - 1 \cdot 6V}{3 \cdot 3\ k\Omega}$$

$$= 2 \cdot 45\ mA$$

If the feed-back loop is closed, so that the radiation from the luminescence diode falls on the phototransistor, the current increases, in relation to the coupling characteristics, to the value I_{F2}. This is calculated as

$$I_{F2} = \frac{V_b - V_{F,TIL31} - V_{CEsat,TIL81}}{R_{V2}}$$

$$= \frac{10\ V - 1 \cdot 6\ V - 0 \cdot 3\ V}{330\ \Omega} = 24 \cdot 5\ mA$$

Then the total forward current I_{Ftot} through the luminescence diode TIL 31 is:

$$I_{Ftot} = I_{F1} + I_{F2} = 2 \cdot 45\ mA + 24 \cdot 5\ mA$$

$$= 26 \cdot 95\ mA$$

Figure 16.5 shows an oscillator circuit with optical feedback. In this, the contact S in *Figure 16.3* has been replaced by a phototransistor TIL 81 opposite the luminescence diode TIL 31. When the phototransistor is irradiated, the GaAs diode is short-circuited. The radiation from the luminescence diode is thus switched off. The phototransistor, which is now not irradiated, increases in resistance, so that a forward current again flows through the GaAs diode. The process is repeated periodically. The frequency of this oscillator lies in the low-frequency range. It is determined by the storage times of the phototransistor and thus depends on the degree of overload of the transistor, which is affected by the coupling characteristic of the coupler (distance). The output signal can be amplified further with an LF amplifier. This circuit is suitable for simple couplers and for alarm installations.

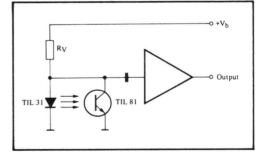

Figure 16.5
Modulator circuit for luminescence diodes using a coupler with optical feed-back

16.2
Sine-wave modulated transmitters with luminescence diodes

The steady-state current of a luminescence

312

diode in sine-wave modulated operation can be produced either from a voltage source through the series resistance R_V or through a constant current source.

16.2.1
Modulated operation with bias voltage sources

If the steady-state current of the luminescence diode is taken from a voltage source, the series resistance R_V is calculated from the formula:

$$R_V = \frac{V_b - V_F}{I_F}$$

In order to avoid modulation distortion, the maximum modulation voltage during the negative half-wave must not reach the lower non-linear part of the characteristic curve $I_F = f(V_F)$. In *Figure 16.6*, the working points A_I and A_Φ for low-distortion modulation of luminescence diodes are plotted.

The luminescence diodes can be modulated by coupling the modulation frequency in parallel with the bias voltage supply. In the circuit b in *Figure 16.7*, additional measures have been taken (choke and filter capacitor) in order to prevent interaction of the modulation voltage of the voltage source which produces the steady-state current.

It is, however, more usual to connect the modulation voltage generator and the

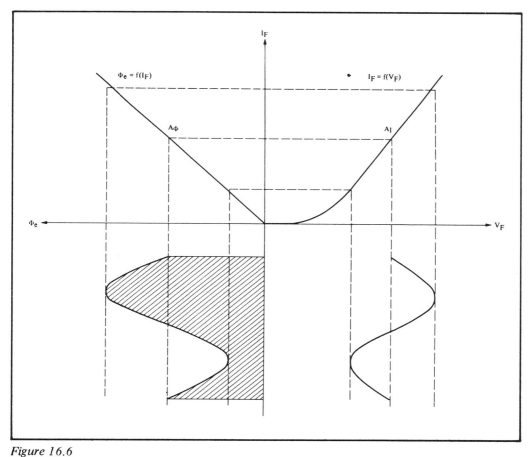

Figure 16.6
Selection of the working point for low-distortion modulation of luminescence diodes

313

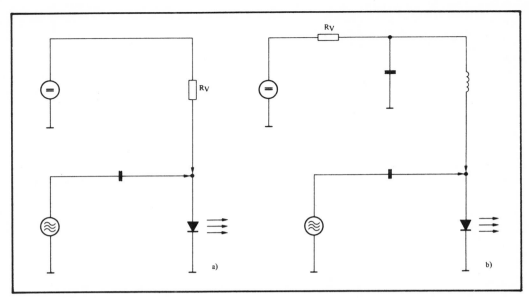

Figure 16.7
Modulator circuits for luminescence diodes (parallel feed)

steady-state supply in series (*Figure 16.8*). This method, with which both functions are combined in one circuit, will be used in almost all the applications to be shown later.

An amplitude-modulated transmitter corresponding to the theoretical circuit in *Figure 16.7* can be constructed with few components. As an example, *Figure 16.9* shows a low-frequency-modulated transmitter for the transmission of music programmes. The transmitter receives the modulation signal from the loudspeaker

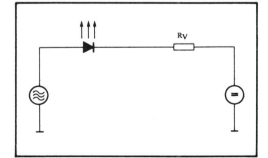

Figure 16.8
Modulator circuit for luminescence diodes (series connection)

output of a radio receiver. The photo-detector can be connected, for example, to the preamplifier of a tape recorder. If additional lenses are fitted in front of the luminescence diode TIXL 27 and the photodiode TIL 81, distances of r > 10 m can be spanned.

In *Figure 16.10*, a series connection of the modulation amplifier and steady-state current source is used. The steady-state current is supplied by the transistor, which works as a current source. The steady-state current is calculated from the formula:

$$I_{F,R} = \frac{V_b \cdot \dfrac{R1}{R1 + R2} - V_{BE}}{R_E}$$

$$= \frac{10\ V \cdot \dfrac{2 \cdot 2\ k\Omega}{2 \cdot 2\ k\Omega + 2 \cdot 2\ k\Omega} - 0 \cdot 7\ V}{220\ \Omega}$$

$$= 20\ mA$$

In *Figure 16.10*, the modulation voltage is fed in to the base of the transistor. Because

314

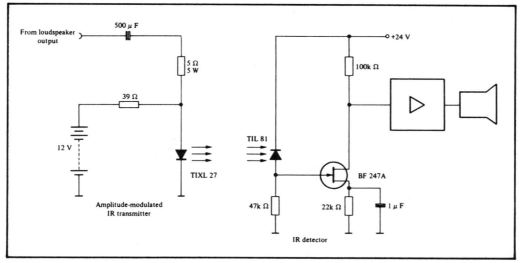

Figure 16.9
Simple IR LF transmission link

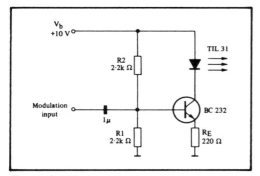

Figure 16.10
In this modulator, the amplifier and steady-state current source are connected in series

of the current feed-back, the alternating current in the transistor is proportional to the modulation voltage.

The circuit in *Figure 16.11*, in which the luminescence diode TIXL 27 is operated with a current $I_F = 300$ mA, is of similar construction. Because of the loss power transistor TIP 33 is used as a driver, and is preceded by an additional emitter follower, in order not to load the modulation source.

Figure 16.12 shows an amplitude-modulated transmitter with the operational

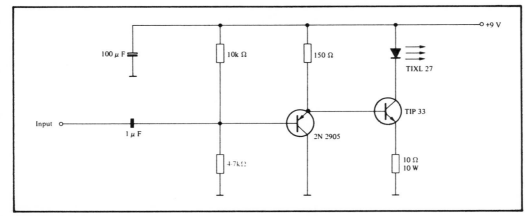

Figure 16.11
Modulator circuit for the GaAs power diode TIXL 27

315

amplifier SN 72741P. The working point is set by the voltage divider R 1, R 2, which at the same time determines the gain. The resistance R 4 determines the steady-state current I_F:

$$I_F = \frac{V_b -}{R4} = \frac{15\ V}{680\ \Omega} = 22\ mA$$

In order to keep the loss power in the operational amplifier low, the greatest part of this current is supplied through the resistance R 3. The operational amplifier permits a modulation variation of approx. ± 10 mA. Thus the necessary modulation voltage V_{mod} at the input of the amplifier is calculated as:

$$V_{modrms} = \frac{10\ mA.680}{\sqrt{2}} \cdot \frac{R2}{R1 + R2}$$

$$= 22\ mV$$

Figure 16.13 shows an HF-modulated transmitter to carry a television picture. The modulator input is driven directly by the HF output signal of a television camera.

The HF power transistor XB 436 works on an oscillator circuit tuned to f_O = 50 MHz. The luminescence diode is connected in series with the oscillator coil. A silicon diode, connected in opposing parallel, protects the GaAs diode from high reverse voltages. It must be noted, that silicon-doped GaAs diodes, such as the TIXL 12 used here, only have a low radiant efficiency at a frequency f = 50 MHz. Of course, the useful radiant power produced is sufficient here, to drive the photodetector.

As the photodetector, the very sensitive Si avalanche photodiode TIXL 56 is used, it has a very high limiting frequency. Its working point must be adjusted so that the high internal "avalanche gain" is utilised with a good signal/noise ratio (see Chapter 9). The avalanche diode TIXL 56 drives a broad-band circuit tuned to f_O = 50 MHz. The parallel-tuned circuit consists of the junction capacitance of the avalanche diode, the circuit capacitance and the coil L 1. The output signal, coupled out at low impedance through L 2, is fed to the 60-Ohm aerial socket of a television receiver switched to Channel 2.

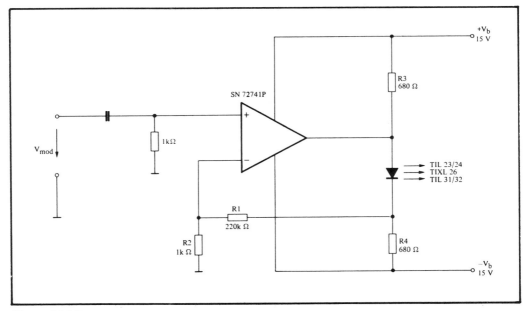

Figure 16.12
Amplitude modulated transmitter with an operational amplifier

316

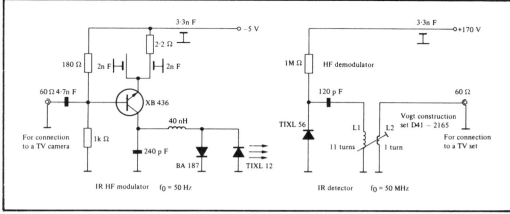

Figure 16.13
Simple IR HF-link for transmission of a TV picture

Figure 16.14 shows an amplitude-modulated IR transmitter, which is used, together with a suitable receiver (see Figure 17.10) as an optical-link telephone. For this reason, the frequency range of the amplifier is designed for a range of 300 . . . 3000 Hz, but can be extended without difficulty, by changing the appropriate components, to a range of 50 . . . 60 000 Hz. The signal coming from the microphone is first amplified with a low-noise transistor amplifier. The voltage gain is calculated from the formula:

$$V_U = \frac{R1 \cdot R2}{R1 + R2} \cdot \frac{I_C}{26 \text{ mV}}$$

$$= \frac{18 \text{ k}\Omega \cdot 100 \text{ k}\Omega}{18 \text{ k}\Omega + 100 \text{ k}\Omega} \cdot \frac{0 \cdot 1 \text{ mA}}{26 \text{ mV}} = 59$$

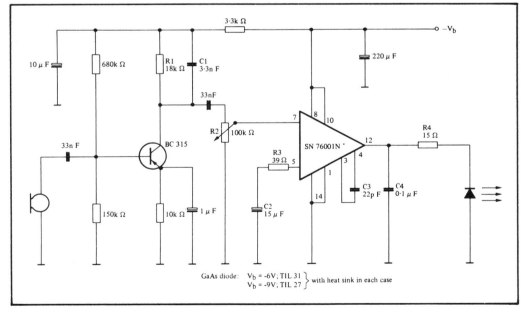

Figure 16.14
Amplitude-modulated IR transmitter with integrated LF power amplifier

317

The upper frequency limit is determined by the time-constant R 1.C 1. The gain of the whole circuit, and thus the modulation level, can be adjusted with the potentiometer R 2. There follows an SN 76001 AN integrated audio power amplifier. The resistance R 3 fixes the gain $V_u = 200 = 46$ dB, while the capacitor C 2 in series with it determines the lower frequency limit $f_u = 300$ Hz. The capacitors C 3 and C 4 compensate the phase and frequency response of the integrated amplifier and thus prevent oscillation of the circuit.

Since the positive input is grounded, the GaAs diodes can be mounted on large-area heat-sinks, without the occurrence of insulation problems. The steady-state current I_{Fo} of the diodes is set by the series

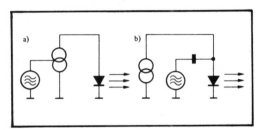

Figure 16.15
Amplitude modulation of the radiation from luminescence diodes with current sources; a) series connection and b) parallel connection of current source and modulation generator

resistance R 4. With an operating voltage of 6 V, for the diode TIL 31, its value is:

$$I_{Fo} = \frac{\dfrac{V_b}{2} - V_F}{R4} = \frac{3 \text{ V} - 1 \cdot 5 \text{ V}}{15\,\Omega} = 0 \cdot 1 \text{ A}$$

and with an operating voltage of 9 V, for the diode TIXL 27:

$$I_{Fo} = \frac{4 \cdot 5 \text{ V} - 1 \cdot 5 \text{ V}}{15\,\Omega} = 0 \cdot 2 \text{ A}$$

In conjunction with a sensitive photodetector (*Figure 17.10*), the range r is more than 10 m, if the TIL 31 is used as the transmitting diode. If additional lenses are fitted to the transmitter and the receiver to focus the beam, several hundred metres can be spanned. In this case, however, exact alignment of the transmitter and receiver is necessary (tripod). As a protective measure against irradiation by direct sunlight, IR filters RG 38 are to be fitted in front of the diodes, since otherwise the focussed solar radiation would destroy the wafer of the photodiode or the luminescence diode.

16.2.2
Modulation operation with constant-current sources

The operation of luminescence diodes with

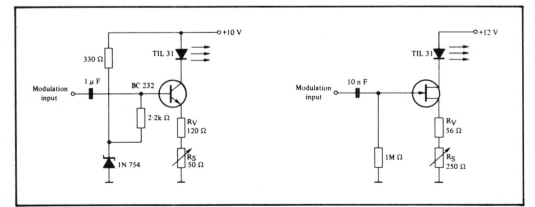

Figure 16.16
Current sources as modulation amplifiers for luminescence diodes

318

current sources has already been described in Chapter 14; therefore it does not need to be discussed further.

Figure 16.15 shows the principle of the modulator circuit in conjunction with the current sources. In *Figure 16.15a*, the current source itself is modulated, so that the diode current varies with the modulation voltage. Circuit b shows the modulation voltage fed in in parallel.

In *Figure 16.16*, constant-current sources work at the same time as modulation amplifiers. The modulation signal is fed in in parallel with the input voltage V_e of the constant-current source. The steady-state current is determined by the resistance R_V, together with the variable resistance R_s.

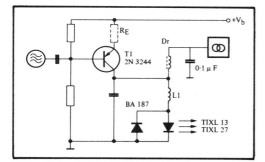

Figure 16.17
HF drive of luminescence diodes

Figure 16.17 shows how high-power luminescence diodes are driven with HF. The transistor T 1 works in Class B. Its load impedance is a parallel tuned circuit located in the collector lead. The GaAs diode is connected in series with the coil of the tuned circuit. During a negative half-wave on the base, a high collector current flows in the tuned circuit. The coil current through the GaAs diode is higher than the current in the transistor, by the Q-factor of the tuned circuit. In linear modulation operation, the working point of the GaAs diode is determined by the constant current. If the GaAs diode is to work in pulsed operation, the constant-current source is omitted. The luminescence diode has to be protected from high reverse voltages. Therefore, a Si diode BA 187 is connected in opposing parallel with it.

As a further example, a sine-wave modulated IR transmitter is shown in *Figure 16.18*. The oscillator consists of a three-stage amplifier with a field-effect transistor in the input stage. Through the feed-back branch (Wien branch), consisting of the RC networks R 1, C 1 and R 2, C 2, the frequency of the oscillator is determined and can be calculated from the formula:

$$f = \frac{1}{2 \pi \sqrt{R\,1.R\,2.C\,1.C\,2}}$$

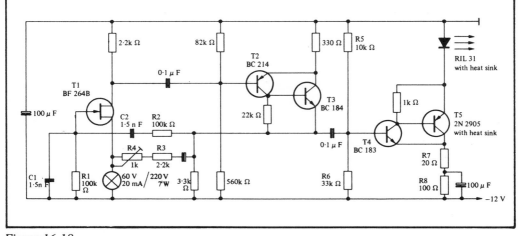

Figure 16.18
Sine-wave-modulated IR transmitter with Wien-Robinson oscillator

319

A non-frequency-dependent feed-back (Robinson circuit), through the resistances R 3, R 4 and the filament lamp, stabilises the amplitude, since with increasing oscillator amplitude the resistance of the filament lamp, and thus the degree of feed-back, increases.

This circuit ensures an excellent frequency and amplitude stability, which is particularly required, if narrow-band-pass amplifiers are used in radiation links, in order to suppress interfering radiation.

A current source, the steady-state current I_F of which is determined by the base voltage R 5, R 6 and the emitter resistances R 7, R 8, serves as the modulator. A steady-state current of

$$I_F = \frac{V_b \cdot \dfrac{R\,6}{R\,5 + R\,6} - V_{BE}}{R\,7 + R\,8}$$

$$= \frac{9\,V\,\dfrac{33\,k\,\Omega}{10\,k\Omega + 3\,k\Omega} - 0.7\,V}{15\,k\Omega + 120\,\Omega}$$

$$I_F = \frac{6.2\,V}{135\,\Omega} = 46\,mA$$

In order to achieve a large modulation excursion with a small oscillator amplitude ($V_{osc\ rms} = 400$ mV), part of the emitter resistance is shunted capacitatively, so that only the resistance R 7 determines the degree of modulation.

$$I_{Fp-p} = \frac{500\,mV.2.\sqrt{2}}{15\,\Omega} = 94\,mA$$

Because of the loss power which occurs, the luminescence diode TIL 31 must be mounted on a heat sink. This is possible without insulation problems, since the positive pole of the supply voltage is grounded. The range is about 5 m and can be increased further with additional lenses.

16.3
Pulse-modulated pulse transmitter with luminescence diodes

For the pulse modulation of luminescence diodes, the two basic circuits shown in *Figure 16.19* are used. With these, the modulation of the diode current can be achieved in the following ways:

1
Periodic interruption of the forward current I_F.

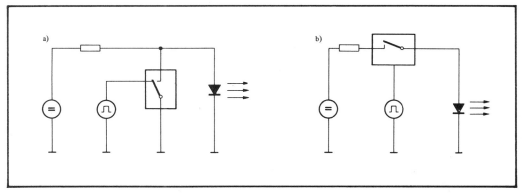

Figure 16.19
Theoretical circuits for the pulse modulation of luminescence diodes; a) pulse modulation by short-circuiting switch in parallel with the luminescence diode, b) pulse modulation by switch in series with the luminescence diode

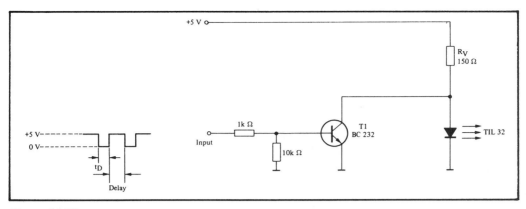

Figure 16.20
Simple pulse modulator circuit to drive luminescence diodes

2
Periodic short-circuiting of a biassed luminescence diode

3
Pulse drive of the modulation amplifer.

4
Discharge circuits with four-layer semiconductors.

As an example, *Figure 16.20* shows a simple modulator circuit for the luminescence diode TIL 32. The bias current I_F only

flows through the GaAs diode during the pulse duration t_D.

$$I_F = \frac{V_b - V_F}{R_V} = \frac{5\ V - 1 \cdot 5}{150} = 23 \cdot 3\ mA$$

During the interval, the forward current I_F flows through the transistor T 1, which is connected in parallel with the GaAs diode. The GaAs diode TIL 32 is then short-circuited. In *Figures 16.21a* and *b*, the base of the amplifier transistor T 1 is driven by the modulation signal. As in *Figure 14.9*,

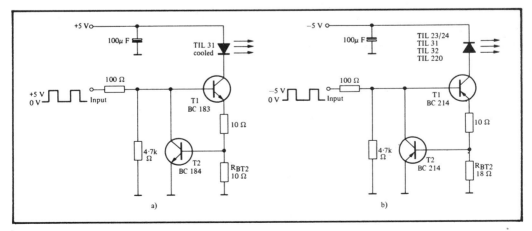

Figure 16.21
Pulse modulators for the control of luminescence diodes; a) with NPN transistors, b) with PNP transistors

321

the transistor T 2 serves to limit the forward current I_F. This is calculated at:

$$I_F = \frac{V_{BE,T2}}{R_{B,T2}}$$

For the generation of high pulse frequencies, discharge circuits with four-layer semi-conductors are suitable. *Figure 16.22* shows an oscillator circuit with the unijunction transistor TIS 43. The capacitor C 1 is charged through R 1. When its potential reaches the response threshold of the unijunction transistor, the resistance of the emitter-base path suddenly falls and the capacitor discharges through this path and the luminescence diode. The peak current I_F can be adjusted with the resistance R 2.

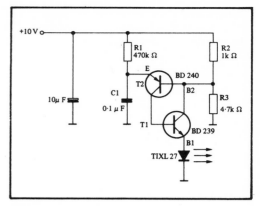

Figure 16.23
Narrow-pulse generator with unijunction equivalent circuit to drive luminescence diodes

Figure 16.23 shows an equivalent narrow-pulse modulator with a unijunction equivalent circuit. In this, the capacitor C 1 is discharged through the emitter-base diode of the transistor T 2, the collector-emitter path of T 1 and the luminescence diode TIXL 27. The maximum pulse current of this circuit is determined by the permissible base current of T 2. If very fast-acting power transistors are used for T 1 and T 2, then with a working voltage of $V_B = +40$ V, a pulse current of $I_{F,p} = 8$ A is achieved with $t_p = 0.5$ μs.

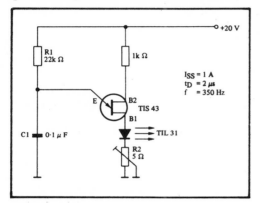

Figure 16.22
Narrow-pulse generator with unijunction transistor for the control of luminescence diodes

Discharge circuits can also be constructed with trigger diodes. *Figure 16.24* shows two narrow-pulse generators with the

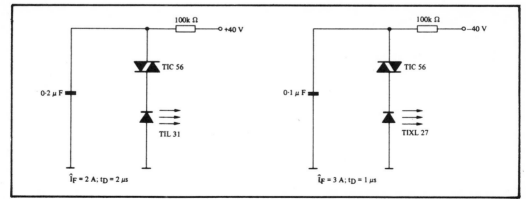

Figure 16.24
Pulse modulators with trigger diode

trigger diode TIC 56. The anode connections of the luminescence diodes are mounted directly on a grounded heat-sink.

Figure 16.25 shows a further narrow-pulse modulator with the thyristor TIC 106. The gate of the thyristor receives the trigger signal from the trigger circuit. The capacitor C 1 discharges through the thyristor and the luminescence diode TIXL 27. Following this, the thyristor turns off, when the current I_V through the resistor R_V is less than the holding current I_H of the thyristor. The capacitor C 1 charges up again. The peak current I_F through the GaAs diode TIXL 27 must not exceed 4 A. The pulse width t_D must then be less than 10 μs and the mark-space ratio must be less than 10%.

Finally, *Figure 16.26* shows a pulse transmitter for the GaAs diodes TIXL 27 and TIL 31. The pulse generator consists of an emitter-coupled multivibrator (T 1, T 2) with short rise and fall times. The driver transistor T 3 drives the power stage T 5, in which a Darlington power transistor is

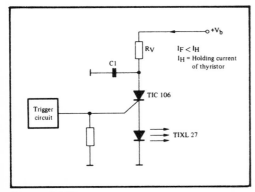

Figure 16.25
Narrow-pulse modulator with thyristor for high pulse currents

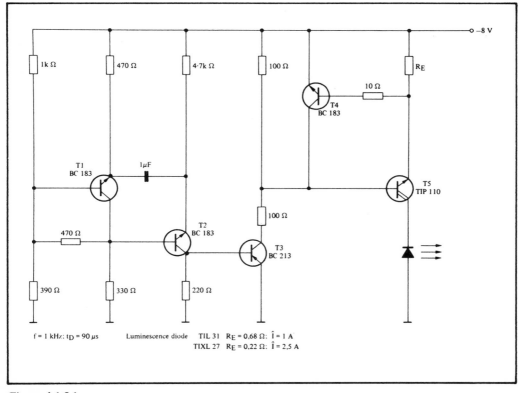

Figure 16.26
IR pulse transmitter with high power rating

323

used, so that a low control power is needed. The transistor T 4 measures the voltage drop on the emitter resistance of the final stage and regulates the current to the pre-selected value. The circuit is again so designed, that the anode connection of the luminescence can be mounted directly on a heat-sink connected to ground. When the TIL 31 is used, the range of the transmitter, without additional lenses, is more than 15 m.

17
Photodetector circuits for modulated radiation

For the demodulation of a modulated optical radiation, phototransistors, photodiodes and photocells can be used.

background radiation can be substantially suppressed by an IR filter (RG 830).

17.1
Circuits with phototransistors

The operation and the working-point conditions of emitter and collector-connected phototransistors have been dealt with in detail in Section 15.3.1.3.

As a first example, *Figure 17.1* shows a sensitive phototransistor receiver. The output signal of the phototransistor T3 is amplified in a two-stage transistor amplifier, with feedback through the resistor R1, to stabilise the working point and the gain.

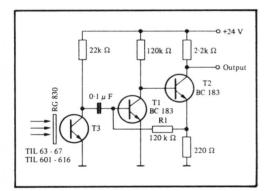

Figure 17.1
Receiver amplifier for pulsed radiation

A very weak incident signal will hardly be amplified at all by the phototransistor, since with very small collector currents this only has a low current gain, while with very large irradiances the phototransistor, and then also the following amplifier, will be overloaded. Therefore this circuit is primarily

Photodetector circuits for modulated radiation

suitable for the reception of pulsed radiation, since distortion (limitation) of the signals is then only of subsidiary importance. Interference effects from

In *Figure 17.2*, a receiver circuit with an operational amplifier is shown. Through the D.C. feedback R1 and R2, the output

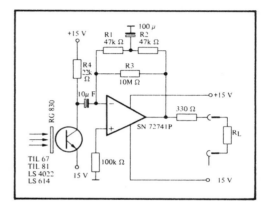

Figure 17.2
Simple phototransistor amplifier for modulated radiation

voltage of the amplifier in the steady-state condition is held at 0 V, while the A.C. voltage gain is determined by the resistance R3. The output voltage is calculated by the formula

$$V_0 = R_3 \cdot I_P$$

where I_P = photocurrent.

The circuit is suitable for the reception of LF-modulated radiation, if the working point of the phototransistor is set by the incident irradiance. Let us dispense with further circuit examples, since firstly *Figure 15.34* has been explained in detail and secondly circuits with photodiodes and photocells are generally more favourable.

17.2
Circuit with the TIL 81 as a photodiode and as a photocell

Photodiodes and photocells are suitable for the demodulation of a modulated optical radiation, even under unfavourable reception conditions. Photodiodes with reverse bias or short-circuited photocells have a linear characteristic $I_P = f(E_e)$. The mode of operation and the setting of the working point for the TIL 81 as a photocell or as a photodiode have been described in Sections 15.3.1.1 and 15.3.1.2.

Figure 17.3 shows the demodulation of a weak useful optical radiation with and without unmodulated interference radiation. In practice, the working points A_1 and A_2 can lie several orders of magnitude apart. Without interference radiation, the lower limit of detection of a radiation source depends on the noise level, or in simplified form, on the dark current of the photodetector.

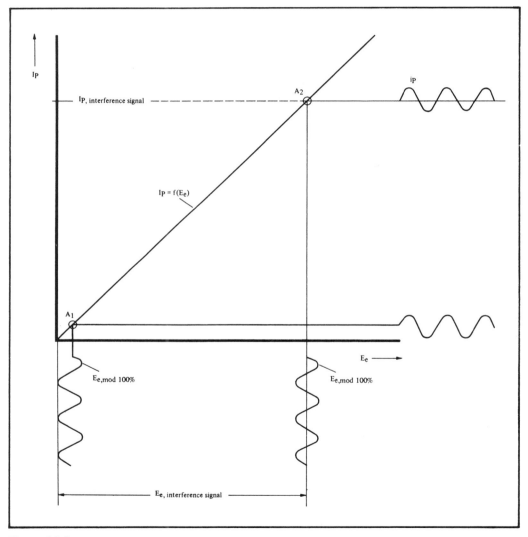

Figure 17.3
Demodulation of a modulated optical radiation

The possibility of evaluation of a weak useful radiation with a simultaneous more intensive interference radiation depends on the required ratio of useful to interference signal. Intensive interference radiations produce high interference photocurrents $I_{P,interf.}$ and thus high input noise.

The sensitivity of a photodetector circuit depends firstly on the absolute sensitivity s of the photodetector and secondly on the load resistance R_A.

The frequency limit is determined by the load resistance R_A and the junction capacitance C_i:

$$f_g = \frac{1}{2\pi \cdot R_A \cdot C_i} \qquad (10.62)$$

The mximum value of the load resistance R_A can be calculated by rearranging the equation (10.62).

$$R_A = \frac{1}{2\pi \cdot f_g \cdot C_i}$$

For pulse operation, the rise time of the photodetector is calculated as:

$$t_r = \frac{0 \cdot 35}{f_g} \qquad (10.61)$$

Figure 17.4 shows the use of FET broad-band amplifiers, driven by the photodiode TIL 81. The high-impedance load resistor R_A is not loaded by the subsequent source follower. In addition, the field effect transistor is suitable for automatic gain control. The control voltage is fed to the gate through the resistance R_G.

As a comparison, Figure 17.5 shows the use of FET broad-band amplifiers, driven by the photocell TIL 81. In comparison with diode operation, the coupling capacitor to the gate of the FET can be omitted in this case. The irradiated cell produces a negative potential on the gate. The shift in the working point of the FET is unimportant, since the maximum no-load photo-voltage is 0·6 V. The gate resistance R_A of the source-follower serves at the same time as the load resistance of the photocell.

The operating mode of the photocell is determined by the value of the load resistance R_A. If very high interference radiation levels are to be expected, the photocell must work in the short-circuited mode, in which case the load resistance must lie between $R_A = 1$ kΩ and $R_A = 5$ kΩ. With low useful and interference radiation levels, the resistance can be

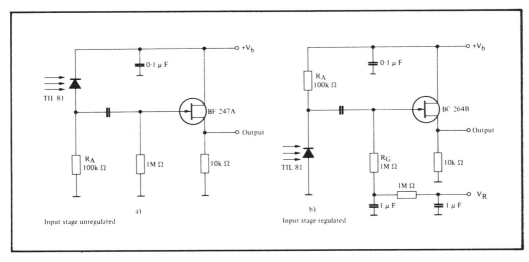

Figure 17.4
Voltage drive of broad-band amplifiers with photodiodes

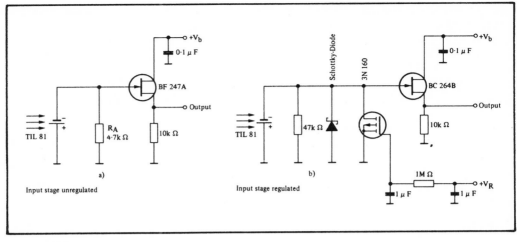

Figure 17.5
Voltage drive of broad-band amplifiers with photocells.

increased to 10 kΩ to 100 kΩ. The interference radiation falling on to the photocell is to be reduced by masks and IR filters. If the load resistance has a value of 1 MΩ, for example, the photocell works in the no-load mode. With medium irradiances, its working point is already in voltage saturation $V_{P,L} \approx 0.6$ V. The useful signal will either be totally suppressed or severely distorted. A remedy is provided by variable load resistances. Against high unmodulated interference radiations, a Schottky diode can be used as an additional variable load resistance. With high irradiances, the working point of the photocell is then determined by the forward voltage of the diode.

With very high and severely fluctuating useful irradiances, it is advisable to use a field-effect transistor, the forward resistance r_{dson} of which is adjusted by a control voltage, as the load resistance (*Figure 17.5b*).

In *Figure 17.6*, two photodetector circuits for modulation frequencies in the LF range are shown. The original FET source-follower has been replaced by a source-connected circuit. The Miller capacitance of the FET reduces the upper frequency limit still further. In these applications, it is advantageous to use low-noise FET types such as BC 264 A - D or BF 805.

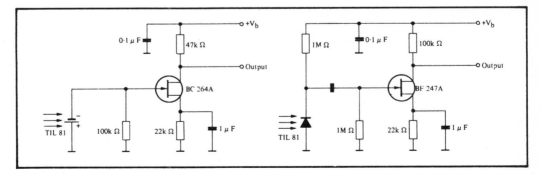

Figure 17.6
Photodetector for modulation frequencies below 10 kHz; (a) with photocell TIL 81, (b) with photodiode TIL 81

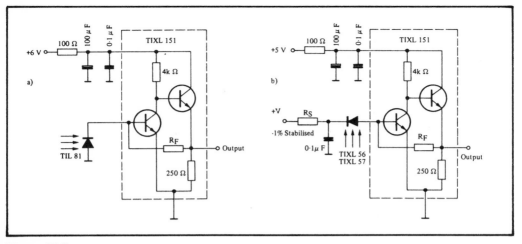

Figure 17.7

Current drive of the broad-band amplifier TIXL 151; (a) with photodiode TIL 81, (b) with avalanche photodiodes TIXL 56 or TIXL 57

Figure 17.7 shows the broad-band amplifier TIXL 151 driven by photodiodes. The base-emitter voltage $V_{BE} \approx 0.6$ V of the input stage can be used as the reverse voltage for the photodiode. Circuit (a) is suitable for use as a measurement receiver. With it, for example, the modulation of the optical radiation from filament lamps, fluorescent tubes or luminescence diodes can be shown on an oscillograph. Also, the frequency response of these radiation sources can be investigated. With circuit (b), optical radiation with sine-wave modulation up to f_{mod} = 40 MHz can be demodulated. The reverse voltage of the avalanche photodiode is to be adjusted exactly for the desired avalanche gain and stabilised to one part per thousand. In this process, the temperature coefficient of the avalanche diode is to be taken into account (see Section 9.1.1.).

With low irradiances, the safety resistance R_S can be omitted. It is to be so designed, that the permissible loss power of the avalanche photodiode is not exceeded with the maximum incident irradiance.

The avalanche photodiode TIXL 56 has the typical *relative* spectral sensitivity of silicon junction photodetectors. Its reverse voltage is approximately V_R = 170 V.

The avalanche photodiode TIXL 57 has the typical *relative* spectral sensitivity of germanium junction photodetectors. Its reverse voltage is approximately V_R = 40 V. In both circuits, the gain of the TIXL 151 is approximately 38 dB.

In *Figure 17.8*, examples of the driving of selective photodetector circuits are shown. The load impedance of the photodetector TIL 81 is a parallel tuned circuit, tuned to the modulation frequency in each case. The Q-factor of the tuned circuit determines the band-width b of the receiver detector circuit.

$$b = \frac{f_{res}}{Q} = R_v \cdot \sqrt{\frac{C_j + C_p}{L}}$$

Rv = Loss resistences in the input circuit.

A high-Q circuit gives reduced band-width and thus reduced noise in the input amplifier. These circuits also have a number of other advantages. Since the tuned circuit only has a high impedance at the resonance frequency, all interference radiations except for the resonance frequency are short-circuited, which also improves the useful-signal/interference ratio. Furthermore, neither the junction capacitance of the photodetector nor the input capacitance of

331

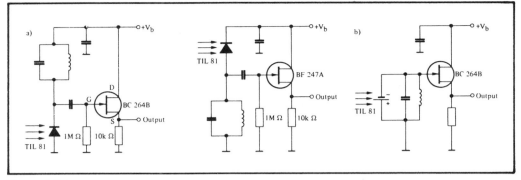

Figure 17.8
Driving of selective photodetector circuits; (a) with photodiode TIL 81, (b) with photocell TIL 81

the amplifier has an adverse effect, since they are included in the capacitance of the tuned circuit.

In selective receivers for LF-modulated radiation, it is advantageous to use active filters (*Figure 17.9*). The photocell TIL 81 works in the short-circuited mode. The subsequent source-follower drives the active filter, which consists of a two-stage amplifier, with feedback through a double-T network, with which the feedback reaches a minimum at the desired frequency. Under

the condition, that $R_1 = R_2 = 2 \cdot R_3$ and $C_1 = C_2 = 0 \cdot 5 \cdot C_3$, the resonant frequency of the filter is calculated by the formula

$$f_{res} = \frac{1}{2\pi \, R_1 \, C_1}$$

With the design values stated in the circuit diagram, a resonant frequency $f_{res} = 1$ kHz is obtained. The filter components should have a tolerance of 1% maximum, in order to obtain a narrow passband. The filter is

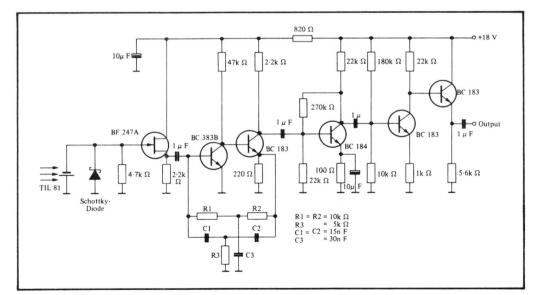

Figure 17.9
Photodetector with active filter

332

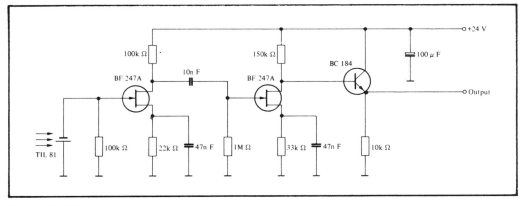

Figure 17.10
Receiver for optical telephony unit

followed by a three-stage amplifier with
emitter-follower output, with a voltage gain
$V_u \approx 3000$, so that even weak signals can
still be evaluated. In order to avoid
interference from fluorescent tubes, it is
advisable to fit an IR filter (Type RG 830)
in front of the photocell.

Figure 17.10 shows a sensitive receiver for
optical telephony units. The photocell
TIL 81 works into a high load resistance
(R = 100 kΩ), so that a relatively large
signal is obtained at the amplifier input.
Care must be taken, by the use of IR filters,
that the cell is not driven, by interference
radiation, into voltage saturation and that
the useful signal is not thus suppressed. The
voltage gain of the three-stage amplifier is
approximately 4000.

The upper frequency limit f_0 is determined
by the load resistance of the photocell, the
internal capacitance of the photocell and
the input capacitance (Miller capacitance)
of the amplifier; in the circuit shown it is
$f_0 \approx 7$ kHz. If a source-follower is
connected in front of the amplifier (see
Figure 17.5a), the upper frequency limit is
increased to 20 kHz, since the effect of the
Miller capacitance of the first amplifier
stage is eliminated. The lower frequency
limit f_u is approximately 350 Hz. It is
determined by the 47 nF capacitors in the
source circuit of the field-effect transistors.

By changing these components, the frequency
limit can be altered in a wide range.

The amplifier shown in *Figure 17.11* can be
used together with the photodetector of
Figure 17.10 as a monitor amplifier. This
circuit can also be used as a measurement
receiver or receiver for A.C. telegraphy.

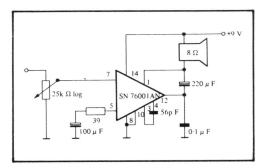

Figure 17.11
Monitor amplifier for connection to the
photodetector shown in Figure 17.10

In *Figure 17.12*, an automatic gain control
is used to compensate for fluctuations in
irradiance. The control amplifier T5 is
driven through the T4 output through a
voltage divider. The output signal taken
from the collector of T5 is rectified with a
peak rectifier. The control voltage thus
obtained is freed of interfering modulation
components with a low-pass filter and then
controls the MOS transistor, which works as
a variable resistance. With an increasing

333

useful signal, the load resistance of the photocell becomes less, so that the cell is not driven into voltage saturation.

The photoreceiver circuits in *Figures 17.9, 17.10* and *17.12* are very sensitive. When evaluating weak signlas, an IR filter should always be placed in front of the photo-detector. Depending on the application and construction, it must be noted that even undesired reflected useful radiations will be evaluated.

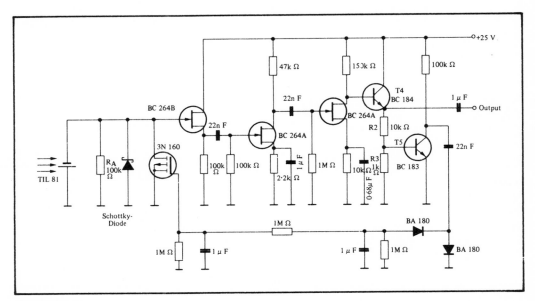

Figure 17.12
Photo-A.C.-telegraphy receiver

334

18
Practical measurement of the photocurrent sensitivity of Si phototransistors

Practical Measurement of the Photocurrent Sensitivity of Si Phototransistors

The user tests optoelectronic components, not only in accordance with the data-sheet test conditions, but also in accordance with his own application specifications. The user's own selection measurements on optoelectric components are usually carried out in the goods-inward inspection. In these, firstly the sensitivity tolerances of the phototransistors delivered are selected into various groups and secondly, criteria related to the application can be taken into account.

Criteria related to the application include:

1
Installation of masks, lenses, objective, filters or light-guides in the beam path between the radiation source and phototransistor.

2
The radiation source used can show a radiation function S_λ which differs greatly from the radiation source stated in the data sheet.

3
The selected working point for the phototransistor test device does not need to correspond with the working point stated under the test conditions in the data sheet (e.g. $V_{CE} = 5$ V). The working point will be determined by the function $I_C = f(V_{CE})$ with an irradiance value related to the application as the parameter and with the selected load resistance.

In many cases, therefore, a test equipment to be built will have almost the same circuit, the same optoelectronic components and an equivalent mechanical construction, corresponding to the system in the finished equipment. With test equipments of this kind, the relative photocurrent sensitivity of phototransistors and diodes is measured. In

the same way, only the relative radiant power of GaAs diodes can be measured.

The values measured with these test equipments cannot necessarily be compared with the data-sheet values, since the data-sheet test conditions do not usually agree with the individual test conditions. Nevertheless, the values found in these measurements are, in most cases, more informative for the user than the pure data-sheet values, since the appropriate construction of the test equipment and the mechanical and optical characteristics of the whole circuit are covered.

18.1
Theoretical circuit of test equipments

Since all tests of this kind involve comparative measurements, the test equipment normally includes a bridge circuit (*Figure 18.1*). For the measurement of the

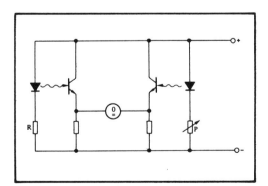

Figure 18.1
Theoretical circuit of a test equipment for the measurement of the photocurrent sensitivity of Si phototransistors

relative radiant power of GaAs diodes, the bridge circuit contains two phototransistors (or photodiodes) with the same photocurrent sensitivity in each case. Since the absolute sensitivity is only of minor importance here, the user can very easily measure two devices with identical characteristics. The resistance R will be so selected, that the same current flows through the test specimen as in the final circuit. The potentiometer P can then very easily be calibrated directly as a percentage of the relative radiant power, since the radiant power is proportional to the current.

For the measurement of the relative photocurrent sensitivity of the phototransistors, the circuit is to be so designed, that both transistors are irradiated with the same radiant power. Instead of the fixed resistance, a potentiometer will then be connected in series with the test specimen and will again be calibrated as a percentage of the relative sensitivity.

Basically, the calibrated values can be measured with high accuracy with the circuit shown. The measurement errors will be primarily determined by the mechanical tolerances of the arrangement.

18.2
Test equipment for measurement of the relative sensitivity of phototransistors.

This test equipment has been developed, to record the squint effect of a phototransistor in a given mechanical arrangement. The squint effect is caused by the fact, that the optical major axis (normal) of optoelectronic components does not coincide exactly with the mechanical axis. This has a particularly adverse effect, when, for example, phototransistors are combined with thin light-guides, since the latter are mounted, for constructional reasons, in the mechanical and not in the optical axis of the phototransistor. Therefore only the sensitivity of the phototransistor in the major axis is decisive for the whole arrangement, while the reception aperture angle, and thus the relative sensitivity of the phototransistor, can be varied over the diameter of the light guide.

Figure 18.2 shows the circuit of the test equipment. It consists of two source-

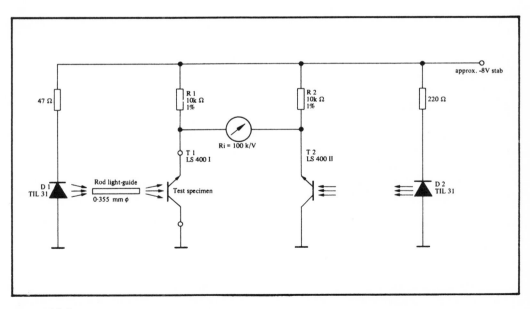

Figure 18.2
Test circuit for measurement of relative photocurrent sensitivity

detector combinations, while the photo-transistor under test, T1, and the reference phototransistor T2 form a bridge circuit with the collector resistances R1 and R2. To suit the application, two GaAs diodes TIL 31 are used as radiation sources. A light guide is located in the beam path between the GaAs diode D1 and the phototransistor T1 under test, as in the practical application.

The bridge circuit compensates the temperature-dependence of the photo-transistor under test, if the same type is selected for both phototransistors T1 and T2. In addition, both phototransistors T1 and T2 should have approximately the same photocurrent sensitivity. The indicating instrument in the bridge shows the deviation of the photocurrent sensitivity of the test specimen from that of the reference phototransistor. The indicating instrument has a high internal resistance. The bridge is balanced by varying the source-detector distance of the reference system. If a changing polarity on the indicating instrument is not desired, the bridge will be set off-balance.

19
Light measurement with Si phototransistors in electronic flash units

19

Light measurement with Si phototransistors in electronic flash units.

The advances in recent years in the miniaturisation of electronic components, combined with a reduction in costs, have led to the production of small but efficient electronic flash units. They thus helped the breakthrough of flashlight photography into the amateur field. But one problem remained: the choice of the correct exposure time.

Determination of the aperture value by the formula: "Aperture = Coefficient/Distance" only gives a very inexact result. Firstly, the exact determination of the distance of the subject is quite complicated. However, a much more important fact, is that with this formula the environmental conditions, such as the reflectivity of the surrounding walls, are not taken into account at all and thus lead, especially in colour photography, to considerable exposure errors. It is therefore worth while to measure the radiation reflected from the subject and to control the emission time of the flash unit from this. This problem can be solved in a simple manner with the aid of optoelectronic components.

19.1
Principle of an electronic flash unit

The supply unit, which consists of an accumulator or a number of dry batteries and the voltage converter, charges the main capacitor to a voltage between 300 V and 500 V. Firing is triggered by the synchronising contact of the camera.

The flash discharge tube is fired through the igniter electrode by a high-voltage pulse from the ignition circuit. In the flash discharge tube, the xenon is partially ionised, so that the discharge of the main capacitor commences. The plasma is excited into intense radiation by very high currents of several hundred amperes. (*Figure 19.1.*)

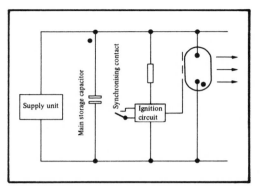

Figure 19.1
Theoretical circuit of a modern flash unit

The various parameters, such as the spectral radiation distribution, luminous flux, luminous efficiency, flash duration and half-value time of xenon-filled flash discharge tubes depend firstly on the construction of the flash tube and secondly on the design of the flash unit. The variation of the flash discharge with time is characterised by the light-emission/time curve. *Figure 19.2* shows a typical emission/time curve with a steep·rise and slow fall. In this curve:

t_1 = Contact or delay time: Time from the triggering of the flash until the commencement of light emission.

t_2 = Start-up time: Time between the triggering of the flash and the time when the half maximum value of the luminous power is first reached.

t_3 = Peak time: Time from the triggering of the flash until the maximum luminous power is reached.

t_4 = Half-value emission time: Period between the first and second times that the half-value of the maximum luminous power is reached. This time is a significant criterion, since,

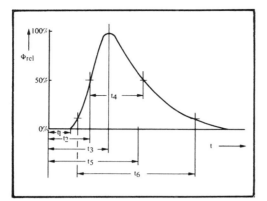

Figure 19.2
Emission/time curve of a flash discharge

together with the maximum luminous power, it determines the photographic efficiency.

t_5 = Mid-point time: Period between the triggering of the flash until the mid-point of the light emission time.

t_6 = Light-emission time: Duration of the quantity of light emitted between the values $\phi_{e,rel}$ = 10% and $\phi_{e,rel}$ = 90%. Other, similar definitions are also common.

The discharge circuit determines the shape of the emission/time curve, mainly with the anode voltage, the capacitance and the total resistance. The total resistance is composed of the internal resistance of the flash tube, the series resistance of the main capacitor and the lead resistances.

The efficiency of a flash discharge lamp is stated as the luminous efficiency. It is the ratio of the total quantity of light produced to the electrical energy consumed.

$$\eta_v = \frac{\Phi . t[lms]}{P . t[Ws]} \qquad (2.35)$$

In amateur flash units, the luminous efficiency η_v lies between 35 lms/Ws and 50 lms/Ws. The luminous efficiency mainly depends on the power and the maximum peak current.

It is calculated from the formula:

$$W = \frac{C . U^2}{2} \qquad (19.1)$$

W = Energy (Ws), C = Capacitance (F or As/V)

Example:
C = 500 μF; V = 360 V; η_v = 40 lms/Ws

$$W = \frac{500 . 10^{-6} \text{ As} . (360 \text{ V})^2}{V \cdot 2}$$

$$= \frac{64 \cdot 83 \text{ AsV}^2}{2 \text{ V}} \approx 32 \cdot 4 \text{ Ws} \qquad (19.2)$$

The quantity of light emitted is:

$$W_v = \Phi . t = \eta . W \qquad (19.3)$$

$$W_v = \frac{40 \text{ lms} . 32 \cdot 4 \text{ Ws}}{Ws} = 1296 \cdot 4 \text{ lms}$$

$$\qquad (19.4)$$

19.2
Types of exposure control

In exposure control circuits, a distinction is made between a switching device in parallel with the flash tube (*Figure 19.3*) and a switching device in series with the flash tube (*Figure 19.4*). The control circuit remains the same in principle in both cases. The switching device in parallel with the flash tube consists of a low-resistance short-circuiting circuit. When the optimum exposure is reached, the remaining energy in the storage capacitor is shorted by the low-resistance parallel circuit, so that the flash radiation is interrupted. In more recently developed flash units, the low-resistance shorting circuit in parallel with the flash tube is replaced by an interrupter circuit in series with the flash tube. In this case, only the energy necessary for the optimum exposure is drawn from the main or storage capacitor. The energy which is not needed remains stored in the main capacitor. For battery

344

and accumulator-operated flash units, this saving of energy gives a larger number of flashes and a more rapid flash repetition rate.

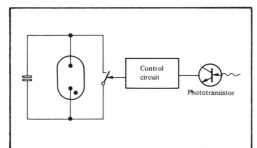

Figure 19.3
Principle of exposure control with a short-circuiting circuit in parallel with the flash-tube

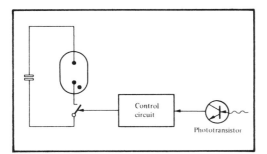

Figure 19.4
Principle of exposure control with an inter-rupter circuit in series with the flash tube

19.3
Circuit of a flash unit

Figure 19.5 shows, as an example, the circuit of an amateur flash unit with a switching device in parallel with the flash discharge tube. With the onset of the flash discharge, the operating voltage for the automatic exposure control is produced at the same time through the capacitor C 3. Thus the automatic exposure control remains unaffected by other flashes. After the flash has been triggered, the phototransistor receives the reflected flashlight. The output signal of the phototransistor is integrated up to a threshold value, the optimum exposure value. The capacitor C 1 serves as the integrating element. The pulse released through the thyristor TIC 47 to the primary side of the ignition transformer Tr 2 is stepped up on the secondary side, so that the quench tube fires. The tolerances on the response sensitivity of the thyristor and on the photocurrent sensitivity of the photo-transistor are eliminated by calibration of the automatic exposure control with R 1. The quench tube, when fired, has a considerably lower resistance than the flash discharge tube. The storage capacitor C 2 therefore discharges through the quench tube, so that the flash emission is interrupted.

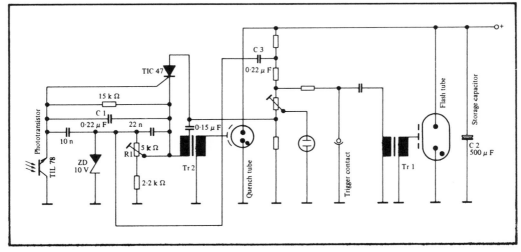

Figure 19.5
Circuit of an amateur flash unit

345

19.4
Si phototransistors for the automatic flash exposure control

Photodetectors for automatic exposure control are selected according to their absolute sensitivity, spectral sensitivity distribution, their sensitivity as a function of the angle of incidence, the turn-on time and sometimes the case dimensions. For reasons of cost, sensitive phototransistors are preferred.

Any proportionality error between the collector current I_C and the irradiance E_e can usually be neglected in practice.

The spatial sensitivity $s = f(\varphi)$ can be adapted to the requirements by an aperture mask. Phototransistors with relatively narrow reception peaks represent an optimum solution, since their sensitivity distribution $s = f(\varphi)$ corresponds approximately to the desired radiation conditions.

In *Figure 19.6*, the relative sensitivity s_{rel} is shown schematically as a function of the angle of incidence for the Si phototransistor TIL 78 with an aperture mask. The measurement area A of the subject illuminated by the flash is covered almost exactly by the reception peak of the TIL 78. Phototransistors with very narrow reception peaks cover the reflected radiation from the measurement area A with greater errors. In addition, the installation tolerances of the phototransistor would have to lie within very close limits. The turn-on time of the phototransistor TIL 78 is less than 1 μs. Thus even very short flash times can be safely dealt with.

The spectral emission distribution of an ideal flash radiation source and the spectral sensitivity distribution of an ideal photo-detector should be linear. Their spectral limits should correspond with the spectral sensitivity limits of the film. In practice, xenon flash tubes have become established

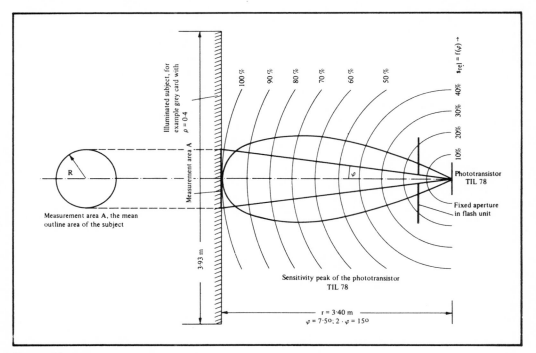

Figure 19.6
Determination of the relative sensitivity s_{rel} as a function of the angle φ of the returned incident flash radiation for the phototransistor TIL 78

as flash radiation sources and silicon photo-transistors as photodetectors.

Figure 19.7 shows the spectral emission distribution $S(XB)\lambda$ of the xenon flash discharge, the spectral sensitivity $s(\lambda)_{TIL\ 78}$ of the phototransistor TIL 78 and the graphic determination of the spectral sensitivity $s(XB)\lambda,_{TIL\ 78}$ of the photo-transistor TIL 78 for the xenon flash radiation XB (seel also Section 9.8). The relative spectral sensitivity $V(\lambda)$ of the eye is shown for comparison. Despite the difference in the spectral sensitivity of the film, the spectral sensitivity $s(XB)\lambda$ is suitable for a relative flashlight exposure measurement, since most substances have almost the same reflectivity in the visible and in the near IR range.

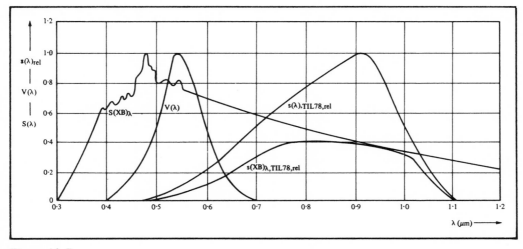

Figure 19.7
Graphical determination of the relative spectral sensitivity of the TIL 78 phototransistor for xenon flash radiation (see Section 9.8). The symbols denote: $S(\lambda)$ = Radiation function of a radiation, $s(\lambda)$ = Spectral sensitivity of the phototransistor TIL 78, $S(XB)$ = Radiation function of the xenon flash radiation, $s(XB)$ = Spectral sensitivity of the phototransistor TIL 78 for the xenon flash radiation, $V(\lambda)$ = Relative spectral sensitivity of the eye

20
Circuits with light-emitting diodes

Circuits with Light-Emitting Diodes

This chapter describes the application of luminescence diodes, the radiation from which lies in the visible range. These components can be used in many ways as optical indicators and are replacing filament and discharge lamps to an increasing extent. A significant advantage of these light-emitting diodes is firstly the low working voltage, as a result of which they can be driven directly by almost all semiconductor components (i.e. transistors, thyristors and integrated circuits). A further advantage to be mentioned, in comparison with filament lamps, is that no current peak occurs at switching on, as is the case with filament lamps because of the low resistance of the cold filament. If the circuit is designed with insufficient care, this very phenomenon can easily lead to its destruction.

20.1
Simple indicators

Luminescence diode can be used in many cases for operational indication in equipments, replacing conventional filament lamps. With low operating voltages, in particular, these diodes consume considerably less power. *Figure 20.1* shows the corresponding circuit. The series resistance R_V can easily be calculated from the formula

$$R_V = \frac{V_b - V_F}{I_F}$$

For the diode TIL 220, with an operating voltage of $V_b = 5$ V and a current $I_F = 30$ mA:

$$R_V = \frac{5 \text{ V} - 1 \cdot 6 \text{ V}}{30 \text{ mA}} \approx 120 \ \Omega$$

In the same way, these diodes can be driven directly from TTL circuits (*Figure 20.2*). Care is to be taken, however, that the

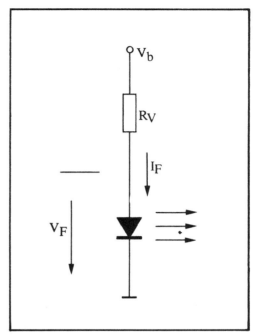

Figure 20.1
Supply voltage indicator

outputs of the T.T.L. can deliver the necessary current. The drivers SN 7416N and SN 7417N, which can deliver an output current up to 40 mA, are particularly suitable for this purpose. The series resistance R_V is calculated from the formula:

$$R_V = \frac{V_{CC} - V_{OL} - V_F}{I_F}$$

In order to indicate changes of state in an equipment as clearly as possible, it is often advantageous, to use indicators of different colours. With the following circuit, the red diode lights up when H-level is present at the input, and the greed diode with L-level (*Figure 20.3*). It must be noted, that green-emitting diodes have a higher forward voltage V_F and thus the value of the series resistance becomes less.

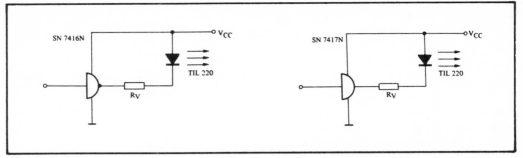

Figure 20.2
Control of luminescence diodes by TTL circuits

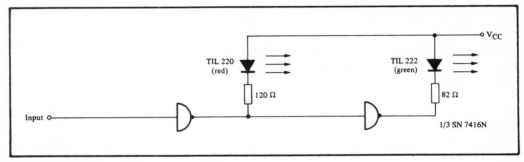

Figure 20.3
Indication with colour change on change of logic level

20.2
Diode tester

The circuit in *Figure 20.4* shows a diode tester, with which the polarity and function can be tested in a simple manner. According to the polarity of the half-wave of the alternating voltage, the current flows either through the green or the red diode. The following four conditions are possible:

1
Neither diode lights up: Open circuit.

2
Red diode lights up: Anode of the test specimen to connection 1; diode in order.

3
Green diode lights up: Anode of the test specimen to connection 2; diode in order.

4
Both diodes light up: Short-circuit in test specimen; diode defective.

20.3
Logic tester

The testing of digital circuits in complete systems often causes difficulty. Several inputs E_n and the corresponding outputs $A = f(E_n)$ must be observed simultaneously, which in most cases is not possible even with multichannel oscillographs.

The logic tester shown in *Figure 20.5*, with which all 16 pins of the integrated circuit to be tested are contacted, is suitable for measurements of this kind. Through a logic circuit, the V_{CC} and GND connections are automatically determined and the actual test circuit is supplied with operating current. At every IC connection there is an amplifier, the input configuration of which corresponds to that of the TTL or DTL circuits. The output of the amplifier dirves a luminescence

352

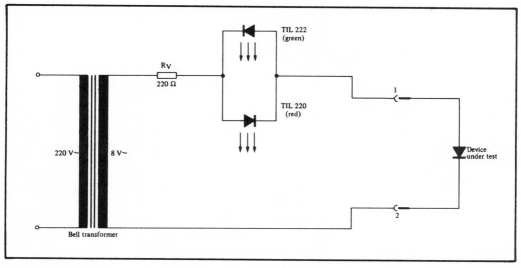

Figure 20.4
Simple diode tester

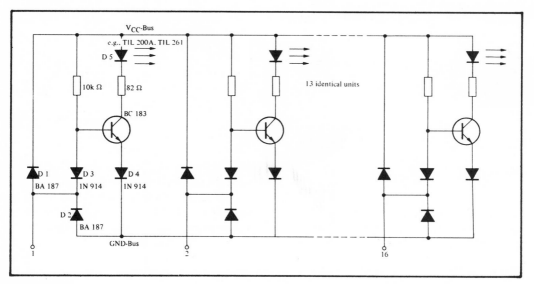

Figure 20.5
Circuit of the logic tester

diode, which lights up, if H-level is present at the amplifier input.

As can be seen from the circuit of the logic tester, the diodes D 1 and D 2 search for the V_{CC} and the GND connections. In considering the mode of operation of the circuit, it is at first assumed, that the V_{CC}

connection carries the highest positive and the GND connection the lowest negative potential. If, for instance, V_{CC} potential (+5 V) is present at connection 16, then the diode D 1 here is conductive and feeds the V_{CC} bus. All other diodes D 1 are connected to the inputs or outputs of the test specimen and, in all cases, a more negative voltage is

353

present at these points, so that these diodes are non-conducting. The diodes D 2, which search for the GND connection and supply current to the GND bus, work in the opposite direction.

Because of the forward voltage V_{F2} of the diode D 2, there is a voltage of 0·7 V, relative to the ground potential of the logic system, on the GND bus. In order that the transistor becomes conductive, the voltage on its base must be at least 2 x 0·7 V more positive (because of the diode D 4). Therefore a voltage $V_I \geqslant 1\cdot4$ V (relative to the GND connection of the circuit being tested) must be present at the cathode of the diode D 3. This value corresponds approximately to the threshold voltage of the DTL and TTL circuits. The design of the other components is very simple. For a diode current $I_F = 10$ mA 10 mA through the luminescence diode TIL 209A, the series resistance is calculated from the formula:

$$R_V = \frac{V_{CC} - V_{F1} - V_{F5} - V_{CEsat} - V_{F4} - V_{F2}}{I_F}$$

$$R_V =$$
$$\frac{5\text{ V} - 0\cdot8\text{ V} - 1\cdot6\text{ V} - 0\cdot3\text{ V} - 0\cdot7\text{ V} - 0\cdot8\text{ V}}{10\text{ mA}}$$

$$= 82\text{ mA}$$

If it is assumed, that the current gain of the transistor $h_{FE(sat)} \geqslant 50$, then the base resistance is calculated as:

$$R_B =$$
$$\frac{V_{CC} - V_{F1} - V_{BE} - V_{F4} - V_{F2}}{I_C} \cdot h_{FE}$$

$$R_B =$$
$$\frac{5\text{ V} - 0\cdot8\text{ V} - 0\cdot7\text{ V} - 0\cdot7\text{ V} - 0\cdot8\text{ V}}{10\text{ mA}} \cdot 50$$

$$= 10\text{ k}\Omega$$

The input current I_{IL}, with which the test specimen is loaded, is thus less than 200 μA,

so that the tester does not represent any significant load.

20.4
Polarity and voltage tester

The test device in *Figure 20.6* permits rapid testing of the polarity of unknown voltage sources. The polarity is indicated by a red and a green light-emitting diode.

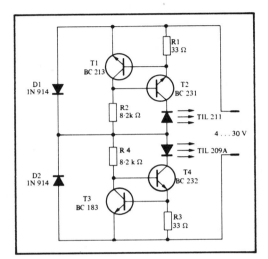

Figure 20.6
Polarity and voltage tester

Such devices have to be required to work in a large voltage range. Therefore it is not possible to stabilise the current for the luminescence diode with a simple series resistance, because with low voltages the diode would only light up weakly or not at all, while with high voltages the current would rise so far that the diode would be over stressed. For this reason, in polarity and voltage testers, the current is regulated to a constant value by a stabilising circuit. For this, the transistor T1 (T3) measures the voltage drop on the resistance R1 (R3). If the voltage becomes greater than 0·7 V, which corresponds to a current of approximately 20 mA, the transistor lowers the base voltage of transistor T2 (T4), so that the current through the luminescence diode remains constant.

354

A separate current source is provided for each polarity, driving a red and a green luminescence diode respectively. The diodes D 1 and D 2 short-circuit the part of the circuit which is not in use because of the polarity. The maximum permissible input voltage is primarily determined by the maximum permissible loss power of the transistors T2 and T4.

20.5
Large-format seven-segment display element

If numerical display units have to be read over long distances, then sometimes the figure size of the display units at present obtainable (e.g. TIL 302) is not sufficient. In this case, a seven-segment display element of any desired size can be constructed from individual diodes. *Figure 20.7* shows the circuit. As the luminescence diode, the type TIL 220 is used, while 5 diodes are connected in series for each segment. A figure height of approximately 60 mm is thus obtained.

Because the diodes are connected in series, the segment current is only 30 mA, which makes it possible to drive this unit from a normal SN 7447AN TTL seven-segment decoder. It is only necessary to provide a 12 V supply for the luminescence diodes.

20.6
Analogue indication of digital values

It is often necessary to indicate signals, which originate from digital measuring instruments, but where the absolute measured value is of no interest, but only the relative value, e.g. the filling level in a tank: Full, ¾, ½, ¼, Empty; or the difference of a motor speed from the nominal value: Positive - Zero - Negative.

Although digital display units are easily read they do not always give the required accuracy. In these cases it is advisable, to display the measured values in analogue form. With a row of luminescence diodes,

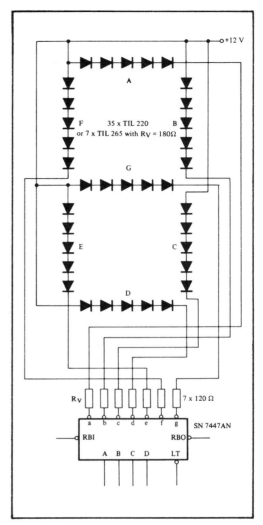

Figure 20.7
Large-format seven-segment display for a figure height of 60 mm

simple scales, which display the measured value in quasi-analogue form, can be constructed. *Figure 20.8* shows a simple circuit, in which a digital 4-bit value is decoded in a 1-from-16 decoder and is then displayed with 16 luminescent diodes. One of the 16 diodes lights up, according to the incoming digital value.

Since the TIL 209A, which already emits sufficiently brightly at 15 mA, is used for the diodes, the SN 74159N is adequate for

355

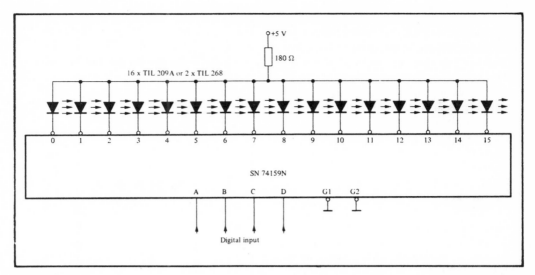

Figure 20.8
Simple analogue indication of digital values

the decoder and driver. Because only one diode ever lights up at a time, only one series resistance, common to all diodes, is used. A less favourable feature of this circuit, however, is that the value read off always has to be compared with a scale, which has to be arranged beside the diodes. In order to avoid this disadvantage, several diodes can be made to light up, according to the value

to be displayed, so that a luminescent strip is produced. In comparison with the circuit in *Figure 20.8*, additional drivers are to be connected between the individual outputs (*Figure 20.9*). Through a "Wired-OR" relationship, these switch all lower-valued outputs to L-level, so that with the input information 8 (= H L L L), the diodes 0 . . . 8 light up.

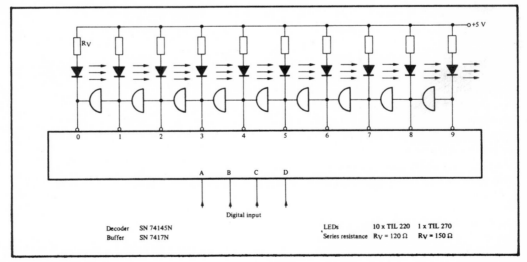

Figure 20.9
Representation of digital values by a light strip

356

Scales can also be constructed, with the zero point in the centre instead of at one end, so that both positive and negative deviations from a nominal value can be indicated. For the representation of negative numbers there are two basic possibilities, which are shown on the two linear scales in *Figure 20.10*. Which of the two numerical representations is selected, depends on the particular application. From the circuit point of view, neither of the two codes has any advantages or disadvantages for the construction of the scales.

Table 20.1 shows the two codes which are used in the circuits in *Figure 20.11* and *Figure 20.12*. In both cases, positive numbers are represented in the same binary code. In both cases, the necessary sign bit is "Low". With code A, negative numbers are characterised by the amount of the number and an H-signal in the sign bit. With code B, on the other hand, the negative number is represented by the unit complement and an H-signal in the sign bit.

In both circuit diagrams, the value of the number is again represented by the length of the luminescent strip and the number of diodes which are lit. In order to be able to distinguish positive and negative numbers easily, positive values are indicated by green and negative numbers by red luminescence diodes. The two circuits only differ in the connection of the outputs, which arises from the significance of the respective decoder outputs in the two numerical codes.

20.7
Analogue measuring instruments with LED indication

Up to the present, electromechanical measurements are indicated almost exclusively in analogue form. These measured values permit a high resolution, which is, however, not required in many cases. Therefore, in the following section, a number of circuits will be described, by means of which analogue measured values are indicated by a row of luminescent diodes

Decimal	Code A				Code B			
	Sign	C	B	A	Sign	C	B	A
−7	H	H	H	H	H	L	L	H
−6	H	H	H	L	H	L	H	L
−5	H	H	L	H	H	L	H	H
−4	H	H	L	L	H	H	L	L
−3	H	L	H	H	H	H	L	H
−2	H	L	H	L	H	H	H	L
−1	H	L	L	H	H	H	H	H
0	L	L	L	L	L	L	L	L
1	L	L	L	H	L	L	L	H
2	L	L	H	L	L	L	H	L
3	L	L	H	H	L	L	H	H
4	L	H	L	L	L	H	L	L
5	L	H	L	H	L	H	L	H
6	L	H	H	L	L	H	H	L
7	L	H	H	H	L	H	H	H

Table 20.1
Code table for negative and positive numbers

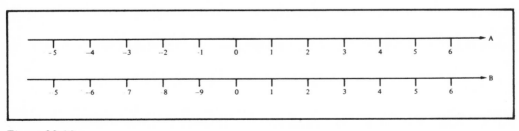

Figure 20.10
Representation of positive and negative numbers

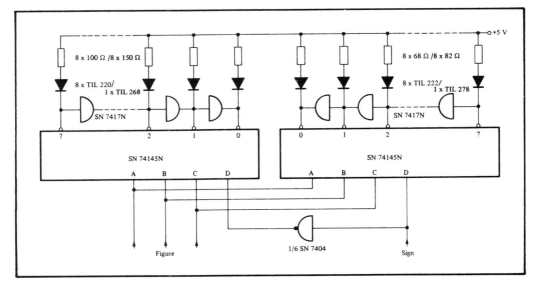

Figure 20.11
Luminescent strip with centre zerc; numerical representation in accordance with code A in Table 20.1

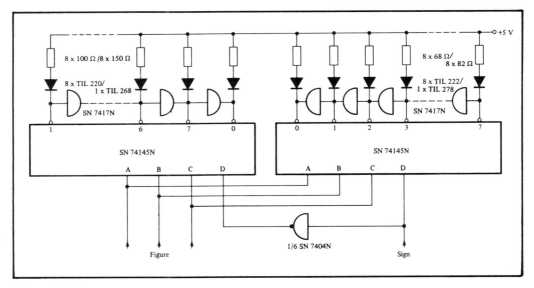

Figure 20.12
Luminescent strip with centre zero; numerical representation in accordance with code B (units complement) in Table 20.1

– while these simulate a scale – similar to the circuits described in the previous section.

It is a requirement of such circuits, that,

according to the voltage (current, resistance) at the input, a number of diodes proportional to the input value lights up or a given diode lights up in a row.

358

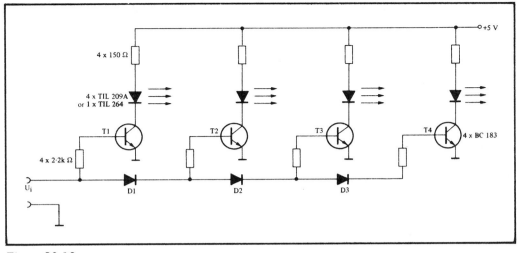

Figure 20.13
Simple voltmeter with luminescence diode indication

Figure 20.13 shows a simple circuit, which is applied in principle — but of course in considerably more extensive form — in the integrated circuits described later. If the input voltage V_i is less than 0·7 V, then all transistors are switched off. Thus none of the diodes light up. If the input voltage exceeds the value 0·7 V (V_{BE} of the transistor T1) the first transistor becomes conductive and the luminescent diode in its collector circuit lights up. Since there are one or more diodes in series with the bases of the other transistors (D1 - D3), these at first remain switched off. With an input voltage $V_i \geqslant 1\cdot4$ V, the diode D1 now becomes conductive, so that the transistor T2 turns on and the second luminescent diode lights up, and so on until with an input voltage $V_i \geqslant 2\cdot8$ V all four diodes light up.

Disadvantages of this circuit are firstly the low input resistance and secondly the inexact switching thresholds, which are determined exclusively by the forward voltages of the diodes and the base-emitter voltages of the transistors.

More suitable for applications of this kind is the integrated level indicator SN 16889P, which has been specially developed for the

purpose. It has a high-impedance input amplifier. The switching thresholds for the individual amplifiers, which drive the luminescence diodes, are stabilised by an internal temperature-compensated reference voltage. The switching thresholds lie at 200, 400, 600, 800 and 1000 mV and can be enlarged as desired by a voltage divider at the input. *Figure 20.14* shows the circuit of this level meter. The operating voltage of this module may lie within the range $V_b = 10$ to 16 V. In order to keep the brightness of the luminescent diodes constant, they are fed from stabilised current sources.

It is useful, to indicate separately when a value exceeds or falls below a given level. This is done in the circuit in *Figure 20.15* in such a way, that when the input voltage falls below the lowest threshold value ($V_i \leqslant 200$ mV), that is, when output 2 switches off, this output is periodically switched on again by feedback, so that the luminescence diode at this output begins to flash.

If the input voltage falls below 200 mV, then output 2 switches off. The voltage at this point rises almost to the supply voltage. Through the 5 MΩ resistor, the 50 μF capacitor is now charged up to 200 mV, so

359

that the output 2 switches on again, until the capacitor has been discharged through the resistances in the input circuit and a new charging process begins.

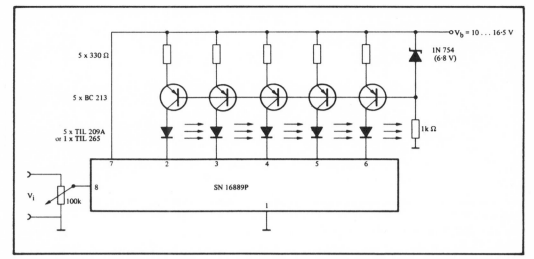

Figure 20.14
Level indicator

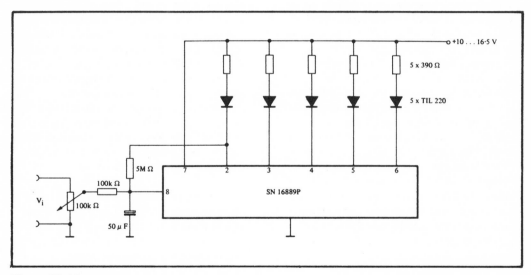

Figure 20.15
Level indicator with flasher circuit to indicate when input falls below a limit

21
Numeric and alphanumeric display units

21.1
Seven-segment display units

In recent times, the so-called seven-segment display units are becoming more and more widespread. In these units the figures or symbols are composed of individual light-emitting bars (segments). With a total of seven segments, the figures 0 to 9 and, for special cases, the letters A, C, E, F, H, J, L, P, S, U can be represented. The basic pattern is formed by an upright rectangle, divided in the middle by a horizontal line. (*Figure 21.1.*)

Display units of this kind are produced by Texas Instruments under the type numbers TIL 302, 303, 304, 312 and 320. The individual segments consist of light-emitting or luminescent diodes. These displays are characterised by high brightness and good legibility, even at flat viewing angles. As a further advantage, the low power consumption should be mentioned. Since the operating voltage is only a few volts (typically 1·7 V or 3·4 V), these units can be used directly with TTL integrated circuits. As decoder and driver, the integrated circuits SN 7446AN (30 V/40 mA output) and SN 7447AN (15 V (15 V/40 mA output), which have an open-collector output configuration, are available (*Figure 21.2*).

There are no problems in the driving of these luminescent diode or LED display systems. For current limitation, it is only necessary to provide small series resistors (*Figure 21.3*). The decimal point can be controlled through a switch, a normal gate or an inverter (SN 7400N, SN 7401N, SN 7404N, SN 7405N). These circuits are capable, without difficulty, of delivering the required diode current of 20 mA, if an output voltage $V_{OL} \geqslant 0\cdot4$ V is permitted,

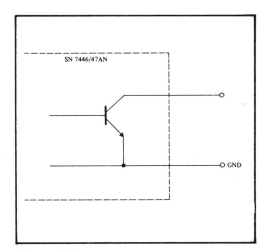

Figure 21.2
Circuit of the output stage of SN 7446AN and SN 7447AN

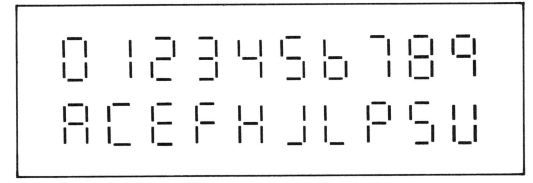

Figure 21.1
Representation of figures and letters with seven-segment display units

which is of no importance for applications of this kind.

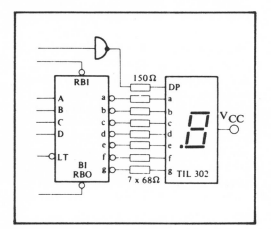

Figure 21.3
Control of display units with light-emitting diodes

The seven-segment decoder drivers also have a number of additional inputs and outputs, with which numerous further functions can be carried out. First, the "Lamp Test" (LT) input should be mentioned. It serves to test the display unit. If this connection is at "Low" level and there is not a "Low" signal at the same time on the "Blanking Input" (BI) connection, all seven segments light up, i.e. the figure 8 is displayed.

The blanking out of figures ("Blanking" and "Zero-Blanking") can be carried out very simply, since the necessary electronics for these functions are already contained in the circuit: If there is a "Low" level at the input "Ripple Blanking Input" (RBI), all output transistors will be switched off, if the inputs A, B, C and D are also "Low", i.e. if a decimal zero is present at the inputs. A further connection "Blanking Input" and "Ripple Blanking Output" (BI/RBO) works simultaneously as an output and an input. If a "Low" signal is applied to this connection through a switch or an IC with open-collector output, the display is extinguished. Thus, for example, it is possible to attach a

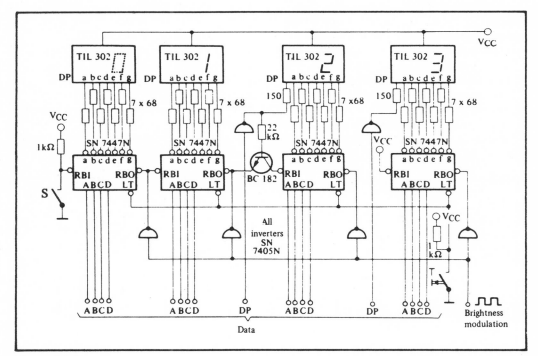

Figure 21.4
Multi-digit display system with seven-segment display units

364

brightness control to this point. If this connection is working as an output, it becomes "Low" if the inputs A, B, C, D and "Ripple Blanking Input" are also "Low". With this function, automatic zero-suppression can be carried out over any desired number of digits.

Figure 21.4 shows the circuit of a display unit, in which the facilities described above are utilised. This system can be used, for example, in a digital frequency-meter. By switching the decimal point, numbers can be displayed in the range from 00·00 to 9999, while zeros before the decimal point are blanked off, so that instead of the number 00·35, only the value ·35 is displayed. Thus reading errors, which easily occur with longer numbers can be avoided. If the switch S is opened, the zero-blanking is switched off. The function of the display units can be tested, with the button T. When the button is pressed, an 8 is displayed in all digits. Through the brightness modulation input, it is possible to control the brightness of the display continuously with a square-wave signal of variable pulse-width. The circuit shown in *Figure 21.5* is a suitable pulse generator.

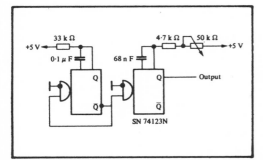

Figure 21.5
Clock generator for brightness control of semiconductor display units

21.2
Multiplex operation of display units

Display units constructed with luminescent diodes can be operated in the multiplex mode without difficulty. For example, with an 8-digit display, the 8 figures to be displayed are then switched on one after another. The switching of the individual figures takes place with such a high frequency, that because of the slow response of the human eye, a stationary image still appears.

Figures 21.6 and *21.7* show the circuit for such a multiplex operation. Despite a somewhat higher component cost, this circuit is hardly any more expensive than 8 individual SN 7446A or SN 7447A decoders. A considerable advantage lies in the fact that, with this method, only 8 data lines and 2 power supply lines are needed between the control and display sections, in contrast to a total of 34 lines with a conventional configuration. This is of particular interest, if long distances have to be spanned between the data source and the display system for remote indications. With displays with more than 8 digits, this method is always advisable, since in these cases the system described here gives price advantages in every case. The circuit works as follows:

A free-running oscillator (SN 7413N) with a frequency of approximately 1 kHz drives the counter (SN 7493AN). The outputs of the latter lead to the 4 multiplexers (SN 74151N), which switch the inputs A1 . . . A8, B1 . . . B8, C1 . . . C8 and D1 . . . D8 in sequence to the lines Y_A, Y_B, Y_C and Y_D, which finally drive the seven-segment decoder SN 7448N. Its outputs are connected with 7 current sources of type SN 75450N (with the decimal point there are 8 current sources), which serve to feed the GaAs luminescent diodes. Because of the low operating voltage (5 V), current limitation by the use of resistors is not advisable, since slight variations in working voltage would result in considerable current variations. The outputs of the counter also lead to the decoder SN 74145N, which, in synchronism with the selected data address in each case, switches on the corresponding display unit (TIL 302) through additional power drivers (BD 736). Each time the counter is advanced, the oscillator also triggers the SN 74122N monostable flip-flop

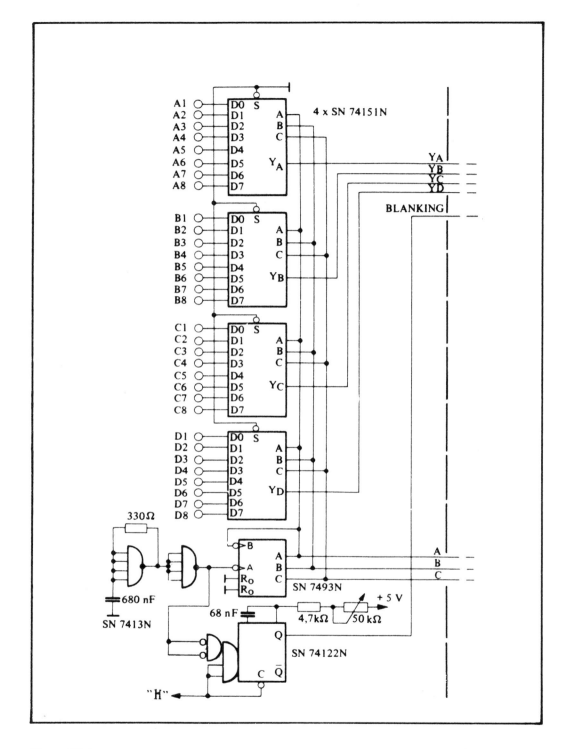

Figure 21.6
Multiplex operation of LED display units, control section

366

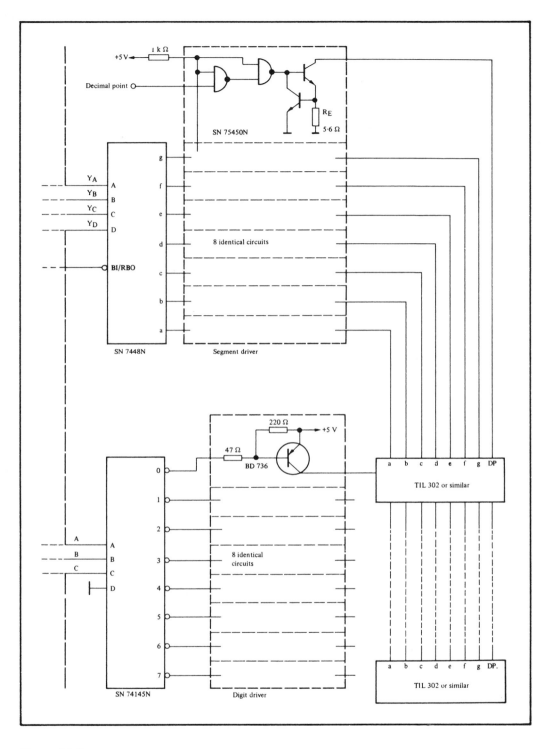

Figure 21.7
Multiplex operation of LED display units, display section

367

which drives the blanking input of the seven-segment decoder. With the 50 kΩ potentiometer, the "On" time of the luminescent diodes can be varied, in order to control their brightness without power loss.

The design of such multiplex circuits will be explained by means of the following example:

A mean segment I_F = 15 mA is required. Since a total of 8 units are driven one after another, the peak current per segment is

$$I_F = \overline{I_F} \cdot 8 = 15 \text{ mA} \cdot 8 = 120 \text{ mA}$$

The current in the segment driver is calculated from the formula:

$$I = \frac{0 \cdot 7 \text{ V}}{R_E}$$

Thus the emitter resistance R_E is obtained

$$R_E = \frac{0 \cdot 7 \text{ V}}{I_F} = \frac{0 \cdot 7 \text{ V}}{120 \text{ mA}} \approx 5 \cdot 6 \ \Omega$$

The digit driver must be capable of producing the current for the seven segments and the decimal point. Thus the digit current works out to:

$$I_D = I_F \cdot 8 = 120 \text{ mA} \cdot 8 = 960 \text{ mA}$$

Therefore, a power transistor, which still has a small saturation voltage, even with a collector current of 1 A, is used as the digit driver.

21.3
Numerical display units with integrated logic

Advances in semiconductor technology are making it possible to construct ever more complex assemblies in the smallest space. Around 1960, to build a counter decade with store, decoder and cold-cathode tube driver, a circuit board of about 150 cm^2 area was needed. With integrated circuits it was possible, from the middle sixties, to solve the same problem with only three integrated circuits: counter SN 7490AN, store SN 7475N, decoder and cold-cathode tube driver SN 74141N. Thus an area of only about 15 cm^2 was needed. A significant advance came with the development of the visible light emitting diodes (VLEDs), which, together with the highly complex logic circuits which are now available, make it possible to combine the assembly described above into a 16-pin Dual-in-Line case. The space requirement is now only about 2 cm^2. At the same time, because of the smaller number of components and connecting leads, the reliability of the whole circuit has increased considerably. The latest development from Texas Instruments are the two highly-complex circuits TIL 306 and TIL 307, which each contain not only the counter, store, and seven-segment decoder/driver, but also the display. The two circuits differ only in the arrangement of the decimal point. While, with the TIL 307, the decimal point is arranged to the right of the figure and with the TIL 306 it is on the left.

The circuit used in the counter has a number of special features which permit universal application of this device. The counter can be operated either asynchronously or synchronously with serial carry or with "carry look ahead". The simplified logical circuit is shown in *Figure 21.8.*

As well as the "Clock" input, the counter has two "Enable" inputs: one for serial carry ("SCEI" = Serial Carry Enable Input) and one for parallel carry ("PCEI" = Parallel Carry Enable Input). A signal at the clock input can only take effect if these two inputs are on "Low". Care is to be taken, that the level at the enable inputs can only be changed when the clock input is "High", since otherwise malfunctions can occur.

Further, the counter has a carry output ("MC" = Maximum Count Output), which becomes "Low" when the counter has reached the figure 9 and if at the same time the serial carry input is "Low". This output

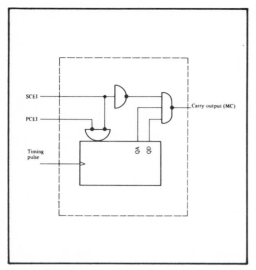

Figure 21.8
Carry circuit in TIL 306/307

delivers the necessary carry to the following decades. Thus there are three possible ways of operating the counter: in asynchronous operation, the output MC is connected to the clock input of the following decade. The enable inputs SCEI and PCEI are always "Low" (*Figure 21.9*). With this arrangement,

the flip-flops in the individual decades work synchronously, while the carries are produced serially. Although the maximum possible counting frequency (approximately 18 MHz) is not affected by this, the response time of the counter is. Since the propagation delay for each counter is approximately 12 ns, with a 4-digit counter it takes about 36 ns, until the clock signal reaches the most significant decade and about 80 ns until this decade has also reached its new state.

In synchronous operation, the enable inputs SCEI and PCEI are connected in each case to the carry output of the preceding decade. The clock signal is fed in common to all counter stages (*Figure 21.10*). With this operating mode, all relevant flip-flops switch simultaneously with the positive-going edge of the clock signal. Thus a very short response time is obtained. The maximum possible counting frequency is, however, limited by the counter length. Before a new clock pulse can be processed, all carry operations must have been dealt with. But since this takes place serially, these signals must run through

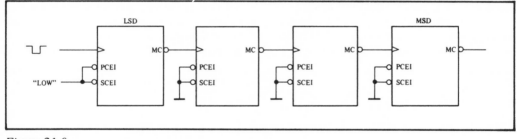

Figure 21.9
Asynchronous counter operation with TIL 306/307

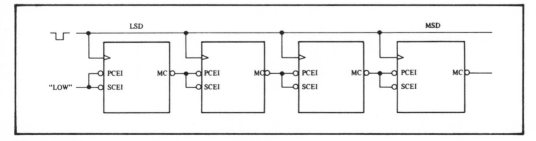

Figure 21.10
Synchronous counter operation with TIL 306/307

369

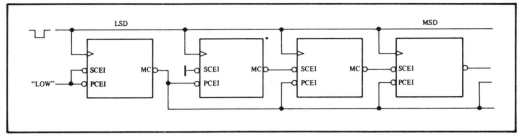

Figure 21.11
Synchronous operation with carry look ahead

all the circuits from the least-significant to the most-significant decade. The time needed for this naturally increases with the counter length and determines the minimum permissible interval between two clock pulses, and thus the maximum counting frequency.

In synchronous operation with carry look ahead (*Figure 21.11*), the clock signal is again fed to all decades in parallel, so that all flip-flops in the counter switch synchronously. The response time of the counter is thus again only about 33 ns and is independent of the counter length. The carry from the first counter is fed in parallel to all other counters through the PCEI inputs. The serial carry enable input SCEI of the second decade is always "Low", so that this decade always switches, when the first

decade switches from 9 to 0. As with all subsequent decades, the carry output is connected to the carry input of the next decade. If the counter now reaches the position 9990, then the serial carries run through the whole counter, for which the time of 9 clock pulses is available, and thus preset the more significant decades. When the counter finally reaches the position 9999, all stages are released ("enabled") by the carry from the first decade. Thus the critical propagation delay is now only determined by the first decade; it is thus again independent of the counter length.

Figure 21.12 shows the complete block diagram of the TIL 306/307. As well as the inputs already mentioned, the counter also has an asynchronous reset input ("Clear"). A "Low" signal at this input sets all flip-

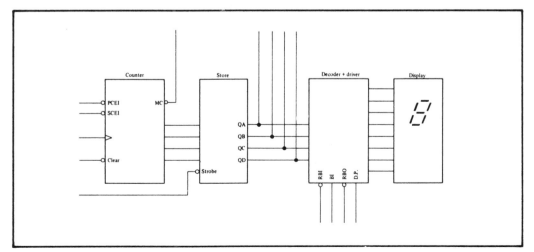

Figure 21.12
Block circuit of the TIL 306/307

370

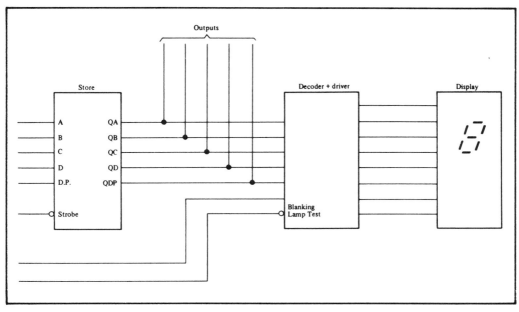

Figure 21.13
Block circuit of the TIL 308/309

flops to zero, irrespective of the signal at the "SCEI", "PCEI" and Clock inputs. If a "Low" signal is applied to the "Strobe" input, the existing counter state is transferred to the store, until the strobe input again becomes "High". The stored number can then be taken off from the outputs QA, QB, QC, QD.

The decoder/driver substantially corresponds in its logical circuit with the SN 7446/47AN, so that is is possible, here too, to construct a circuit for zero blanking without additional expense. The "Blanking" input is not, however, combined with the "Ripple-Blanking" output, so that brightness control is simpler to carry out than in *Figure 21.4.* Finally, the driver for the decimal point has also been integrated, so that the latter can be switched on or off with a standard TTL signal.

The TIL 308/309 (*Figure 21.13*) each contains a store, which, as well as the four input bits A, B, C, D, also stores the signal for the decimal point. The outputs of the store are brought out through the connections QA, QB, QC, QD and QDP. During a "Low" signal at the "Strobe" input of the store, the latter accepts the data at the inputs A, B, C, D and DP. A "High" signal at the "Blanking" input switches off the display, while a "Low" signal at the "Lamp Test" input causes all diodes in the display section to light up.

21.4
Monolithic display elements

As well as the above-mentioned display devices, in which the segments consist of individual GaAs diodes or combinations of these, Texas Instruments also manufacture monolithic display units, which, because of their small dimensions, are particularly intended for pocket calculators, digital watches and similar applications. Numerous circuits are available to drive these units, which bear the type designation TILD 100. Each circuit always contains several Darlington amplifiers, which can deliver output currents up to several 100 mA with only small input currents.

371

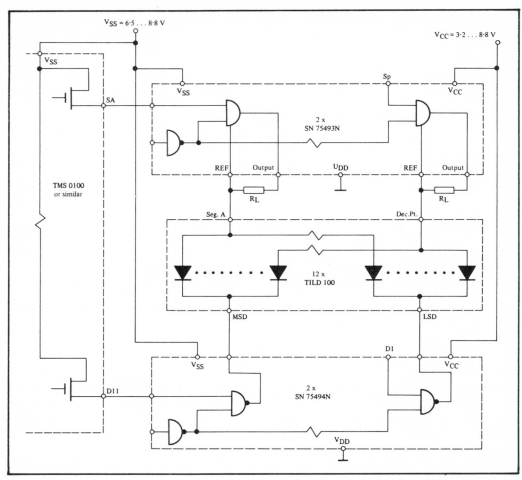

Figure 21.14
12-digit display unit for pocket calculator with the TILD 100 monolithic display devices

Figure 21.14 shows the circuit for a 12-digit display unit in pocket calculators, with which all circuits which have a multiplex seven-segment output and can deliver an output current of 0·5 mA, such as the TMS 0100, for example, can be used as the calculator circuit. As the digit driver, two SN 7549N are used. The segment drivers SN 75493N form a special feature. With these, the output current is not set by a simple series resistance; this would result in the segment current and thus the brightness being greatly dependent on the working voltage. Instead, each output of this circuit has an additional reference input, which measures the output current through an external resistance R_L and regulates it to a constant value. In the circuit shown, the segment current should be about 10 mA. The corresponding resistance R_L is then calculated from the formula:

$$R_L = \frac{0·7 \text{ V}}{I_{seg}} = \frac{0·7 \text{ V}}{10 \text{ mA}} \approx 68 \text{ }\Omega$$

With the circuit shown here, the supply voltage V_{CC} for the display unit can then fluctuate between 3·2 V and 8·8 V, without variation of the brightness of the luminescent diodes.

372

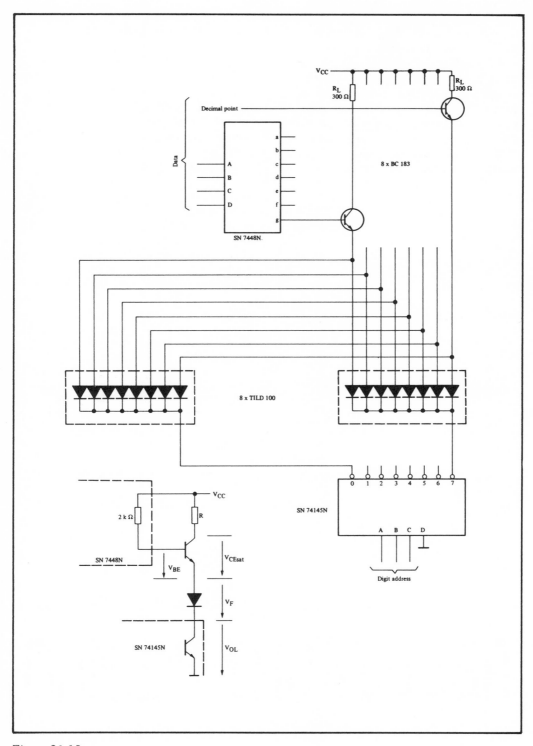

Figure 21.15
Display unit TILD 100 driven with TTL circuits

373

But the TILD 100 display devices can also be used together with TTL circuits (*Figure 21.15*). As the multiplexer, the same circuit as in *Figure 21.7* is then used. As the digit driver, the demultiplexer SN 75145, which is capable of delivering an output current of 80 mA, is used. As segment drivers, NPN transistors of type BC 183, which are driven directly from the seven-segment decoder SN 7448N, are used. The segment current is calculated here from the formula:

$$I_{seg} = \frac{V_{CC} - V_{OL} - V_F - V_{CEsat}}{R_L}$$

$$+ \frac{V_{CC} - V_{OL} - V_F - V_{BE}}{2 \text{ k}\Omega}$$

Then, for a segment current of 10 mA:

$$10 \text{ mA} = \frac{5 \text{ V} - 0.5 \text{ V} - 1.6 \text{ V} - 0.3 \text{ V}}{R_L}$$

$$+ \frac{5 \text{ V} - 0.5 \text{ V} - 1.6 \text{ V} - 0.7 \text{ V}}{2 \text{ k}\Omega}$$

$$10 \text{ mA} - 1.1 \text{ mA} = \frac{2.6 \text{ V}}{R_L}$$

$$R_L = \frac{2.6 \text{ V}}{8.9 \text{ mA}} \approx 300 \text{ }\Omega$$

21.5
5 x 7-point matrix display units

The display units so far described were only capable of representing figures. By advances in semiconductor technology, it is now possible, to produce devices which are suitable for the display of any desired letters, figures or symbols. In these, the desired character is formed from a matrix of 5 x 7 points (*Figure 21.16*).

In such a display unit (e.g. TIL 305), every point in a matrix is represented by a light-emitting diode. In each case, all cathodes of a line and all anodes of a column are connected together, as is shown in *Figure 21.17*.

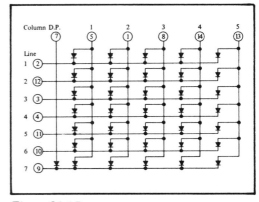

Figure 21.17
Arrangement of light-emitting diodes in display unit TIL 305

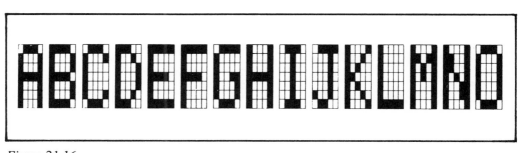

Figure 21.16
Example of the representation of the letters A to O in a 5 x 7 point matrix

With this circuit arrangement, of course, multiplex operation is again necessary for the drive. *Figure 21.18* shows the complete circuit with which all letters, figures and a number of special characters can be displayed. The form of the display is shown in *Figure 21.19*.

An oscillator, constructed with the Schmitt trigger SN 7413N (f = 50 kHz), (*Figure 21.18*), drives the counter SN 7493AN, in which only the first three flip-flops (outputs A, B, C) are used. Through the decoder/driver SN 7445N, the cathodes of the luminescent diodes are now selected a line at a time. At the same time, the associated column data is read out through connections J1, J2, J3 of the constant store TMS 2501NC and is fed to the display unit through 5 NPN transistors (e.g. type BC 183). Selection of the total of 64 possible characters is made through the leads J4 to

J9 and CS1, the individual characters being coded in accordance with the UASCII code.

Construction of a multi-digit display using the circuit shown in *Figure 21.18* is not economical, since in this case a multiplex control, including the character generator, is necessary for each digit of the display. A more suitable circuit is described below for a 16-digit display (*Figure 21.20*). It consists of a store (16 characters of 6 bits each), comparator, clock generator, character generator TMS 2501, a driver stage, the decoder, character and line counters, a timer stage and the control and synchronisation.

The position, in which the character corresponding to the data is to be written, is marked by the address. The strobe signal causes the synchroniser to deliver a write signal to the store. With a clock frequency of 1 MHz and a 100 μs pulse from the timing

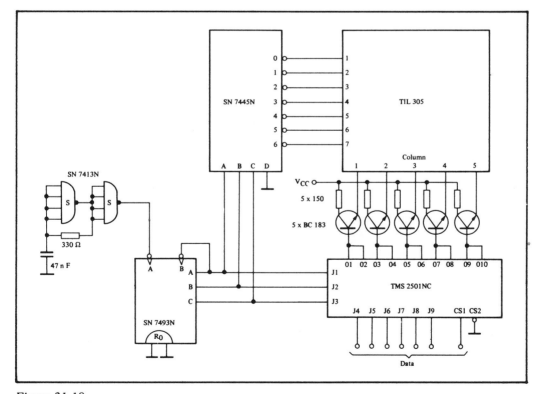

Figure 21.18
Drive circuit for a 5 x 7-point matrix, using the TIL 305

375

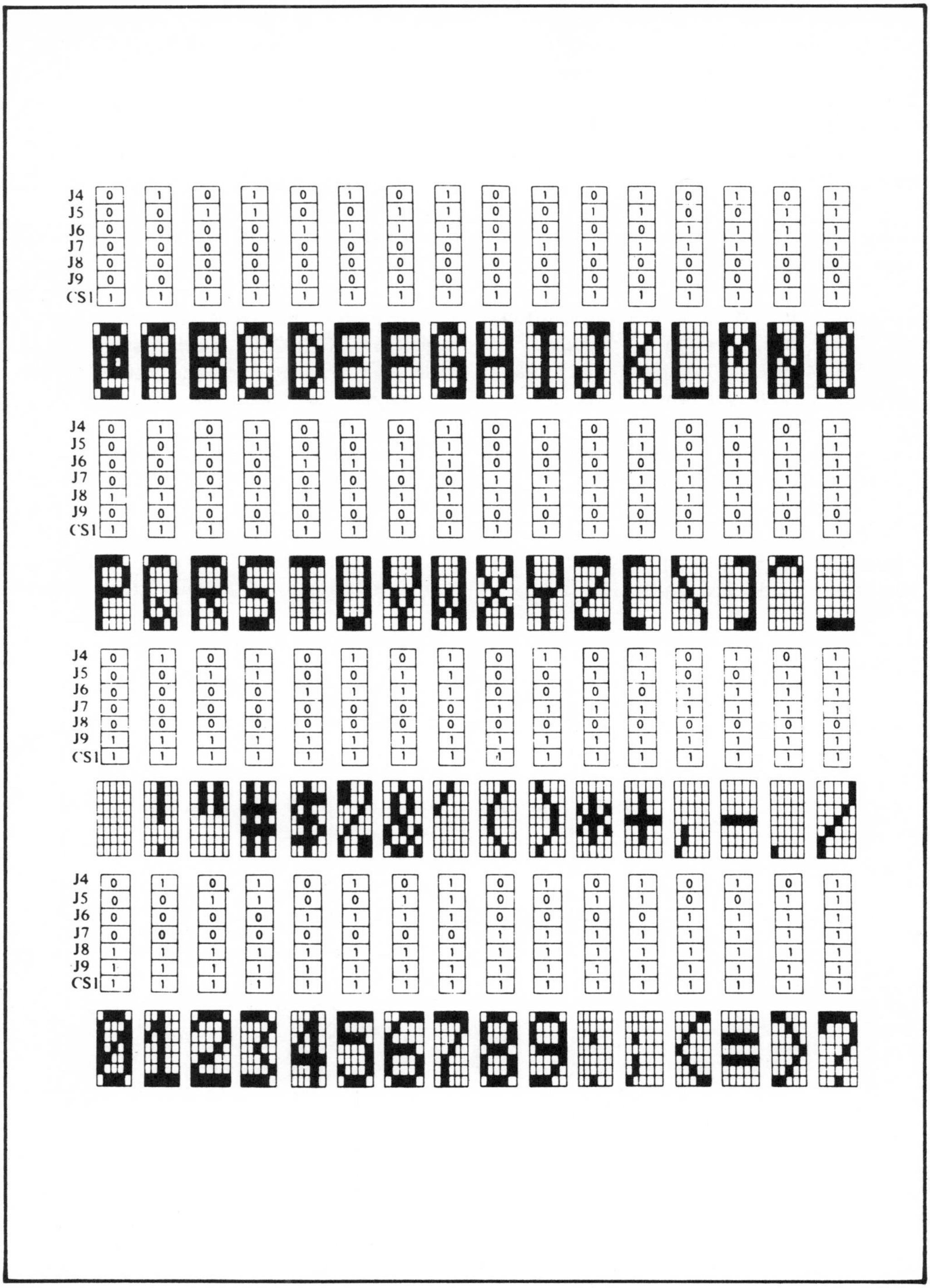

J4	0	1	0	1	0	1	0	1	0	1	0	1	0	1	0	1
J5	0	0	1	1	0	0	1	1	0	0	1	1	0	0	1	1
J6	0	0	0	0	1	1	1	1	0	0	0	0	1	1	1	1
J7	0	0	0	0	0	0	0	0	1	1	1	1	1	1	1	1
J8	0	0	0	0	0	0	0	0	0	0	0	0	0	0	0	0
J9	0	0	0	0	0	0	0	0	0	0	0	0	0	0	0	0
CS1	1	1	1	1	1	1	1	1	1	1	1	1	1	1	1	1

(characters: @ A B C D E F G H I J K L M N O)

J4	0	1	0	1	0	1	0	1	0	1	0	1	0	1	0	1
J5	0	0	1	1	0	0	1	1	0	0	1	1	0	0	1	1
J6	0	0	0	0	1	1	1	1	0	0	0	0	1	1	1	1
J7	0	0	0	0	0	0	0	0	1	1	1	1	1	1	1	1
J8	1	1	1	1	1	1	1	1	1	1	1	1	1	1	1	1
J9	0	0	0	0	0	0	0	0	0	0	0	0	0	0	0	0
CS1	1	1	1	1	1	1	1	1	1	1	1	1	1	1	1	1

(characters: P Q R S T U V W X Y Z [\ ] ^ _)

J4	0	1	0	1	0	1	0	1	0	1	0	1	0	1	0	1
J5	0	0	1	1	0	0	1	1	0	0	1	1	0	0	1	1
J6	0	0	0	0	1	1	1	1	0	0	0	0	1	1	1	1
J7	0	0	0	0	0	0	0	0	1	1	1	1	1	1	1	1
J8	0	0	0	0	0	0	0	0	0	0	0	0	0	0	0	0
J9	1	1	1	1	1	1	1	1	1	1	1	1	1	1	1	1
CS1	1	1	1	1	1	1	1	1	1	1	1	1	1	1	1	1

*(characters: (space) ! " # $ % & ' ( ) * + , - . /)*

J4	0	1	0	1	0	1	0	1	0	1	0	1	0	1	0	1
J5	0	0	1	1	0	0	1	1	0	0	1	1	0	0	1	1
J6	0	0	0	0	1	1	1	1	0	0	0	0	1	1	1	1
J7	0	0	0	0	0	0	0	0	1	1	1	1	1	1	1	1
J8	1	1	1	1	1	1	1	1	1	1	1	1	1	1	1	1
J9	1	1	1	1	1	1	1	1	1	1	1	1	1	1	1	1
CS1	1	1	1	1	1	1	1	1	1	1	1	1	1	1	1	1

(characters: 0 1 2 3 4 5 6 7 8 9 : ; < = > ?)

Figure 21.19
Code table and representation of characters

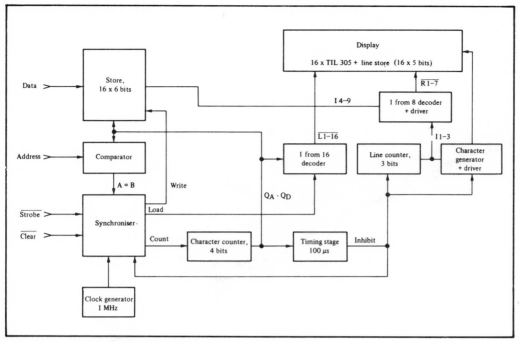

Figure 21.20
Block circuit of a multi-digit alphanumeric display with TIL 305

stage, the strobe signal must be present for at least 116 μs, in order to be able to be evaluated. The circuit is so designed, that the 7 lines of the 16 display elements controlled are switched on one after another, the associated column information is output from the character generator and temporarily stored in the drive.

During a clock cycle, the synchroniser delivers a write signal (only in conjunction with a strobe signal), a load and a count signal, one after another. With the load signal, a signal is given to the drive through the decoder and the column information corresponding to the particular character is stored. The counting pulse advances the character counter and the load process is repeated. After 16 counting pulses, the whole of the column information has been read out, the timing stage is triggered and the inhibit signal (100 μs) switches on the relevant line of the display. Following this, the line counter is advanced and the process described is repeated for the next line. Because of the divider ratio of the line

counter, a mark-space ratio of 1:8 is obtained for the display.

If a character is to be erased, this can only take place during the read-out cycle. When the input address corresponds with the address of the character counter, the comparator gives the signal A = B to the synchronisation. If the strobe signal is present at the same time, the information for the new character, at the data inputs, is taken into the store with the write signal.

With the clear signal, the whole display can be erased. This is done by writing the character "Space", with which none of the LEDs are lit, into all 16 store spaces (D_{1-6} = 0 0 0 0 0 1). The clear signal must be applied throughout a complete read-out cycle, that is, for a minimum of 116 μs, in order to overwrite the whole contents of the store.

The whole circuit of this multi-digit alphanumeric display is as shown in *Figures 21.21* to *21.23*.

377

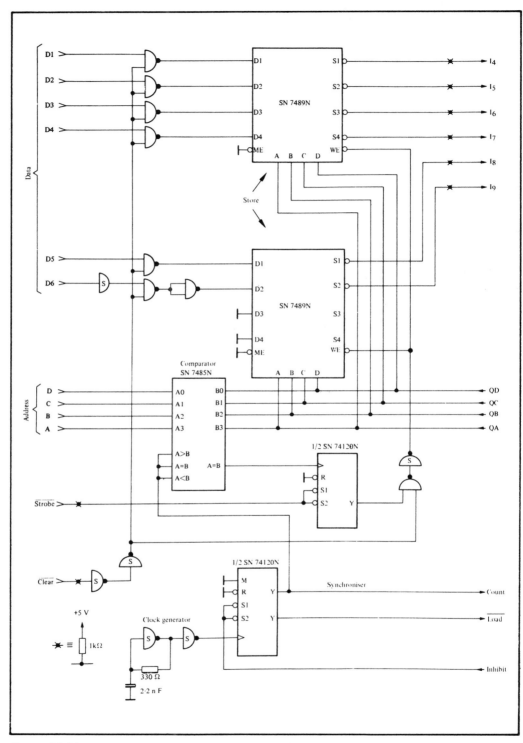

Figure 21.21
Store, Comparator, Synchroniser and Clock Generator

378

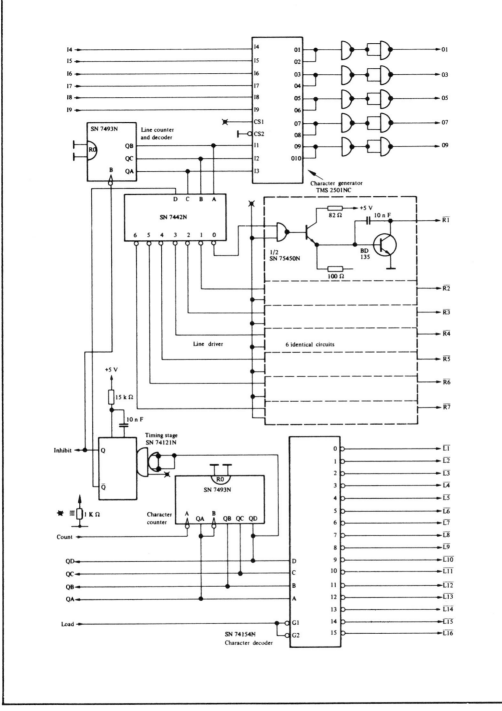

Figure 21.22
Character generator, line and column drive

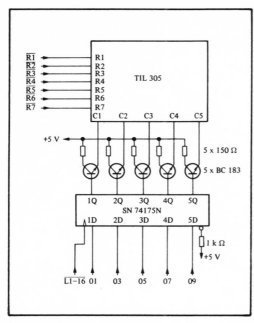

Figure 21.23
Drive for the TIL 305

22
Direction-dependent photocell units

Direction-dependent photocell units

Direction-dependent photocell units make it possible to recognise the direction of motion of moving or rotating objects. With these photocell units, forward-or-reverse counters can then be used to evaluate the signals. Their applications include rotational speed measurement with detection of direction, counting people with detection of direction, counting of goods on conveyor belts, frame counters in film cutting machines, tape counters in audio tape recorders, measurement of angles with the aid of incremental angular position transmitters, etc.

22.1
Principle of operation

Direction-dependent photocell units contain two individual optoelectronic couplers. The

photodetectors T1 and T2 are arranged as shown in *Figure 22.1*, staggered with respect to one another. During a switching operation, the photodetectors T1 and T2 operate in sequence, but overlapping in time. Thus it is possible for the direction to be recognised by the following logic. *Figure 22.1* illustrates the correct sequence of switching of the staggered photodetectors T1 and T2. Depending on the application, the photodetectors are operational either when irradiated or when the irradiation is interrupted. A direction-dependent photocell unit for short ranges (up to 10 cm), needs a GaAs diode with built-in lens, e.g. types TIL 31 or TIL 24, as the radiation source and two silicon phototransistors with built-in lenses, e.g. types LS 400, LS 600 or TIL 81, as the detectors.

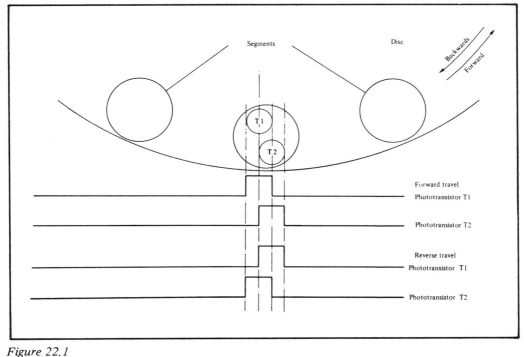

Figure 22.1
Correct switching sequence of the staggered phototransistors

22.2
Direction-dependent counter

Figure 22.2 shows an optoelectronic counter, which can take the place of mechanical counters, for example in tape recorders. A separate tape roller has transparent windows. The radiation from the GaAs diodes TIL 24 can only fall on the phototransistors LS 4022 through the transparent windows. Depending on the spacing of the windows, the tape length can be counted in centimetres or other units. Depending on the direction of tape travel, the phototransistors T1 and T2 are turned on in sequence and overlapping in time. During forward running, the phototransistor T1 operates earlier than the phototransistor T2, and conversely, during reverse running T2 operates before T1. The output signals from the two phototransistors are converted by the Schmitt triggers U 1A and U 1B into TTL-compatible signals.

The pulse diagram in *Figure 22.3* shows the signal at the output U 1A leading the signal at output U 1B during forward travel and the signal at the gate output U 1B leading the signal at the gate output U 1A during reverse travel. The signals from the two

Schmitt triggers are differentiated in each case. The negative-going pulses are inverted by U 2A and U 2B respectively and drive the NAND gates U 2B and U 2D. During forward travel, the gate U 2B is turned on by the signal from U 1B, so that the up-down counter is driven by the forward counting pulses. The gate U 2D remains off through the low level signal at U 1A. During reverse travel, the gate U 2D is unblocked by the high level signal at the output U 1B, so that the reverse counting pulses drive the reversible counter. The SN 74192 counts in the 8421 BCD code and drives the optical display TIL 308. On the SN 74192, the unused "Load" input is connected through 1 kΩ to +5 V. The "Carry" output, for cascading the forward counting pulses, drives the forward-counting input of the next decade. The "Borrow" output, to cascade the reverse-counting pulses, drives the reverse-counting input of the next decade.

The circuit in *Figure 22.2* only works satisfactorily if the phototransistors T1 and T2 are turned on and off consistently on every counting operation. If, for example, the phototransistors are first only turned on in the sequence T1, T2, corresponding to the

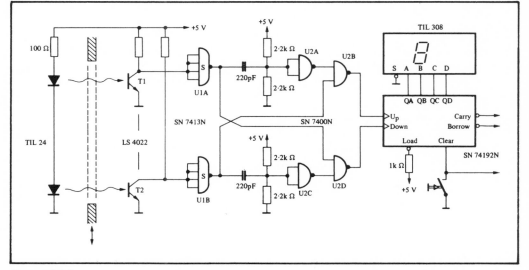

Figure 22.2
Circuit of a direction-dependent counter

384

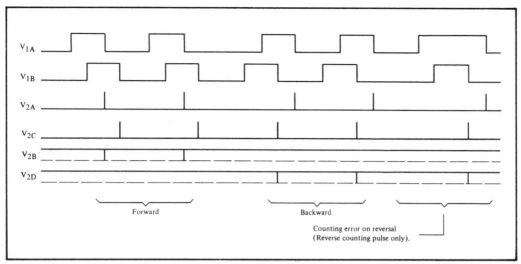

Figure 22.3
Pulse diagram of the direction-dependent counter

forward direction, and following this they are turned off in the reverse direction, then although the original tape-length condition has been restored, a reverse counting pulse is obtained. In the counting of tape lengths, this case, which occurs infrequently, can be disregarded.

22.3
Direction-dependent optoelectronic coupler

The circuit in *Figure 22.4* no longer has this defect. It is therefore suitable for use in incremental angular position transmitters. The phototransistors are in a staggered

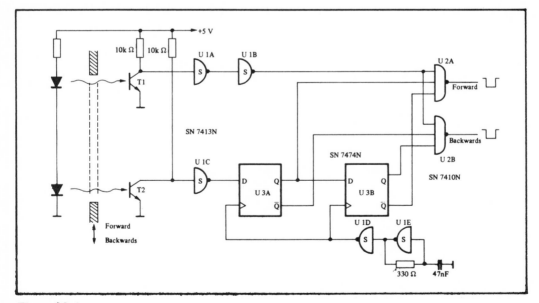

Figure 22.4
Improved evaluation circuit for direction-dependent optoelectronic couplers

385

arrangement, for example, as in *Figure 22.1*. As amplifiers and to produce the necessary fast pulse edges to drive the subsequent TTL circuits, each phototransistor is followed by a Schmitt trigger (U 1A and U 1C). The T1-signal is inverted again by the inverter U 1B and is connected to one input of each of the gates U 2A and U 2B.

The T2-signal is divided by two subsequent flip-flops U 3A, U 3B into time-delayed signals. Through the appropriate logical relationship of the flip-flop output signals Q and $\overline{Q}$ with the T1-signal, the negative-going forward-counting pulses are obtained at the output of the gate U 2A and the negative-going reverse counting pulses at the output of the gate U 2B. The flip-flops receive their clock signal from the oscillator, consisting of the two Schmitt triggers U 1D, E, with $f \approx 25$ kHz.

When the transistor T2 is turned on, the output of the Schmitt trigger U 1C is at a high level. With the next clock pulse, the flip-flop U 3A is set, and the following pulse sets the flip-flop U 3B. When the phototransistor T2 is turned off, the flip-flops are reset in the same order. Thus two states of the two flip-flops, which characterise the turning on and off respectively of the phototransistor, are obtained:

$$\text{Turn-on} = Q_{U3A} \cdot \overline{Q_{U3B}}$$

$$\text{Turn-off} = \overline{Q_{U3A}} \cdot Q_{U3B}$$

Since the output gates U 2A, B only allow signals to pass, if the phototransistor T1 is turned off, the following relationships apply:

$$\overline{\text{Forward}} = T1 \cdot Q_{U3A} \cdot \overline{Q_{U3B}}$$

$$\overline{\text{Reverse}} = T1 \cdot \overline{Q_{U3A}} \cdot Q_{U3B}$$

Accordingly, a pulse, the width of which corresponds to the cycle time of the oscillator (in this case 40 μs), is obtained in each case at the output of the gate. This arrangement ensures, that no false signals are given on reversal of direction: if the reversal takes place from forward to reverse, when T1 has not yet turned on, first a forward signal is given and then a reverse signal after the reversal of direction, so that the total in the subsequent counter is unchanged.

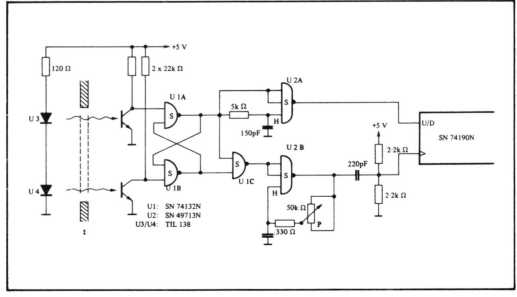

Figure 22.5
Digital rotary control knob

386

22.4
Digital rotary knob

A further application for direction-dependent optoelectronic couplers is the "digital control knob", which is used together with a counter to generate digital set-point values. In contrast to decade switches, the digital control knob has the advantage that the numerical range which is of interest can be traversed continuously and that there are no steps on carrying from one decade to the next, which can never be completely avoided with the usual decade switches.

The heart of this digital control knob, in *Figure 22.5*, is the oscillator U 2B, the frequency of which can be varied over a wide range by means of the potentiometer P. Because the start and end of the potentiometer track are connected together, the oscillator generates the lowest frequency when the potentiometer is in the central position. A disc with a cut-out is mounted on its spindle, so that when the slider is in the central position, both phototransistors are turned on and thus both outputs of the Schmitt triggers U 1A, B are at the logic high level. Through the gate U 1C, the oscillator is thus switched off. If the potentiometer is now turned in one or the other direction, first one of the two phototransistors is covered. Through the feedback between the gates U 1A and U 1B, this "Latch" adopts a stable position and stays in this condition, even when the second phototransistor is covered. Now, the further the potentiometer is turned from the centre position, the higher the frequency of the oscillator becomes and the faster the counter advances. Through the gate U 2A, the position of the flip-flops U 1A, B is transmitted to the counter and thus the required counting direction is set. The RC network at the input of the gate U 2A delays the negative edge at the output by about $0.5\ \mu s$ and thus prevents an error in the counting direction, if the potentiometer is turned back to the central position.

23
Optoelectronic rangefinder

The optoelectronic rangefinder described here was originally developed for film cameras. A connected control circuit then had to set the appropriate focal distance automatically for the measured distance of the subject. For the circuits described below, after appropriate modifications, there are numerous other possible applications. To mention one further application, for example, a highly-sensitive proximity switch in alarm systems. This equipment can equally well be used as a highly-accurate level detector in silos or tanks, while it should be emphasised, that the measurement takes place without contact with the contents.

23.1
Phase measurement as a measuring principle

This short-range rangefinder, the principle of which is illustrated in *Figure 23.1*, is suitable for measuring the distance from fixed and moving objects in the range from 1 m to 15 m. The phase-measurement principle is applied. The emitter delivers modulated infra-red radiation, which is optically focussed onto the object. The radiation is modulated with a crystal-controlled oscillator with an accurate frequency of 4433 kHz (this frequency was chosen, because low-priced crystals are available). As the IR source, a GaAs luminescence diode is used.

In the IR receiver, which forms part of a fixed assembly with the source, the IR radiation is first demodulated, to recover the modulation frequency. A fast-acting Si photodiode serves as the photodetector. According to the propagation time of the IR beam, the distance travelled by which is equal to twice the distance r to the object, the received signal shows a lagging phase displacement φ. Since, through the relationship with the velocity of light, c, φ is proportional to r, the measurement of the distance r can be obtained directly from the measurement of φ.

For the propagation time: $t = 2\dfrac{r}{c}$ (23.1)

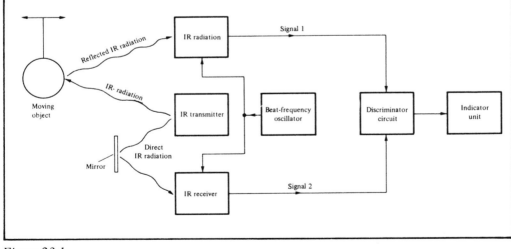

Figure 23.1
Construction principle of an optoelectronic rangefinder

and for the phase displacement:

$$\widehat{\varphi} = 2\pi f \cdot t \qquad (23.2)$$

Therefore: $\widehat{\varphi} = \dfrac{4\pi r \cdot f}{c} \qquad (23.3)$

At a frequency of 4·433 MHz and a distance of 10 m, the phase displacement obtained is:

$$\widehat{\varphi} = \frac{4\pi \cdot 10 \text{ m} \cdot 4{\cdot}433 \cdot 10^6 \text{ s}^{-1}}{3 \cdot 10^8 \text{ ms}^{-1}}$$

$$= 0{\cdot}5911 \cdot \pi = \varphi = 106° \qquad (23.4)$$

Because of the circuit selected, phase measurement is possible up to an angle $\varphi = \pi \ (\cong 180°)$. At a frequency of 4,433 MHz, the maximum measurable distance r_{max} is then

$$r_{max} = \frac{\widehat{\varphi} \cdot c}{4\pi \cdot f} = \frac{\pi \cdot 3 \cdot 10^8 \text{ ms}^{-1}}{4\pi \cdot 4{\cdot}433 \cdot 10^6 \text{ s}^{-1}}$$

$$= 16{\cdot}9 \text{ m} \qquad (23.5)$$

The measured result is completely independent of the signal amplitude. The signal received by the photodetector and then demodulated is amplified, so that phase comparison with the modulation signal is possible. At the same time, exact measurement of the phase displacement is only possible, if the stability of the circuit fulfils strict requirements.

By using the method, which is often used in phase meters, of superimposing an auxiliary frequency from a second oscillator, the received electrical signal is transposed to a lower frequency while retaining the same phase displacement φ. For reasons of stability, this beat-frequency oscillator is also a quartz crystal oscillator. Its frequency is 80 Hz higher than that of the transmitter. A difference frequency of 80 Hz is then formed in the mixer stage. Following the mixer stage, this low frequency is amplified further with relatively simple low-frequency amplifiers and is freed of interfering noise

components with non-critical and economical RC filters. To improve the interference rejection further, the signal is limited in its amplitude in subsequent limiter stages. This is permissible, because the distance information is present in the time-displacement of the zero crossing point voltage of the signal and not in its amplitude. Simple and cheap transistor circuits are adequate for limiting, while, because of the low frequency used, the transistor storage time does not yet have an interfering effect.

Furthermore, through the frequency conversion from 4·433 MHz to 80 Hz, the system accuracy is significantly increased. A phase difference which is still detectable, of $\Delta\varphi = \pm1/4°$ at 80 Hz corresponds, from equation (23.2), to a time displacement of

$$\Delta t = \frac{\Delta\varphi}{360° \cdot f}$$

$$\Delta t = \pm \frac{1}{4 \cdot 360 \cdot 80 \text{ s}^{-1}} \approx \pm 8{\cdot}6 \ \mu s$$

$$(23.6)$$

Since the phase displacement is carried over in the frequency transposition, it corresponds to a change in the propagation time of the high-frequency signal of

$$\Delta t = \pm \frac{1}{4 \cdot 360 \cdot 5 \cdot 10^6 \text{ s}^{-1}} \approx \pm 138{\cdot}8 \text{ ps}$$

$$(23.7)$$

According to equation (23.1), this change in propagation time is equal to a range tolerance of

$$r = \pm \frac{138{\cdot}8 \cdot 10^{-12} \text{ s} : 3 \cdot 10^8 \text{ ms}^{-1}}{2}$$

$$\approx \pm 2 \cdot 10^{-2} \text{ m} \qquad (23.8)$$

In order that the phase comparison can be carried out with at least the same accuracy, a reference signal with good time stability must be available.

Figure 23.2 shows the block diagram of the rangefinder. Through the circuit selected, a reference signal of high stability is produced. Two IR receivers of identical construction are used. Each receiver has a photodetector at its input. The IR radiation from the transmitter is directed through an optical lens system on to the object to be measured and is fed, as the reflected component, delayed by the propagation time, to the IR receiver 1. In front of the photodetector, this also has a lens system with a narrowly focussed radiation characteristic, directed at the target. A small part of the transmitter radiation is taken off before or after the transmitter lens system and fed by a short, direct path to the IR receiver 2. This can be done by deflection of the beam at the inner wall of the equipment housing or by a small reflecting mirror.

This method with two IR receivers eliminates the effect of interfering phase shifts in the stages of the transmitter, since the paths of the signals for receivers 1 and 2 pass through identically-constructed stages and have the same phase shift.

The signal photocurrent from the photodiode and the original current from the beat-frequency oscillator meet at the HF bandpass and pass together through the HF amplifier to the mixer stage, in which the product of mixing – the difference frequency – is produced. Since the two frequencies have little difference between them, the effect of phase shift on both signals is practically equal, so that the low-frequency signal is formed without a recognisable phase error. Since the following LF amplifiers are of similar construction, residual phase displacements, which may occur in the RC band-passes and in the LF amplifier, cancel each other out between signals 1 and 2.

The last assembly in the rangefinder is the discriminator circuit. This has the function of an exclusive-OR gate. As shown in the truth table

Signal 1	Signal 2	Output
L	L	L
L	H	H
H	H	L
H	L	H

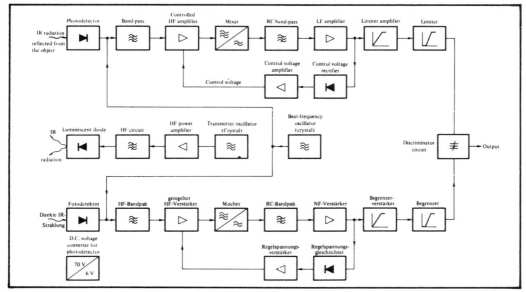

Figure 23.2
Block diagram of the rangefinder

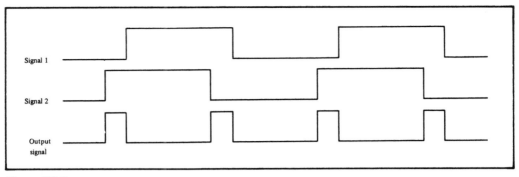

Figure 23.3
Phase comparison of the two received signals

a high level signal appears at the output, whenever the signals 1 and 2 are different from one another. *Figure 23.3* shows two signals 1 and 2, limited to square-wave form, while signal 1 has a phase delay in relation to signal 2. The output signal shown below has a pulse width which is equal to the time displacement between signals 1 and 2. Both rising and falling edges are used to form the output signal. For this reason, the maximum phase displacement which can be evaluated is 180°. The relationship between the output pulse width and the distance is linear.

An appropriately calibrated moving coil instrument can indicate the distance directly in metres. By integration with an RC

network, an analogue voltage, the value of which is a measure of the distance, is obtained. For control purposes, the output signal can be fed directly to a comparator stage, in order to control a servo-motor.

23.2
Practical circuit of the rangefinder

In *Figure 23.4*, the IR transmitter circuit is shown, with its design values. A frequency-stabilised quartz crystal oscillator drives the modulator stage T16 through the emitter follower T15. The parallel-tuned circuit in the collector lead of T16 is tuned to the oscillator frequency of 4·433 MHz. The

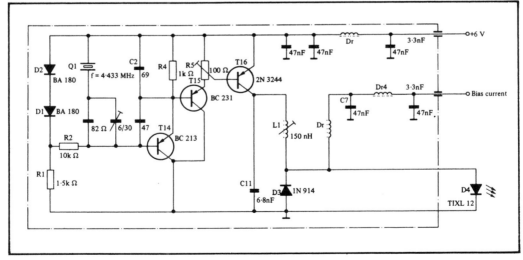

Figure 23.4
IR transmitter circuit of the optoelectronic rangefinder

394

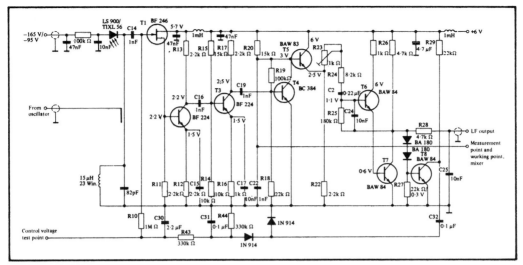

Figure 23.5
IR receiver of the optoelectronic rangefinder

working point of the modulator stage is set
with the potentiometer R5. The GaAs diode
is connected in series with the coil L1 of the
tuned circuit. Thus the modulation current
(coil current) is higher than the total current
in the leads by the Q factor of the tuned
circuit. GaAs diodes have a low breakdown
voltage. Therefore the TIXL 12 is protected
against excess negative potentials by a
1N 914 silicon diode, connected in opposing

parallel. The GaAs diode can be operated
with or without bias current, as desired.

Direct and indirect cross-talk from the
transmitter to the receiver is prevented by
various measures. The transmitter is well
screened to prevent electromagnetic pick-up.
In the same way, the main receiver, the
references receiver and the auxiliary
oscillator are each well screened as individual

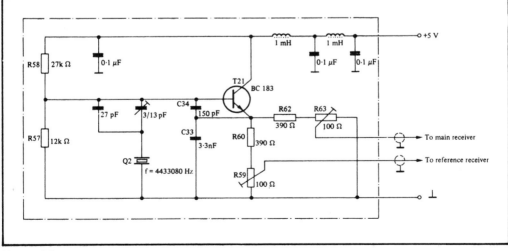

Figure 23.6
Circuit of the auxiliary oscillator

395

modules. The supply leads inside and outside the modules contain effective HF filters. In general, all leads are kept short and some are screened. Each module has only one ground connection. Thus undesirable earth loops can be simply avoided throughout the system.

Figure 23.5 shows the receiver circuit for the main and reference receivers. The difference between the two receivers lies in the fact that the main receiver contains a highly sensitive Si photo-avalanche diode and the reference receiver contains a Si photo-diode with low dark current and a high limiting frequency. The reverse voltage of approximately 165 - 170 V, applied to the photo-avalanche diode, must be adjusted to the most favourable signal/noise ratio. In addition, the reverse voltage is stabilised, at least with an accuracy of 0·1%.

The load impedance of the photodiodes in each case is a parallel-tuned circuit, tuned to the modulation frequency. The input stage

of the subsequent HF amplifier is an FET source-follower. With this, the damping of the tuned circuit is still relatively slight. The working point of the FET is determined by the control voltage, according to the received signal. For this, the positive half-waves of the LF output signal are amplified by T8. From the T8 output signal, the control voltage is obtained through a peak-path rectifier. The time constant of the sub-sequent control voltage filter chain is so designed, that large changes in the reception level, occurring in relatively rapid succession, are well compensated for by the control voltage. Otherwise, the mixer stage T4 will be overloaded. The transistor T4 in the mixer stage is the low-flicker-noise type BC 384. The LF amplifier contains a band-pass tuned to the mixer output of 80 Hz.

Figure 23.6 shows the auxiliary oscillator circuit. The phase angle can be controlled with the 100 Ω potentiometers.

Finally, in *Figure 23.7*, the evaluation section is shown. The LF output signals

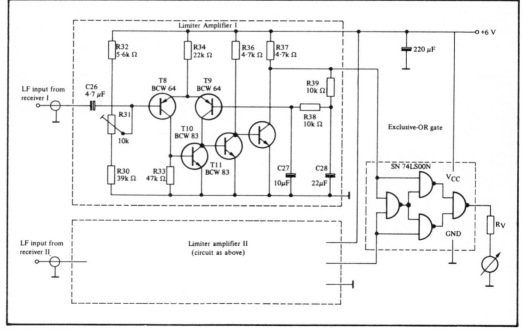

Figure 23.7
Evaluation section of the optoelectronic rangefinder

from the main receiver and the reference receiver are formed into symmetrical square-wave pulses in the relevant limiter amplifiers. The operating voltage is stabilised with a value of 5 V. The limiter amplifiers are TTL-compatible, because a TTL circuit, connected as an exclusive -OR gate can carry out the phase comparison. The measured distance is indicated by a moving-coil instrument.

24
Data transmission with optocouplers

24.1
Interference on transmission links

The transmission of data in a digital system over long distances proves to be very difficult, if low error rates are to be achieved. The reasons for this are interference voltages, which are induced in the transmission line and introduces errors to the useful signal.

The simplest forms of interference to deal with are those which originate from adjacent signal leads. The interference voltage produced through cross-talk is about 15% of the signal voltage, if twisted pairs are used. Even with a large number of pairs of leads in the cable harness, cross-talk increases to a maximum of 20%, because only lines in the immediate vicinity of the affected line determine the cross-talk level. Since the interference rejection of a logic system is about 30-45% of the signal amplitude, no problems are to be expected on this score.

Interference, which originates from other sources, such as relays, motors, etc. is far more difficult to deal with. In this case, the interference voltages reach amplitudes of up to several kilovolts. If it is assumed, that these interference voltages are induced from a cable running parallel to the signal line, a cross-talk level of about 15% can again be expected. These voltages are then large enough in any case, to interfere with the data transmission, even with signal voltages of 12 V or more. Further, it must be noted, that the cross-coupled interference voltages are now present on the relatively low line impedance (approximately 100 Ω), so that the interference energies can very easily destroy the transmitter or receiver circuits, if special protective measures are not taken. Since undisturbed data transmission is impossible under these circumstances, the signal leads should always be screened, in order to keep the effects of interference sources, external to the system, as small as possible.

Different earth potentials between transmitter and receiver are the third source of interference. As a basic rule, a balanced transmission line would help guard against this (push-pull output stage in the transmitter and differential amplifier in the receiver input). Of course, the available integrated circuits for symmetrical operation are only capable of suppressing common-mode voltages up to about 15 V. In practice, however, considerably higher potential differences occur between transmitter and receiver, so that these circuits can only be used to a limited extent. The only possibility, of safely suppressing interference of this kind, consists of electrical isolation of the transmitter from the receiver.

For this, optoelectronic couplers are excellently suitable. On the one hand, they are capable of isolating potential differences up to several 1000 V between the input and output. Of course, the practicable value is only a few hundred volts, since because of the small spacings between conductor tracks on the printed circuit, higher voltages cannot be permitted in most cases, unless special constructional forms such as the coupler TIL 109 are used. On the other hand, optoelectronic couplers can be driven directly from integrated circuits and can themselves drive integrated circuits directly, so that relatively simple and low-priced interface circuits are produced.

24.2
Construction and characteristics of optoelectronic couplers

24.2.1
Current transfer ratio

Optoelectronic couplers, or optocouplers for

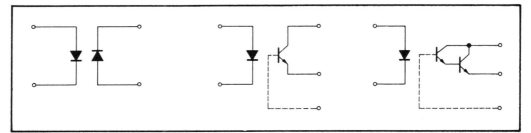

Figure 24.1
Circuits of optoelectronic couplers

short, contain a luminescent diode as the radiation-emitting part and a photodiode or phototransistor (Darlington transistor) as the radiation detector (*Figure 24.1*).

Optocouplers, which only contain a photo-diode as the detector, are not commonly used. For application, in which this operating mode is necessary, couplers in which the base connection of the photo-transistor is brought out are generally used, and the collector-base junction is used as a photodiode. The emitter of the transistor is then unconnected.

The current transfer ratio I_F/I_P is of the order of approximately 0·001, i.e. for an input current of 10 mA (luminescent diode), about 10 μA of output current is obtained from the photodiode or photocell (this is a typical value). The actual current transfer or conversion ratio is taken from the relevant data sheet.

The phototransistor amplifies the photodiode current according to the current gain B(h_{FE}) of the transistor. The current gain of the transistors is about 50−500, so that in this operating mode a current transfer ratio of about 0·1−0·5, corresponding to 10−50%, is obtained. Darlington photo-transistors, with current gains of about 10^4, give a current transfer ratio of from · 1 to 10.

24.2.2
Mechanical construction

Figure 24.2 shows, in simplified form, the mechanical construction and the resultant parasitic coupling capacitance between input (luminescent diode) and output (photo-transistor). Because of the method of construction, high insulation resistances, of typically 10^{11} Ω between input and output, can be achieved, so that in practice the actual

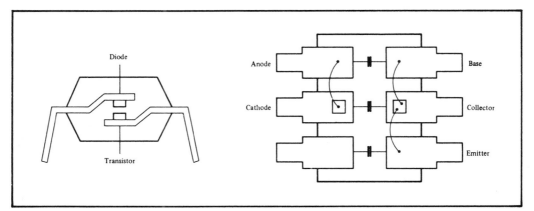

Figure 24.2
Construction of optoelectronic couplers in the Dual-in-Line package

402

insulation resistance is primarily determined by the circuit board material and the surface contamination which inevitably occurs in service.

However, the coupling capacitances between the luminescent diode and the phototransistor are more likely to limit performance in normal operation. Although the capacitance, depending on the coupler type, is only between 0·5 and 2 pF and is thus considerably less than the coupling capacitance of other isolating devices (line transformers, etc.), under extreme conditions it can still have a disturbing effect. In this respect, the capacitance between the anode connection of the luminescent diode and the base connection of the phototransistor is particularly critical, while the effect is increased still further by the capacitance due to the conductor tracks on the printed circuit.

The base connection is the most sensitive part of the coupler, since interference currents coupled in here reappear at the output, amplified by the current gain factor. If the greatest possible interference suppression is required, it is therefore advisable to use couplers, in which the base connection of the phototransistor is not brought out.

Under these conditions, photodiode operation with a subsequent highly sensitive amplifier is not advisable, since in this case even small interference voltages between the input and output of the optocoupler can interfere with the following circuit.

24.2.3
Dynamic performance

The switching data given in the optocoupler data sheets must always be taken into consideration in connection with the relevant measurement circuits. The latter are selected, so that they have little or no influence on the coupler, and so that in fact only the performance of the coupler itself is

measured. In practice, the subsequent circuit reacts on the optocoupler, so that in this case the data sheet values can only be used with reservations to determine the switching performance of the circuit.

24.2.3.1
Photodiode operation

The switching performance of the optocoupler in this operating mode is generally

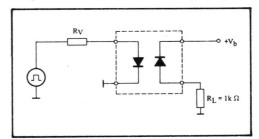

Figure 24.3
Measuring circuit for the dynamic performance of optocouplers (photodiode operation)

measured with a load resistance $R_L = 1\ k\Omega$ (*Figure 24.3*). The delay times are then determined by the following parameters:

a
The rise time of the radiation from the luminescent diode.

b
The junction capacitance of the photodiode (it should be noted that the capacitance of the diode depends on the applied reverse voltage).

c
The value of the load resistance, which, together with the capacitance of the diode, detemrines the time constant of the output circuit.

Since photodiode operation will only be used, if short switching times are required, significantly higher load resistances will not be used in practice, in order to keep the effect of the photodiode capacitance and the capacitances of the subsequent circuit as small as possible. Thus, in this case, the speed of the subsequent circuit can be determined directly from the data-sheet values. For the TIL 103 optocoupler, for example, rise and fall times for the diode current of t_r, t_f = 150 ns are stated. Thus the duration of a cycle at the maximum input frequency is T = 2 . t_r, t_f = 2 . 150 ns, which corresponds to a frequency of about 3 MHz, a value which can be achieved without difficulty in practice.

24.2.3.2
Phototransistor operation

Figure 24.4 shows the measurement circuit to determine the switching times in photo-transistor operation. Basically, the statements

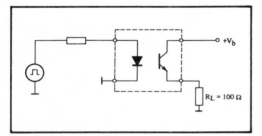

Figure 24.4
Measuring circuit for the dynamic perform-ance of optocouplers (phototransistor opera-tion)

made in the previous section also apply here, since the phototransistor can be visualised as a combination of a photodiode and an NPN transistor. In addition, however, the following points must be noted. The photodiode capacitance is in parallel with the collector-base junction and thus acts as a Miller capacitance. With the collector voltage $V_b > 10$ V used in the measuring circuits, firstly this capacitance is relatively small and secondly the voltage gain of the measuring circuit, with a load resistance R_L = 100 Ω, is only small. The effective Miller capacitance C_M is calculated from the formula:

$$C_M = C_{CB} . (V_U + 1)$$

In practice, however, load resistances of several kilohms are mostly used, as a result of which the voltage gain of the circuit is 10 to 50 times greater than in the measuring circuit. Secondly, the transistor is turned on as far as saturation; but with these small collector-base voltages, the capacitance of the photodiode (= collector-base capacitance) rises further by a factor of 3 to 4, which further increases the rise and fall times (*Figure 24.5*). This explains why trans-mission frequencies of only about 5 to 10 kHz are achieved with simple circuits.

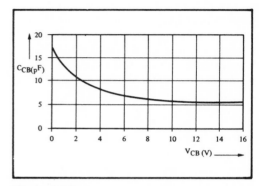
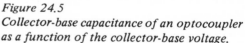

Figure 24.5
Collector-base capacitance of an optocoupler as a function of the collector-base voltage.

24.3
Simple transmission links

Optoelectronic couplers are easily interfaced with TTL. i.e. they can be driven directly from TTL circuits and are in turn capable of driving TTL circuits directly. *Figure 24.6* shows two circuits, which are adequate for many applications.

To determine the output current of the gate and thus the current in the luminescent diode, in *Figure 24.6*, the internal circuit of the gate must be used. *Figure 24.7* shows the equivalent circuit, in which the resistances R_{L1} and R_{L2} represent the line resistances.

404

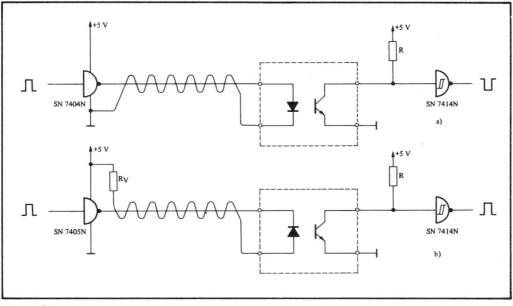

Figure 24.6
Simple transmission links

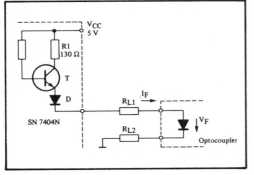

Figure 24.7
Circuit for determination of the output
current

The current is calculated from the formula:

$$I_F = \frac{V_{CC} - V_{CEsatT1} - V_D - V_F}{R_1 + R_{L1} + R_{L2}}$$

If a twisted pair, each of 0·4 mm diameter,
is used as the line, then with a line length of
100 m, the resistance $R_{L1} = R_{L2} = 14\ \Omega$.
Thus the input current I_F of the optocoupler is:

$$I_F = \frac{5\ V - 0·3\ V - 0·7\ V - 1·2\ V}{130\ \Omega + 14\ \Omega + 14\ \Omega}$$

$$\approx 17·7\ mA$$

For type TIL 117, with a minimum current
transfer ratio of 50%, an output current
I_{OL} = 9 mA is then obtained.

In order to take account of the process
variations which affect the components
within the SN 7404N, of fluctuations in
working voltage and of ageing of the optocoupler, the calculation is continued with
only half this value (I_{OL} = 4·5 mÁ). The
resistance R in *Figure 24.6a* is selected to be
as low as possible, in order to achieve short
switching times. Thus:

$$R = \frac{V_{CC}}{I_{OL} - I_{IL7414}} = \frac{5\ V}{4·5\ mA - 1·2\ mA}$$

$$= 1·5\ k\Omega$$

The component values for the circuit shown
in *Figure 24.6b* can be calculated accordingly.

405

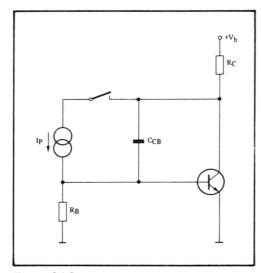

Figure 24.8
Equivalent circuit for the time-constant components in the optocoupler

24.4
Improvement of the switching performance of optocouplers

The time which primarily determines the maximum possible transmission frequency, is the turn-off time of the phototransistor. While the Miller capacitance $C_M = C_{CB}$ $(V + 1)$ is charged up relatively rapidly during turn-on by the photodiode current I_P (*Figure 24.8*), when the transistor is turned off, this capacitance must be discharged through the high-resistance base-emitter junction, which can take up to about 100 μs unless additional measures are taken. If, on the other hand, a resistance R_B is connected in parallel with the base-emitter junction, then an additional current, which discharges the Miller capacitance, is delivered from this path. Naturally, the lower the value of this resistance, the more it shortens the switching time.

On the other hand, it also diverts a part of the photocurrent I_P and thus reduces the current transfer ratio of the coupler. With a base-emitter voltage $V_{BE} = 600$ mV on the phototransistor, a photocurrent $I_P = 10$ to 20 μA and a resistance $R_B = 100$ kΩ, up to

60% of the photocurrent will be diverted to ground and thus the current transfer ratio will be reduced by the same percentage. Against this, there is a reducation of the turn-off time by about 50%, which often makes up for this disadvantage.

It is true that this very simple circuitry improves the switching performance of the phototransistor, but the actual cause – the Miller capacitance – is not affected. This is only possible, if the voltage gain of the circuit is reduced. A low-value load resistance R_C (approximately 100 Ω) would however, also reduce the output amplitude to the same extent. A circuit which avoids these disadvantages is the cascode circuit, well known from high-frequency work, although it was used there for other reasons.

In this case, the phototransistor works into the low input impendance (approximately 20-100 Ω) of the base-connected transistor T1, the working point of which is set by the diode D. Thus the voltage gain of the photo-transistor falls to very low values (1 to 10). At the same time, its collector-base voltage is prevented from falling below 2 V and thus causing an excessive rise in the collector-base capacitance (see *Figure 24.5*), which also improves the switching performance. This operating mode of the phototransistor is very similar to the measurement conditions stated in the data sheets, so that a direct comparison can be made from these values on the performance of the circuit in *Figure 24.9*.

The subsequent transistor T2 provides an adequate output signal so that TTL circuits can be driven directly. Since this transistor ensures short storage times and short rise and fall times, even in saturated operation, a Schmitt trigger is unnecessary in most cases.

The mode of operation of the circuit in *Figure 24.10* corresponds to that in *Figure 24.9*. By using the SN 75450N interface circuit, however, the number of discrete components is considerably reduced. The circuits in *Figure 24.9* and *24.10* permit

406

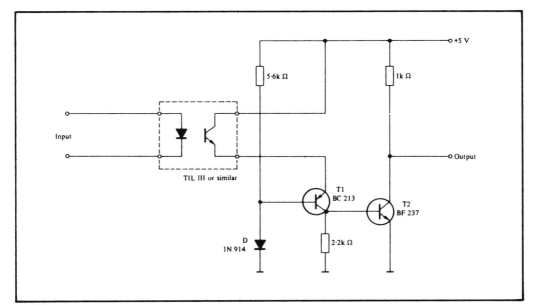

Figure 24.9
Cascode circuit to improve the switching performance of optocouplers

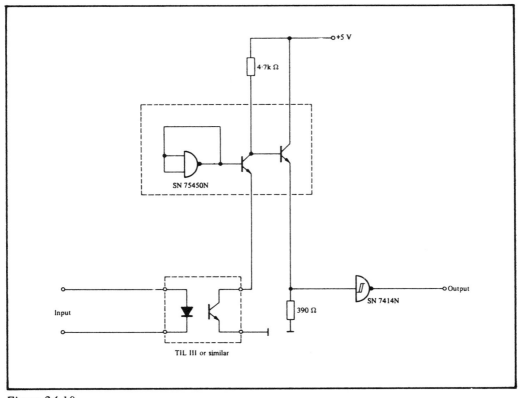

Figure 24.10
Cascode circuit with the SN 75450N interface circuit

407

maximum transmission frequencies of 100-300 kHz and are thus about 10 times faster than the circuits in *Figure 24.8*.

On the other hand, the circuit in *Figure 24.11* permits transmission frequencies of up to about 100 kHz (typically 50 kHz). Here, the phototransistor works into the relatively low input resistance of an emitter-connected transistor (approximately 1 to 5 kΩ). Because of the input characteristic of the amplifier, the collector-base capacitance of the phototransistor is again prevented from becoming excesssively large due to the small collector-base voltages. An additional resistance between base and emitter allows the Miller capacitance to be rapidly discharged.

24.5
Optocouplers in the photodiode mode

As was shown in the previous section, the physical characteristics of the photo-transistor limit the maximum transmission frequency. If only the photodiode of an optocoupler is used, the undesirable effects of the Miller capacitance and the storage time of the phototransistor are avoided. The frequency limit of these circuits can then be

moved up into the range above 1 MHz. Since the photodiode only delivers small currents, but on the other hand only low values of load resistance may be used to avoid large time-constants, the received signal must be interfaced to the subsequent logic circuits by highly sensitive amplifiers.

For this, the comparator SN 72710 or similar circuits are particularly suitable. (*Figure 24.12*). In this case, the load resistance of the photodiode is 1 kΩ. A bias current, fed in through a high resistance, defines a specific working point in the quiescent condition, this bias current being chosen to be about half the value of the opposing current of the photodiode ($I_F \approx 40\ \mu A$).

At the same time, a number of measures have been taken here, firstly to prevent line reflections and secondly to protect the circuit from damage due to coupled interference voltages. This is easily accomplished at the end of the line, i.e. at the input to the optocoupler. The luminescent diode can handle aperiodic forward current pulses of up to 1 A. With a 100-Ω line, this corresponds to an interference voltage of 100 V. A clamp diode D3 is provided, solely to limit negative voltages.

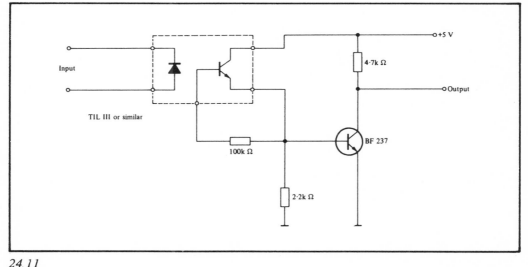

24.11
Amplification circuit for transmission frequences up to approximately 100 kHz

408

In the case of the line driver, however, further measures are necessary. The diodes D1 and D2 limit the interference voltage and the resistance R3 limits the current to non-critical values. The BAV 24 diode, which is used, permits aperiodic pulse currents up to 4 A. Together with the resistance R3, interference voltages on the line, up to 400 V, can then be safely suppressed. In the case of the circuit used as a line-driver, however, further measures must still be taken. With the above-mentioned currents, the forward voltage of the clamp diodes is more than 1 V. However, the voltage at the output of an integrated circuit must not become more negative than the component substrate, since otherwise damage can occur. In the case of the SN 75450 driver, the substrate is brought out separately. A voltage of -2 to -15 V is applied to this connection, in order to avoid an incorrect polarity. Finally, the resistances R1 and R2 protect the output transistors from excessively large collector currents (> 300 mA).

The driver circuit used delivers a no-load voltage $V_{OH} \geqslant 4$ V, while the output impedance is only a few ohms, both at High and Low level. With the aid of the resistance R3, the internal resistance of the driver is then matched to the line impedance, so that line reflections are safely prevented.

Naturally, the measures described here can be used successfully with the circuits stated previously.

24.6
Duplex operation with optocouplers

With long transmission distances, the cable costs play a substantial part. Since the data generally has to be passed in both directions between two stations, it is worth while to do this over the same line. Parallel connection of the two stations — as is generally usual with TTL circuits — is not possible here, since the luminescent diodes in the optocouplers have a very low resistance and therefore do not permit definite current sharing in the two receivers. It is, however, possible to connect the two stations in series, so that they handle a common current. The two logic levels are then represented by turning the current on or off.

Figure 24.13 shows the circuit for such a transmission system. In the rest condition, the two enable inputs are at low level. The current source formed by the transistor T1 now sends a current of about 20 mA through the phototransistor, which is also turned on, in coupler I and the luminescence diode in coupler II. For a data transmission from

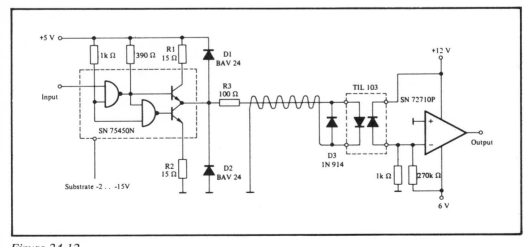

Figure 24.12
Data-transmission link for transmission frequencies in the MHz range

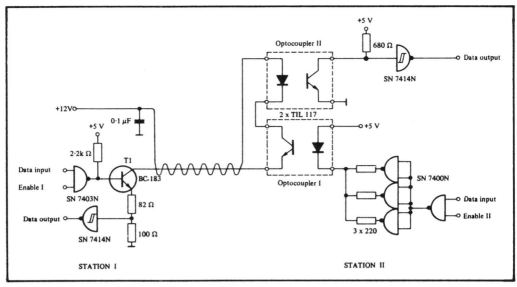

Figure 24.13
Duplex data transmission with optocouplers

station I to station II, the input enable I is switched to a high level. The information at the data input now modulates the current in the line. The coupler II transmits this to the Schmitt trigger, at the output of which the information is again available in TTL-compatible form. Data transmission from station II to station I takes place by interruption of the current path by the coupler I (the input enable I must of course be "High"). The voltage at the emitter of the transistor T1 thus collapses. Finally, the signal is coupled out again through a Schmitt trigger.

With a line length of 1000 m, this system permits transmission frequencies up to 10 kHz. Here, the limiting frequency is not so much determined by the speed of the coupler, as by the slow level changes, originating from line reflections, on the line, since the latter is not correctly terminated at its two ends.

24.7
Common-mode suppression of optocouplers

The advantage of the high common-mode

suppression of optocouplers is mainly due to the low coupling capacitance between the GaAs luminescent diode and the photo-transistor. The coupling capacitance depends on the insulation material, the distance between the luminescent diode and the phototransistor and also on the base area of the transistor.

Investigation of common-mode suppression for the optocoupler was undertaken by means of the test circuit shown in *Figure 24.14*. Voltage pulses, with different amplitudes and rise times are applied to the input of the coupler. The rates of rise of voltage (dv/dt) at which the flip-flop is set are considered as the criterion for common-mode suppression. The measurements were carried out with an input diode current $I_F = 0$ and with an I_F value, which corresponds to the typical $I_{C(on)}$ value of the data book in each case. In order to eliminate the effect of the brought-out base in types TIL 102, 111 and 113, these connections were clipped off.

With the TIL 102, 108 and 109 optocouplers no reaction occurred with voltage pulses with an amplitude of 1000 V and rates of

410

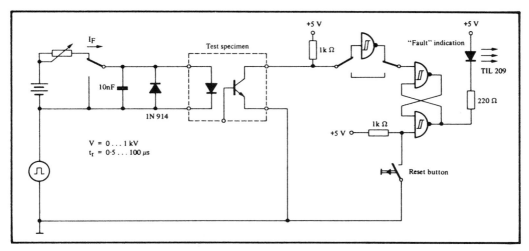

Figure 24.14
Test circuit for determination of the common-mode suppression of optocouplers

rise of 400 V/μs, either when the current was or was not flowing.

On the other hand, the TIL 113 coupler reacts, because of its large current transfer ratio, even to voltage pulses of low amplitude and slope. Measurements without current were discontinued at values of V = 180 V and dv/dt = 1 V/μs.

When carrying current, the couplers TIL 111 and TIL 113 showed no reaction up to the values V = 1000 V and dv/dt = 500 V/μs.

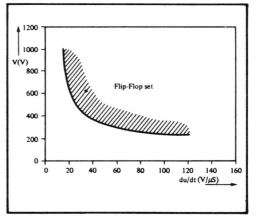

Figure 24.15
Common-mode suppression of the opto-coupler TIL 111 when not carrying current

411

25
Light exposure switch for photographic enlargers

Light exposure switch for photographic enlargers

The amateur photographer determines the optimum exposure time for photographic papers, either by test strips or by means of an exposure meter. In an exposure switch, the exposure measurement is coupled to a time switch for switching the lamp in the enlarger on and off. The time constant of the timing circuit is varied according to the irradiance falling on the photodetector. Simple exposure switches work with a trigger circuit, e.g. a monostable multi-vibrator. The timing circuit usually consists of an electrolytic capacitor and a photo-resistor. In this case, either the charging or discharging of the electrolytic capacitor through the irradiated photoresistor serves as a measure of the necessary exposure time.

The two components which determine the time have several disadvantages. The capacitance of electrolytic capacitors is temperature-dependent. In addition, they have a relatively large leakage current. Photoresistors are also temperature-dependent, they show fatigue and ageing effects and their function $R = f(E_e)$ is not exactly linear. Also, photoresistors with a high dark resistance must be selected. Generally, these have a very slow response. Therefore low irradiances (small lens apertures) are used, so that an exposure time of about 5 to 10 seconds is obtained. Some of the above-mentioned disadvantages of simple exposure switches can be avoided with modern semiconductor components.

25.1
Principle of construction of a light exposure switch with Si photodiodes

Figure 25.1 shows the block diagram of the exposure switch. The exposure process is prepared by pressing the normally-closed push-button. After the button is released, the Schmitt trigger receives a D.C. voltage level which is high in comparison with its threshold value; this causes the light source

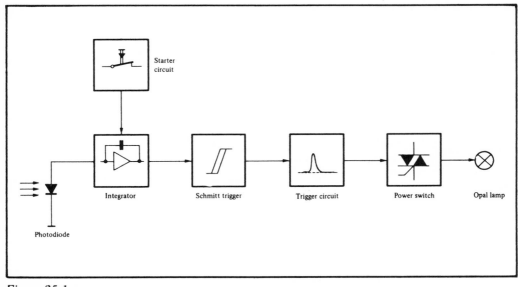

Figure 25.1
Block circuit of an exposure switch

in the enlarger to be switched on through the switching amplifier and the triac. The enlarger projects an image of the negative on to the photosensitive paper. The photodiode receives the radiation reflected from the paper. The output voltage of the integrator decreases in proportion to the light exposure of the photodiode. After the optimum exposure, it reaches the turn-off level of the Schmitt trigger. The Schmitt trigger terminates the exposure.

The maximum distance r from the photodiode to the projected image is determined by the reception peak of the diode. The photodiode should only evaluate the central picture content A. Evaluation of darkened or over-exposed areas is then prevented. In a greatly simplified form, the solid angle Ω can be determined with the half-value points of the reception peak. Here, the angle φ is the half aperture angle of the photodiode.

$$\Omega = 2\pi \, (1 - \cos \varphi) \; \Omega \qquad (3.4)$$

$$r = \sqrt{\frac{A}{\Omega}} \qquad (25.1)$$

The effectively evaluated area A is greater, since the aperture angle only takes account of the decrease in sensitivity down to 50%. For photodiodes with very narrow reception peaks, this error is negligible, since the effectivity evaluated area lies within the illuminated image. The photodiode TIL 81 satisfies this condition.

The reception characteristic of photodiodes with large reception peaks is narrowed, either by an aperture mask or an external lens. With an aperture mask, a more uniform sensitivity is obtained over the solid angle which is of interest. At the same time, the absolute sensitivity of the photodiode decreases according to the degree of masking.

When the enlargement scale is changed, the evaluated solid angle should remain constant, in proportion to the projected image size. The distance r is then to be corrected accordingly.

25.2
The timing system

The photodiode, the normally-closed push-button and the integrator represent the timing system of the exposure switch. If a photodiode with a reverse bias voltage is irradiated, a photocurrent flows. The photo-current is independent of the applied reverse voltage. It is calculated with the equation

$$I_p = s \, . \, E_e \qquad (9.26)$$

The capacitor is charged or discharged by the diode current. The voltage on the capacitor, which is proportional to the stored charge, serves as the evaluation criterion.

Figure 25.2 shows the circuit principle of the timing system. The amplifier is connected as an integrator. Therefore the capacitor lies between the drain and gate of the 2N 3822 N-channel junction FET. The photodiode TIL 81 is connected between gate and ground. The normally-closed push-button is located between the storage capacitor, the FET load resistance and the drain. In the quiescent condition, a voltage $V_{DS} \approx 0$ V is present at the gate.

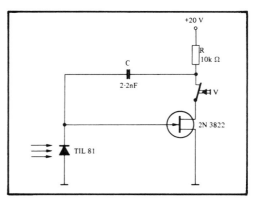

Figure 25.2
Circuit of the timing network

If the drain circuit is interrupted with the button U, the storage capacitor charges up through the load resistance R and through the parallel-connected diodes, the photo-diode and the gate-source diode, until it

416

reaches the working voltage. When the contact U is closed, the voltage at the drain falls, since the transistor is now conducting again.

The decrease in voltage at the drain is transmitted with negative potential, through the capacitor C, to the gate. The gate voltage V_{GS} falls and reduces the drain current until no further voltage change takes place at the drain. The negative gate voltage serves as a reverse voltage for the photodiode TIL 81. If radiation now falls on the photodiode, the capacitor C will be discharged by the photocurrent. A subsequent emitter follower (see *Figure 25.3*) reduces the output impedance of the integrator. The voltage on the load resistance R now decreases linearly with the exposure. The exposure error is calculated from the sum of the dark current $I_{D,25}$ and the residual gate current $I_{GSS,25}$:

$$I_{Leak} = I_{D,25} + I_{GSS,25} \qquad (25.2)$$

Because the ambient temperature in the darkroom is between $20^{\circ}C$ and $25^{\circ}C$, as required for the temperature-sensitive photographic baths, the leakage currents of the semiconductors also only have to be taken into account at this temperature. The maximum dark current of the photodiode TIL 81, for the measurement conditions $V_{CB} = 10$ V at $t_U = 25^{\circ}C$, is about $I_{D,25,max} = 10$ nA. The typical values are two to three orders of magnitude lower. For the measurement conditions $V_{CB} = 3$ V at $t_U = 25^{\circ}C$, typical values of $I_{D,25,typ} = 20$ pA to 50 pA are obtained. For the measurement condition $V_{GS} = -30$ V at $t_U = 25^{\circ}C$, the maximum residual gate current of the 2N 3822 FET is about $I_{GSS,25,max} = 100$ pA For the measurement condition $V_{GS} = -3$ V at $t_U = 25^{\circ}C$, the typical value is $I_{GSS,25,typ} = 15$ pA. Thus, in accordance with equation (25.2), the leakage current I_{Leak} is obtained as

$$I_{Leak} = 50 \text{ pA} + 15 \text{ pA} = 65 \text{ pA}$$

The measurement condition $V_{CB} = 3$ V or $V_{GS} = -3$ V was chosen, because the typical pinch-off voltage of the 2N 3822 FET is about $V_{GS,typ} = -1$ V to -3 V. The capacitance of the storage capacitor is determined by the shortest exposure time needed, t, and by the photocurrent sensitivity of the photodiode TIL 81. The capacitance of the capacitor C was determined experimentally at $2\cdot2$ nF. The time constant, determined by the leakage currents, is calculated in accordance with the following equation:

$$t_{Leak} = \frac{C \cdot U}{I_{Leak}} \qquad (25.3)$$

$$t_{Leak} = \frac{2\cdot2 \cdot 10^{-9} \text{ As} \cdot 3 \text{ V}}{65 \cdot 10^{-12} \text{ A} \cdot \text{V}} = 101 \text{ s}$$

With a permissible exposure error of 10%, the exposure time must therefore not exceed t = 10 s. The value t_{Leak} increases to 25 minutes, if the photodiode TIL 81 and the FET 2N 3822 are selected for their leakage currents. In simplified form, the residual gate current of the FET without the photodiode connected can be calculated in accordance with equation (25.2). For this, it is necessary to measure the discharge time t_1 due to the gate residual current in a well-insulated circuit. The discharge time t_2 is measured with the photodiode installed but completely darkened. By means of the time difference $t_1 - t_2$, the dark current of the photodiode can be calculated. It is better to measure leakage currents with an electrometer. If the FET is selected for minimum gate leakage current, then the BF 805, BC 264B and BC 264C can also be used. If, in addition, the pinch-off voltage is also restricted to the range between $V_{GS} = 1$ V and 3 V, the BF 245 FET can also be used.

Even more severe requirements can be satisfied with the photodiode TIXL 80; it has a more favourable ratio of photocurrent to dark current. In this case, the capacitor is enlarged accordingly.

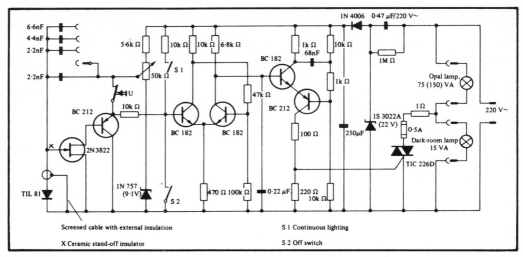

Figure 25.3
Circuit of the light exposure switch

25.3
Practical circuit of the light exposure switch

Figure 25.3 shows the circuit of the light exposure switch with the TIL 81 photodiode. The performance of the timing circuit is considerably affected by insulation resistances. This problem will be eliminated, if all connection of the gate circuit are taken to a ceramic support or if the printed circuit (on glassfibre reinforced epoxy material) is sealed with insulating spray varnish and baked out. A screened cable is to be used as the lead to the photodiode.

The exposure time is adjusted to the sensitivity of the photographic paper by R1. This varies the charging voltage of the storage capacitor. Switching of the capacitance of the storage capacitor permits variation of the exposure in wide ranges.

Through the emitter-coupled Schmitt trigger, the integrator drives an oscillator, which fires the triac during the exposure time and thus switches on the opal lamp.

The 15 W darkroom lamp is connected in series with the opal lamp of the enlarger. In the stand-by condition, the darkroom lamp is lit, while the opal lamp still remains dark. In the exposure condition, the darkroom lamp is short-circuited by the triac and the opal lamp is thus switched on. Thus a false exposure through the darkroom lamp is prevented. The light can be switched on continuously with the switch S1. S2 serves as an emergency switch. The supply voltage is produced through a capacitor as a series impedance, is rectified with the IN 4006 diode and is stabilised with a 22 V Zener diode.

418

26
Optoelectronic couplers as switches for analogue signals

26.1
Semiconductor switches and potential isolation

Optoelectronic couplers, have found a wide field of application in digital techniques for the transmission of digital signals, while at the same time the transmitter and receiver are electrically isolated from one another. In many cases, however, a switching device, in which the control and switch sections are electrically isolated from one another, is also needed for analogue signals. This problem cannot be solved with normal bipolar transistors, since the control circuit (base-emitter path) has a physical connection with the switching circuit (collector-emitter path). Better isolation is achieved with MOS field effect transistors, in which, because of the extremely high input impedance of the transistor, a quasi-isolation between the control and switching sections is achieved. The disadvantage is that the potential difference between the gate and source connections affects the forward resistance of the transistor, i.e. the voltage to be switched must lie in a certain potential range in relation to the control voltage.

Optoelectronic couplers as switches for analogue signals

With the optoelectronic coupler, the conditions are considerably more favourable. The phototransistor, which works as a switch, is controlled by the radiation emitted by the GaAs diode, so that there is no longer a physical connection between the control and switching sections. Therefore, potential differences of 1000 volts or more can exist between the control and switching sections. The low coupling capacitance, of only 1 pF, between the diode and transistor, should also be emphasised. The semiconductor switches just mentioned are compared in *Figure 26.1*.

26.2
The phototransistor as a switch in the optocoupler

26.2.1
Steady-state performance

A switch is required to have a blocking resistance in the open condition which is practically infinitely high, while the

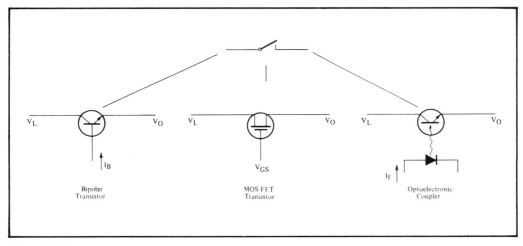

Figure 26.1
Comparison of various semiconductor switches

conduction resistance in the closed condition should be as small as possible. The required conduction resistance depends on the particular application. If only small currents are to be switched in high-resistance circuits, a conduction resistance of a few hundred Ohms is often of no importance.

The blocking resistance of a phototransistor is determined by the collector-emitter leakage current I_{CEO}. It is about 1 nA with a collector-emitter voltage of $V_{CE} = 10$ V and 25°C ambient temperature. This corresponds to a blocking resistance of 10^{10} Ω.

It is, of course, considerably more difficult to achieve a low conduction resistance in the phototransistors in optocouplers than with normal bipolar transistors. In the stated applications, the latter are operated with a base current of several milliamperes, so that the control current is several orders of magnitude greater than the load current.

Because of the low efficiency of the optocoupler, these conditions cannot be achieved here. The effective base current in the phototransistor is only a few times 10 μA. The result of this is that the forward resistance of the switch only reaches a few tens or a hundred Ohms, but in many cases this is of subsidiary importance. As an example, *Figure 26.2* shows the forward characteristic of a phototransistor in an optocoupler. Each curve relates to a different forward current I_F in the diode which emits the radiation. In inverse operation, the transistor very quickly comes into the current saturation range, because of its low current gain, so that in this case only currents of a few microamperes can be switched with reasonable efficiency. In *Figure 26.3*, the forward resistance is shown as a function of the diode current. If a low conduction resistance is required both for positive and negative currents, two photo-transistors can be connected in opposing parallel (*Figure 26.4*). It should be noted,

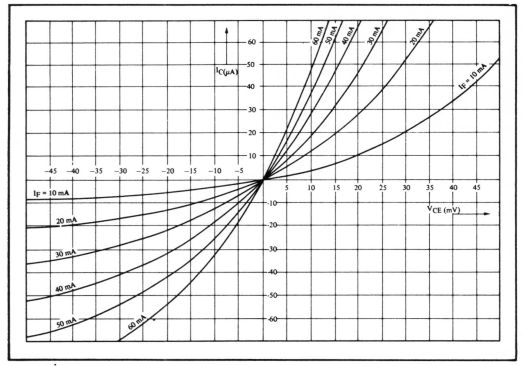

Figure 26.2
Characteristic of the phototransistor in the TIL 111 optocoupler

however, that in this case the breakdown voltage of the switch will be determined by the breakdown voltage of the base-emitter diode of the phototransistors and only amounts to about 7 V.

26.2.2
Dynamic performance

In the applications described here, the turn-on and turn-off times of the phototransistor will be affected less by the transit frequency of the transistor than by the collector-base capacitance (in inverse operation the emitter-base capacitance) and the internal resistance of the circuit (*Figure 26.5*). The latter determines the apparent voltage gain V of the transistor and thus also the actual effective capacitances.

The effective Miller capacitance is calculated as follows:

a
for normal operation:

$$C_M = (V_N + 1) . C_{CB}$$

b
for inverse operation:

$$C_M = V_{NI} + 1)C_{EB}$$

In order to achieve a high radiation sensitivity of the phototransistor, it must have a large base area. This leads automatically to a large collector-base capacitance, which lies approximately between 20 and 100 pF. Since the transistors still cause a

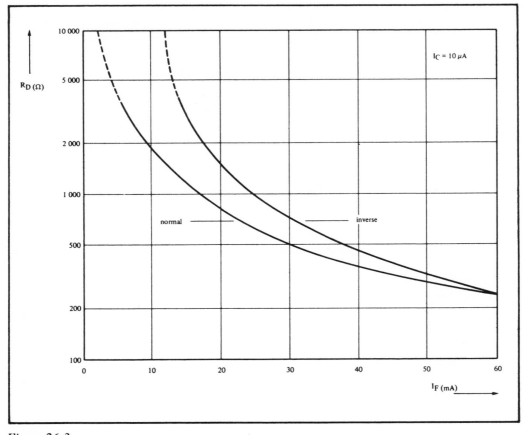

Figure 26.3
Conduction resistance of the phototransistor

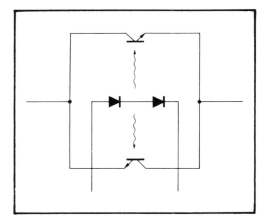

Figure 26.4
*Anti-parallel connection of phototransistors
to reduce the conduction resistance*

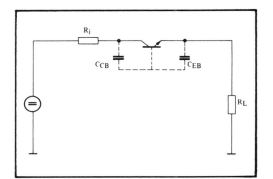

Figure 26.5
*Equivalent circuit of a phototransistor for
dynamic operation*

very large voltage gain in high-resistance circuits, the effective Miller capacitance C_M reaches values of several nanoFarads. This then leads to switching times, which can amount, in the least favourable case, to several milliseconds. The conditions are considerably more favourable during inverse operation of the phototransistor. Here, firstly, the emitter-base capacitance is considerably less than the collector-base capacitance, Secondly, because of the low current gain of the transistor in this operating mode, a smaller voltage gain is obtained and thus a considerably smaller effective Miller capacitance.

The switching times were determined with the measurement circuit shown in *Figure 26.6*. The results are summarised in *Table 26.1*.

As was to be expected, the switching times of the phototransistors in inverse operation are considerably less than the values in normal operation. Therefore, if short switching times are required, it is advisable to operate the phototransistors in reverse. If both positive and negative voltages are to be switched, two transistors are to be connected in series with opposite polarity. Since one transistor then always works in reverse, short switching times are always obtained, irrespective of the polarity of the voltage.

	$V_i = 1$ V		$V_i = -1$ V		$V_i = 1$ V[1]		$V_i = 1$ V[2]	
R [kΩ]	10	100	10	100	10	100	10	100
t_{on} (μs)	1·3	1·2	–	2·0	1·3	1·0	–	1·4
t_{off} (μs)	130	810	–	21	130	830	–	20

[1] Two transistors connected in opposing parallel
[2] Two transistors connected in opposing series

Table 26.1
Switching times of phototransistors (TIL 111)

424

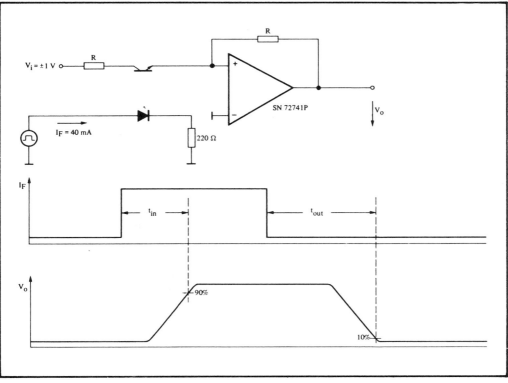

Figure 26.6
Measurement circuit for determination of switching times

26.3
Use of optocouplers in a digital voltmeter

Digital voltmeters are replacing analogue
voltmeters to an increasing extent. The
advantage of these instruments is, that the
measured results are presented in a form,
which can be processed further by data-
logging systems or process computers without
intermediate treatment, which is of great
importance for process automation.

It is often an important requirement with
measuring instruments of this kind, that the
analogue input and the digital output are
isolated from one another, because the
measurement source and the digital signals
have different reference potentials. There is
often a danger, that the measured values will
be falsified or the digital output signals will
be interfered with through transient currents
through undefined earth loops. Such errors

can be eliminated by isolation of the
measurement section from the evaluation
section.

In the following paragraphs, a simple digital
voltmeter, in which isolation is achieved
between the measuring and evaluation
sections in a very simple way, by use the of
optocouplers as switches for analogue signals
and as transmission devices for analogue and
digital signals.

26.3.1
Measurement method used

The unknown voltage is measured by the
"Dual-slope" method. In this, the voltage
V_x is integrated over a given time t_1
(*Figure 26.7*). After the time t_1 has elapsed,
the voltage V_x is switched off and a reference
voltage V_{Ref} is switched to the integrator.

425

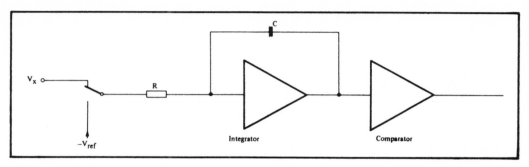

Figure 26.7
Theoretical measuring circuit of a digital voltmeter

Thus the capacitor C is discharged again. When the output voltage of the integrator reaches zero, the comparator operates. *Figure 26.8* shows the variation of the voltage.

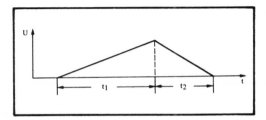

Figure 26.8
Variation of voltage with time at the integrator

The following relationships apply:

$$V_{c_1} = \frac{1}{RC} V_x \cdot t_1$$

and

$$V_{c_2} = \frac{1}{RC} (-V_{Ref}) t_2$$

With $V_{c_1} + V_{c_2} = 0$, then:

$$\frac{1}{RC} V_x t_1 + \frac{1}{RC} (-V_{Ref}) t_2 = 0 \text{ or}$$

$$V_x = \frac{t_2}{t_1} V_{Ref}$$

As can be seen, the external time constant of the integrator (R and C) has no effect on the accuracy of the measurement. Since only the ratio of the times t_1 and t_2, but not their absolute value, is of interest, this source of error can also be easily eliminated. An oscillator, which serves as a time-base, only has to have adequate short-time constancy. The absolute value of the frequency does not affect the accuracy of the measurement. On the other hand, the reference voltage V_{Ref} has a direct influence of the measured result. Therefore, high-stability Zener diodes with low temperature coefficients must be used to generate it.

In the above calculation, the performance of the operational amplifier in the integrator has been disregarded. However, the operational amplifiers which are currently available have such small offset voltages and currents and such an excellent temperature response, that errors of this kind can be neglected in the circuit described here.

26.3.2
Practical circuit of the digital voltmeter

26.3.2.1
Analogue section

As can be seen from the overall circuit in *Figure 26.9*, the analogue section consists of the integrator U1, the comparator U2, the

426

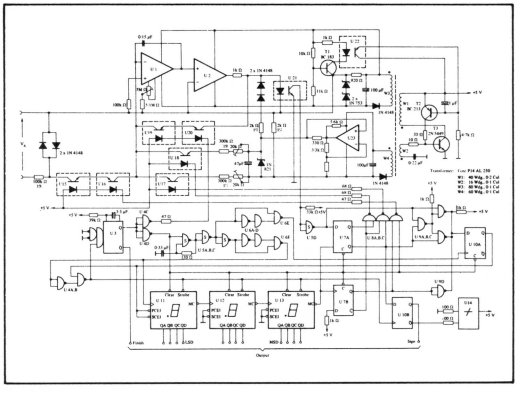

Figure 26.9
Digital voltmeter, overall circuit

reference voltage source and a total of four analogue switches, which are formed by phototransistors. During the time t_1, throughout which the switch U14/16 is closed, the unknown voltage V_x reaches the integrator U1 through a 100 kΩ resistor and charges up the feedback capacitor. Since both positive and negative voltages have to be switched here, two phototransistors are arranged with opposite polarities, so that one transistor is always operated in the inverse mode and thus short switching times are ensured. The offset voltage of the operational amplifier is balanced with the 5 MΩ potentiometer and thus the zero point of the voltmeter is set. the subsequent operational amplifier U2 serves as the comparator. The amplifier has no feedback, so that it works with its full no-load gain and switches over suddenly when the voltage at its input passes through zero. The output signal from the comparator is coupled out

through the optocoupler U21 and is fed to the digital section. During the time t_2 the capacitor is discharged again, either by a positive or negative reference voltage, depending on whether the voltage V_x was negative or positive. The reference voltage is produced by a temperature-compensated Zener diode, which is fed with a constant current from the operational amplifier U23. The mid-point of the voltage divider across this Zener diode is at zero potential, so that a negative reference voltage is obtained at the anode and an equal positive reference voltage, of about 3·1 V, is obtained at the cathode. The full-range value for positive and negative input voltages is balanced with the two 20 kΩ potentiometers. After completion of the measurement, the integrator is reset to zero through the switch U19/20. Since the comparator is involved in the discharge of the integration capacitor, its offset voltage is compensated.

427

26.3.2.2
Digital section

When considering the functional sequence in the digital section, it is best to start from the fact, that the monostable U3 has been set at the end of the last measurement. Thus the output of the gate U4C is at a low level and the switch U19/20 is turned on (see *Figure 26.10*). At this time, the analogue section is at rest. After about 100 ms, the monostable resets and thus initiates the next measurement. The voltage V_x is switched to the integrator through the gate U8C. At the same time, the clock generator U5A, B, C starts and,

delivers, at the gate outputs U6E and U6F, two pulse sequences, displaced by about 180° to one another, with a frequency of approximately 100 kHz. The counter, consisting of U11, 12, 13 now counts from 000 to 999. Through the subsequent carry, the flip-flop U7B is set and the voltage V_x is disconnected from the integrator. At the same time, the carry signal sets the flip-flop U7A according to the polarity of the output voltage of the integrator, detected by the comparator. Thus the appropriate reference voltage is switched through the gate U8A or U8B to the integrator. When the output of the integrator again reaches zero voltage, so

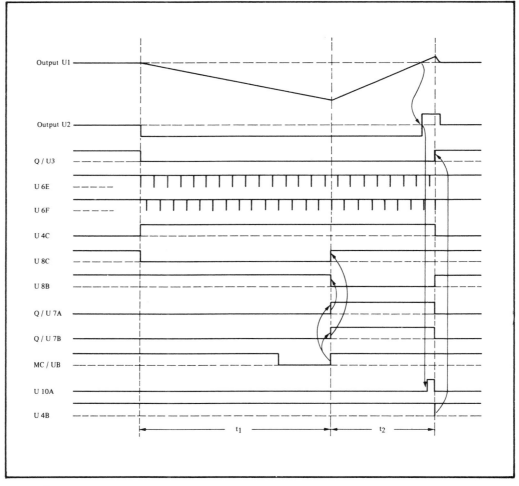

Figure 26.10
Pulse diagram of the digital voltmeter

that the comparator switches over, there is then a high level signal at the D-input of the flip-flop U10A.

The next clock pulse sets the flip-flop, so that the counter is stopped through the enable inputs. The next clock pulse transfers the contents of the counter to the store. At the same time, the position of the flip-flop U7A is interrogated and is transferred to the flip-flop U10B. The signal just mentioned also triggers the monostable U3. Now the "Ready" signal carries a low level and thus reports that the new measured value is ready at the output of the instrument. At the same time, the counter and flip-flop are set to their initial state through the gate U4D.

26.3.2.3
Voltage converter

As already mentioned, the requirement existed, for the analogue and digital sections to be electrically isolated from one another. In the control of the analogue switches and the return signal from the comparator, this was achieved by the use of optocouplers. In order to make it possible for the digital voltmeter to be operated from only one 5 V supply, a voltage converter has been provided, to deliver the supply voltages for the operational amplifiers and the reference voltage. Blocking converters are voltage converters with a good efficiency, but their output voltage is very dependent on the load. Therefore, stabilisation must be provided. A conventional stabiliser circuit at the output of the converter cannot be used, since this would unnecessarily reduce the efficiency of the circuit.

It is more economical, to regulate the output voltage of the converter by varying the mark-space ratio on the transistor T3. For this, the transistor T1 measures the deviation of the output voltage from the voltage of the reference diode (2 x 1N 753). The transistor T2, the collector current of which determines the mark-space ratio in the converter and thus the output voltage, is then driven through the optocoupler U22.

Basically, optocouplers are only suitable with restrictions for the transmission of analogue signals, since firstly the temperature-dependent output current is not exactly proportional to the input current; and secondly there is a variation in the current transfer ratio from device to device. In the application described here, however, this only has a slight effect on the performance of the circuit, if care is taken that the control gain is high enough, even in the worst case, to ensure the required stability of the output voltage.

The components used in the digital voltmeter are summarised in *Table 26.2*.

U 1:	SN 72307P	U 8:	SN 7437N
U 2:	SN 72741P	U 9:	SN 7403N
U 3:	SN 7412N	U 10:	SN 7474N
U 4:	SN 7437N	U 11-13:	TIL 306
U 5:	SN 74132N	U 14:	TIL 304
U 6:	SN 49703N	U 15-22:	TIL 111
U 7:	SN 7474N	U 23:	SN 72741P

Table 26.2
Parts list of the integrated circuits and optocouplers used

26.4
D.C. voltage amplifier with chopper

In instrumentation, D.C. voltage amplifiers are often needed, to amplify very small signals, originating, for example, from thermocouples, so that they can be processed further by subsequent equipment. Generally, severe requirements are imposed on these amplifiers, such as a) high input impedance or low input current, in order not to load the signal source, b) defined gain, c) low zero drift, d) constant parameters with variations in temperature and operating voltage.

With conventional D.C. amplifiers, these requirements can usually not be met, or can only be met at great expense. Therefore it is often simpler, to chop the D.C. voltage signal, then amplify it with a conventional A.C.

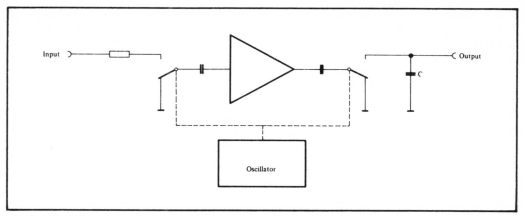

Figure 26.11
Theoretical circuit of a chopper amplifier

voltage amplifier and finally rectify it again. The principle of such an amplifier is illustrated in *Figure 26.11*. The amplifier concerned here is a conventional A.C. voltage amplifier. The desired gain is set by an internal current or voltage feedback. The no-load gain of the circuit should be at least 10 times greater, so that variations in operating voltage and temperature do not affect the performance of the amplifier. An oscillator controls the switches at the input and output. At the input, the direct voltage to be measured is chopped, so that a square-wave voltage, the amplitude of which corresponds to the voltage to be measured, is obtained. At the output end, the amplified square-wave voltage is rectified again with a switch and is smoothed with the capacitor C.

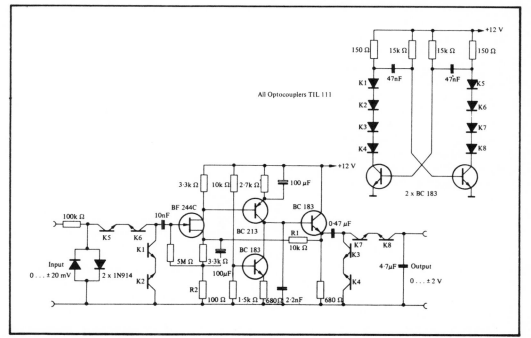

Figure 26.12
D.C. voltage amplifier with chopper

Figure 26.12 shows the complete circuit of a simple amplifier with a chopper. In the first stage, a field-effect transistor ensures a high input impedance. The second stage has a current source, formed by a transistor, as its load impendance. In this way, a gain of 2000 is achieved with an amplifier with only two stages. The amplified signal is fed out through an emitter follower. The resistances R1 and R2 form a feedback loop, so that the gain of the amplifier is reduced to 100. At the same time, the working points of the transistors are stabilised by this. For the switches, the phototransistors in opto-couplers are again used, with two transistors connected in series with opposing polarity in each case, in order to achieve short switching times, both with positive and negative input voltages. A multivibrator with a frequency of approximately 1 kHz drives the diodes in the optocouplers. The 150 Ω resistors in the collector leads limit the diode current to about 40 mA. It should be noted, that the feedback capacitors are not connected to the collectors, but to the anode of the diodes. This ensures, that the base-emitter junctions of the transistors are not driven into breakdown.

26.5
Line tester

The very many cable connections which are needed in electrical system, necessitate a simple test system, with which the cables can be tested for defects, such as short-circuits between two conductors and open circuits. The following report describes an instrument, with which multicore cables can be tested in a simple manner.

26.5.1
Testing principle

In the testing of cables, the following defects must be reliably detected:

a
Short-circuit between two cores,

b
Open-circuit in a core,

c
Cross-over of two cores.

With an appropriate test circuit, the last defect can be detected together with that mentioned under b).

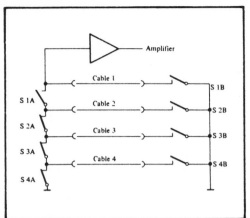

Figure 26.13
Theoretical circuit of the tester

Figure 26.13 shows the theoretical circuit of the tester, while for the sake of simplicity, the circuit for four conductors only is illustrated. Testing commences with all switches SnA closed, except for the switch S1A. The switches SnB are open. Now, if there is a short between conductor 1 and any other conductor, the input of the amplifier is at ground potential. This voltage level characterises the defect named. Then the switch S1B is closed. If conductor 1 has continuity, then again the amplifier input is at ground potential. If conductor 1 has an open circuit or − what produces the same defect − if two conductors have been wrongly identified at one end of the cable, the amplifier input is not at ground potential, which identifies the fault in this case. Then the switch S2A is opened; all other switches SnA are closed and the test for conductor 2 is carried out as described above. Thus, for testing a four-core cable, the switch diagram shown in *Table 26.3* is obtained.

431

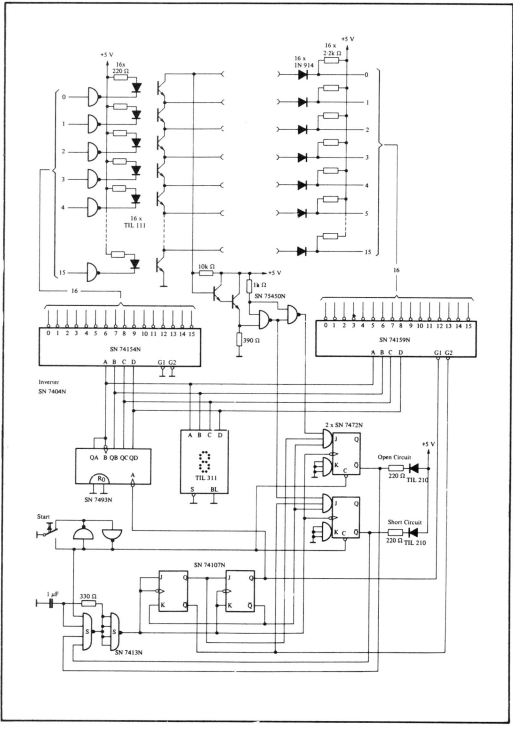

Figure 26.14
Complete circuit of the line tester

432

Switch position (x = Switch closed)								Test	
								Short circuit	Open circuit
S1A	S2A	S3A	S4A	S1B	S2B	S3B	S4B	Conductor	
	x	x	x					1	
	x	x	x	x					1
x		x	x					2	
x		x	x		x				2
x	x		x					3	
x	x		x			x			3
x	x	x						4	
x	x	x					x		4

Table 26.3
Switch diagram

26.5.2
Practical circuit of the line tester

Figure 26.14 shows the circuit of a cable tester, which tests multicore cables with a maximum of 16 cores. By appropriate extension, cables with more than 16 cores can also be tested. A start-stop oscillator (SN 7413N) with a frequency of about 3 kHz triggers the two flip-flops (SN 74107N), which control the two tests (short and open circuit).

First, the cable is tested for short-circuit, the decoder SN 74159N is unblocked and switches one core to a low potential and thus effects the test for open circuit. The SN 74450N amplifies the signals coming from the test specimen and drives the two flip-flops (SN 7472N). If a fault is detected, the corresponding flip-flop is set by the clock signal and the fault is indicated with a luminiscent diode (TIL 210). The oscillator is thus stopped, and the number of the defective core can be read off on the TIL 311 display. This device works in the hexadecimal code, i.e. the figures 0 to 9 are shown in the conventional way, while the numbers 10 to 15 are denoted by the letters A to F (*Figure 26.15*). This has the advantage, that only a one-digit display is necessary to represent 16 states and the complex and thus expensive code conversion into the BCD code is eliminated.

By actuating the "Start" button, the fault indication is cleared and the oscillator is started. The SN 7493 counter is advanced by one and selects the next core to be tested through the demultiplexer (SN 74154N and SN 74159N).

Optocouplers are used as potential-free switches. They ensure that, under no circumstances is the switching element

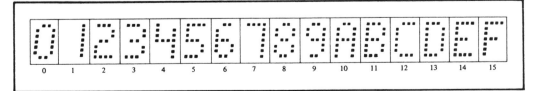

Figure 26.15
Display format of the TIL 311

433

(phototransistor) affected by the control part (luminescent diode). Also, they have the advantage, that they can be driven directly from TTL circuits.

The demultiplexer SN 74159 is used as the switch which connects the corresponding cable core during continuity testing to ground potential. It must be noted, however, that every output of this module draws a leakage current, in the "Off" condition, of 50 μA max. The sum of these currents (maximum 800 μA) could, in some circumstances, falsify the measurement and cause the amplifier (SN 75450N) to operate. Therefore this current is diverted to the supply line through resistors. Diodes in the individual leads ensure reliable decoupling.

Index

438